Membrane Transport
of Calcium

Membrane Transport of Calcium

Edited by

ERNESTO CARAFOLI

Laboratory of Biochemistry, Swiss Federal Institute of Technology (ETH), Zurich, Switzerland

1982

ACADEMIC PRESS

A Subsidiary of Harcourt Brace Jovanovich, Publishers

London   New York
Paris   San Diego   San Francisco   São Paulo
Sydney   Tokyo   Toronto

ACADEMIC PRESS INC. (LONDON) LTD.
24/28 Oval Road,
London NW1

United States Edition published by
ACADEMIC PRESS INC.
111 Fifth Avenue
New York, New York 10003

British Library Cataloguing in Publication Data
Carafoli, E.
    Membrane transport of calcium.
    1. Calcium–Physiological effect–Addresses, essays, lectures
    2. Cells–Permeability–Addresses, essays, lectures
    I.   Title
    615'.01524   QP913.C2

ISBN 0-12-159320-7

LCCCN 81-68980

Printed in Great Britain at the Alden Press, Oxford

List of Contributors

Mordecai P. Blaustein, *Department of Physiology, University of Maryland School of Medicine, Baltimore, Maryland 21201, U S A*

Felix Bronner, *The University of Connecticut Health Center, School of Dental Medicine, Farmington, Conn. 06032, U S A*

Ernesto Carafoli, *Laboratory of Biochemistry, Swiss Federal Institute of Technology (ETH), 8092 Zürich, Switzerland*

Giuseppe Inesi, *Department of Biochemistry, University of Maryland, School of Medicine, Baltimore, Maryland 21201, U S A*

P. G. Kostyuk, *A.A. Bogomoletz Institute of Physiology, Bogomoletz str. 4, 252601 GSP, Kiev-24, U S S R*

Leopoldo de Meis, *Instituto de Ciências Biomédicas, Departamento de Bioquimica, Centro de Ciências da Saúde, Universidade Federal do Rio de Janeiro, Cidade Universitária, Ilha do Fundão, 21.910, Rio de Janeiro, Brasil*

Mark T. Nelson, *Department of Physiology, University of Maryland School of Medicine, Baltimore, Maryland 21201, U S A*

Barry P. Rosen, *Department of Biological Chemistry, University of Maryland School of Medicine, Baltimore, Maryland 21201, U S A*

H. J. Schatzmann, *Universität, Veterinär-Pharmakologisches Institut, Länggass-Strasse 124 3000 Bern, Switzerland*

Preface

It is a lucky coincidence that this book should appear almost 100 years after the publication of the landmark contribution of Ringer on the role of calcium in the contraction of heart. This book is devoted essentially to the messenger function of calcium, a field which has now become immensely popular, and which has undoubtedly been opened by the remarkable observations made by Ringer in 1883.

An appropriate way of celebrating an anniversary, then. But also a way of telling what a long way the field of calcium-related research has come since the days — not too old, in fact — when the overwhelming interest of workers in the area was in the structural role of calcium. Against the background of this overwhelming interest stood a few scattered observations which suggested completely different possibilities. As is so frequent in science, most of them had come ahead of the times, and their implications were clear only to a handful of remarkably foresighted pioneers. At a meeting I attended in 1975, I remember Annemarie Weber quoting one of them, L. V. Heilbrunn, as once saying (or writing) *"Kalzium macht alles"*. A remarkable comment, indeed. It may have sounded grossly exaggerated at the time it was made, but we can see today instead how close it was to reality.

In retrospect, it is difficult to trace back the present phenomenal explosion of knowledge and interest in the field of calcium biochemistry to a single factor. Whether it was the findings and the development of concepts in the field of nervous conductance, or the progresses made in the molecular mechanisms of endocrinology. Whether it was the rapid advances in the field of membrane transport of calcium, and the early work on sarcoplasmic reticulum. Or the discoveries in the area of high-affinity calcium ligation, with their implications for ionophoric transport and muscle contraction. Or was it, last but not least, the recent discovery of calmodulin? Reasonably, each one of these developments played a role, and it was a fortunate happening that situations favourable to important breakthroughs developed simultaneously in so many parallel directions. Not surprisingly, the effect has been autocatalytic, and has been responsible for the atmosphere of great excitement the area enjoys today.

One obvious corollary of the messenger function of calcium is the necessity of controlling very precisely its activity in the intracellular environment, where the targets of the messenger function are located. So important is the messenger function, indeed, that a considerable number of systems for transporting calcium across biomembranes have been developed during evolution. These are independent systems, molecularly and mechanistically different from one another. Their multiplicity is probably a unique case in biology, and underscores the importance of calcium as a messenger. Using simple words, one could say that cells cannot afford significant fluctuations in the activity of calcium, and thus employ all possible ways and means to prevent it. The transport of calcium across biomembranes becomes thus a central component, perhaps *the* central component, of calcium biochemistry and physiology. One may object that this statement, which is reflected in the subject matter of this book, is somewhat restrictive, since it does not pay attention, at least in the traditional way, to the biochemistry of mineralization. This may well be so. But it certainly reflects the most exciting developments and the brightest prospects in the field.

The essays collected here review in detail the calcium transporting systems which are known today to operate in eukaryotic and prokaryotic membranes. Written by leading experts, they cover general properties, molecular aspects, mechanisms and function. The histories of the various calcium transport processes, however, have not been uniform. Out of necessity – the purification of the carrier proteins has generally come late – mechanistic and functional studies have frequently preceded molecular approaches. Thus, an important point in the book is the comparison of developments, both technological and conceptual, in neighbouring areas. Ideally, seasoned specialists as well as novices ought to find in it cross-fertilizing leads for future research endeavours. It is hoped that in the end significant advances in this beautifully exciting field may result.

October 1981   *Ernesto Carafoli*

Contents

1

Penetration of Calcium Through the Membrane of Excitable Cells

P. G. KOSTYUK

A. A. Bogomoletz Institute of Physiology, Academy of Sciences of the Ukrainian SSR, Kiev, USSR

GENERAL ASPECTS OF THE PROBLEM

All cells maintain high gradients of Ca^{2+}, Na^+ and K^+ ions across the surface membrane but it is the Ca^{2+} gradient which is the most effectively maintained, due to powerful calcium binding mechanisms in the cytoplasm. This suggests that, if membranes possess ion-conducting molecular structures, Ca^{2+} ions may be highly effective carriers of electric current across the membrane and may be instrumental in recharging membrane capacity and in triggering active cellular reactions.

The classical work of Hodgkin and Huxley (1952) on perfused squid giant axons was the start of a number of investigations which showed the existence of molecular structures which allow Na^+ or K^+ ions to pass as a result of lowering of the membrane potential (membrane depolarization). The properties of electrically operated sodium and potassium channels have since been shown to be similar in many excitable membranes. Special 'grating'' groups which can open or close a path in ionic channels for ionic movement under influence of external energy provide an exceptionally effective mechanism for the rapid release of potential energy accumulated by the cell in the form of Na^+ and K^+ gradients.

In 1953 Fatt and Katz showed in crayfish muscle fibres that a propagating action potential can occur in an extracellular medium even if sodium ions and hence sodium inward current are absent. The inward current necessary for membrane depolarization is due to divalent cations (Fatt and Ginsborg, 1958). Calcium-dependent action potentials in sodium-free solution could be obtained only if potassium conductance by the membrane was blocked by tetraethylammonium (TEA); however, this treatment could be avoided if extracellular calcium was replaced by Sr^{2+} or Ba^{2+} ions, which appear to be more potent carriers of charges across the membrane than do Ca^{2+} ions.

More detailed information on the role of calcium ions in the electrical excitability of muscle fibres has been obtained by Hagiwara and his colleagues working on giant fibres of *Balanus nubilis* (up to 2 mm in diameter) (Hagiwara and Naka, 1964; Hagiwara *et al.*, 1964; Hagiwara and Nakajima, 1966). They showed that full-scale calcium-dependent action potentials in *Balanus* muscle can always be elicited by injecting calcium-chelating agents into the cell to cause an artifical decrease in the intracellular concentration of free calcium. The amplitude of action potentials in the concentration range of 20–100 mM extracellular calcium increased proportionally to the logarithm of $[Ca^{2+}]_o$, with a slope of about 29 mV for a 10-fold concentration increase. This dependence was close to the theoretical one for a calcium electrode according to the Nernst equation:

$$E_{Ca} = \frac{RT}{nF} \cdot \ln \frac{[Ca^{2+}]_o}{[Ca^{2+}]_i} \tag{1}$$

If at the peak of the action potential the membrane resembles the calcium electrode in its properties, then at two different concentrations of extracellular calcium, with its constant intracellular content, the difference between the peak amplitudes will be:

$$E_2 - E_1 = \frac{RT}{nF} \ln \frac{[Ca^{2+}]_2}{[Ca^{2+}]_1} \tag{2}$$

and at a temperature of 22° C this difference must be 29.5 mV with a 10-fold change of external calcium concentration. A good correlation between the theoretical and the experimental value was observed.

However, interpretation of these results was difficult. As mentioned, Hagiwara and colleagues showed that creation of a very low concentration of free calcium inside the cell (below 8×10^{-8} M) is a necessary condition for generation of the action potential in *Balanus* muscle fibre. This should correspond to extremely high positive values of the calcium e.m.f. (about $+ 150$ mV), which exceeds by several times the observed values of the membrane potential during calcium spikes. This discrepancy can be accounted for if we assume:

(1) that intracellular calcium concentration near the membrane considerably exceeds its concentration in the bulk of the cell; (2) the potential of the calcium spike is determined not only by calcium conductance but also by potassium conductance and is dependent on the I_{Ca}/I_K ratio. In any case, the fact that the amplitude of the muscle fibre action potential spike is strongly dependent in sodium-free solutions on extra- and intracellular calcium concentration, clearly indicate the role of calcium ions in recharging the membrane.

Direct measurements of transmembrane ionic currents by a voltage clamp technique were carried out by Hagiwara *et al.* (1974). These authors' reversal potentials proved to be considerably lower than the expected values, probably because calcium and other transmembrane currents (potassium) were superimposed. When special measures were taken to suppress potassium conductance the equilibrium potentials shifted to more positive values (Keynes *et al.*, 1973). However, assessment of such potentials could be only approximated because direct current–voltage measurements at high positive membrane potentials proved to be virtually impossible and the values were extrapolated.

These authors confirmed that calcium ionic channels of muscle membrane allow other divalent cations to pass, namely Sr^{2+} and Ba^{2+}. In such cases, the quantitative changes in the amplitude of the action potential were also complex. The dependence of the action potential peaks on the logarithm of ion concentration did not show saturation effects with increasing concentration of Ba^{2+} ions, whereas such saturation was revealed with a change in Ca^{2+} concentration. In Ba^{2+}-containing solutions this dependence was nonlinear, its maximal slope being considerably higher than the theoretical value. Hagiwara (1973) proposed a model which could explain the following properties of calcium currents: (1) saturation of the current with increasing calcium concentration; (2) competition between calcium ions and other divalent cations (and also with some trivalent cations).

According to this model, the calcium ion, in order to pass through the membrane, must first be adsorbed by a certain membrane receptor, x:

$$Ca_o + x \underset{k_2}{\overset{k_1}{\rightleftharpoons}} Ca\,x \overset{k_3}{\longrightarrow} Ca_i + x \tag{3}$$

Since the total number of receptors, x, is constant at adsorption equilibrium:

$$K_{Ca} = \frac{([x] + [xCa])[Ca]_o}{[xCa]} \tag{4}$$

where K_{Ca} is the dissociation constant for calcium ions and receptor. From the equation (4) it follows:

$$[xCa] = \frac{[x]}{1 + K_{Ca}/[Ca]_o} \tag{5}$$

According to the model, $I_{Ca} \sim [xCa]$, and $I_{Ca}^{max} \sim [x]$. Then

$$I_{Ca} = \frac{I_{Ca}^{max}}{1 + K_{Ca}/[Ca]_o} \tag{6}$$

where I_{Ca}, I_{Ca}^{max}, K_{Ca} in the general case are potential-dependent values.

The quantitative description of this model completely fits the Michaelis-Menten equation (in the case of equilibrium) and the Langmuir isotherm. Equation (6) describes the saturation of calcium current with increase in external calcium concentration. If cation M^{2+} is present in the medium and binds to the same receptor with dissociation constant K_M but does not penetrate the channel, the equation which describes the calcium current will be:

$$I_{Ca} = \frac{I_{Ca}^{max}}{1 + (1 + [M]_o/K_M)K_{Ca}/[Ca]_o} \tag{7}$$

If I_{Ca} is the calcium current in the absence of other ion species and I'_{Ca} is the current in the presence of such species, then the ratio between these currents will be:

$$\frac{I_{Ca}}{I'_{Ca}} = \frac{1}{1 + [M]_o/K'_M} \tag{8}$$

where $K'_M = K_M(1 + [Ca]_o/K_{Ca})$.

This model does not take into consideration the potential-dependence of the measured values. Such dependence requires correction for the corresponding shifts in potential characteristics when K_{Ca} is determined. It can be achieved, for example, by creating a high concentration of impermeable divalent cations (e.g. 100 mmol Mg^{2+} ions) in external solution; however, these experimental conditions are largely artificial and can result in error.

As follows from the results of Hagiwara and Takahashi (1967), a series of divalent cations and one trivalent ion, lanthanum, can be arranged in the following order according to their ability to bind competitively to a presumed receptor in the calcium channel:

$$La^{3+}, UO_2^{2+} > Zn^{2+}, Co^{2+}, Fe^{2+} > Mn^{2+} > Ni^{2+} > Ca^{2+} > Mg^{2+}$$

The proposed model also allows evaluation of the relative ability of calcium channels to pass the above ions. So, for currents measured in equimolar calcium and barium solutions from the equation (7), one obtains:

$$\frac{I_{Ca}}{I_{Ba}} = \frac{I_{Ca}^{max}}{I_{Ba}^{max}} \cdot \frac{C + aK_{Ba}}{C + aK_{Ca}} \qquad (9)$$

where $a = 1 + [M]_o/K_M$, $C = [Ca]_o = [Ba]_o$.

The ratio between maximal currents in this equation was called the "mobility factor" and the second term the "affinity factor". In the absence of blocking ions ($a = 1$) the following sequence of permeabilities has been obtained: $I_{Ca}:I_{Sr}:I_{Ba} = 1.0:1.05:1.3$ (Hagiwara et al., 1974).

In 1956, from investigations of action potentials in mammalian heart muscle, it was observed that the plateau phase of these potentials is much less sensitive to removal of Na^+ ions from external solution than is the fast initial spike (Coraboeuf and Otsuda, 1956). Similar observations have also been made on frog heart (Niedergerke and Orkand, 1966). It became obvious that in heart muscle fibres, unlike nerve fibres, a considerable inward Ca^{2+} current appears during excitation. The detection of an increase in this current on application of adrenaline was of great importance. This conclusion was made first on the basis of the finding that adrenaline increases the amplitude of the action potential plateau in cases where sodium channels are blocked by tetrodotoxin (Stanley and Reuter, 1965). Increase in calcium influx was then confirmed by measuring ^{45}Ca entry into the muscle (Reuter, 1965). Counter to this, blockers of calcium currents (lanthanum, manganese, cobalt, nickel) caused a decrease in the amplitude of the plateau (Hagiwara and Nakajima, 1966). Results obtained in experiments with voltage clamping confirmed the presence of separate systems of sodium and calcium conductance in heart muscle fibre membrane; in sodium-free solution a slow inward current dependent on the external concentration of Ca^{2+} has been found (Reuter, 1967). This current was also seen in Na-containing solutions but in such cases it was complicated by inward flow of Na^+ ions (Vitek and Trautwein, 1971). Data have been obtained on the conductance of calcium channels in heart muscle for Ba^{2+} and Sr^{2+} ions, as well as on the measurable conductance for Mg^{2+} ions contrary to the results obtained in other membranes (Kohlhardt et al., 1973). Sodium and calcium components of the inward current in heart muscle fibre membranes have been partially separated by verapamil or its derivative D-600, which suppresses calcium conductance competitively (Kohlhardt *et al.*, 1972).

The reversal potential of the slow calcium inward current in heart muscle proved to be surprisingly low — about + 50 mV. A conditioning depolarization resulted in a small shift to more positive membrane potential values (for several millivolts — Bassingthwaighte and Reuter, 1972). Such low values of the reversal potential observed for calcium current in the heart muscle were explained by the low selectivity of the corresponding channels, which may pass Na^+ and K^+ ions together with Ca^{2+} ions. Rations P_{Ca}/P_{Na} and P_{Ca}/P_K were supposed to be $\sim 1:0.01$ (Reuter and Scholtz, 1977).

Studies on replacing extracellular ions and on the effect of specific blockers of ionic channels on the amplitude, rate of rise and duration of electrical responses in mammalian smooth muscle membrane gave largely similar results, despite individual differences in action potentials. In all smooth muscles investigated so far the action potentials are, on the one hand, highly sensitive to changes in concentration of external Ca^{2+} ions and to the blocking effect of Mn^{2+}, Co^{2+}, La^{3+} and D-600; on the other hand, they are resistant to the blocking action of TTX (Bülbring and Tomita, 1969; Anderson *et al.*, 1971; and others). Various smooth muscles are capable of generating action potentials in Na-free solution for a long period of time, their amplitude being dependent on the concentration of external Ca^{2+} ions (Holman, 1958; and others).

These findings led to the conclusion that calcium conductance in electrically excitable membranes of smooth muscle cells is not only an essential factor but the main mechanism of the generation of a propagating action potential.

Because of technical difficulties only a few measurements have been carried out on the analysis of ionic currents by voltage clamping in muscle fibres of taenia coli, uterus and portal vein (Kumamoto and Horn, 1970; Anderson *et al.*, 1971; Mironneau, 1974; Kao and McCullough, 1975; Inomata and Kao, 1976; Bury and Shuba, 1976). The results obtained in these experiments do confirm the above conclusion; however, the authors failed to perform a detailed analysis of calcium currents.

All these data called for a more thorough analysis of the question as to whether the mechanism of generation of the action potential in the nerve fibre also involves the activation of calcium conductance of the membrane together with sodium and potassium conductances. Hodgkin and Keynes (1957) investigated ionic fluxes in the squid using labelled isotopes. They found that $0.006\,pmol/cm^2$ of Ca^{2+} ions actually enter the fibre in the course of generation of one nerve impulse. This quantity is negligible compared with inward-going Na^+ ions $(4\,pmol/cm^2)$. Obviously, in the nerve fibre Ca^{2+} ions are capable of transferring only a small proportion of the charge necessary for generation of the nerve impulse.

More detailed investigation of this question became possible due to the discovery of aequorin, a protein that emits light in the presence of traces of ionized calcium (Shimomura et al., 1962). By introducing this protein inside the axon, Hodgkin and his colleagues identified two phases of calcium entry (Baker et al., 1970; 1971a, b; 1973b). The first, fast, phase was comparable with sodium inward current in its time dependence and was correspondingly blocked by TTX. Therefore, Baker et al. concluded that this portion of calcium current is carried through sodium channels due to the imperfect selectivity of the latter. Later, it was possible to measure directly the selectivity of sodium channels and to show that the ratio between sodium and calcium conductance in sodium channels is close to 100 (Meves and Vogel, 1973).

The second, slow, phase of calcium entry in the axon was dependent on external concentration of these ions and was characterized by the presence of a region of negative resistance in the current–voltage curve. At constant depolarization, calcium entry during this phase decreased from a maximal value to a considerably lower level. The time course of this process was approximately exponential with a time constant of the order of several seconds (Baker et al., 1971c). The slow phase of calcium entry into the fibre was insensitive to TTX and TEA but could be suppressed by the action of Mg^{2+}, Mn^{2+} and Co^{2+} on the outer surface of the membrane, as well as by the more active derivative of verapamil, D-600 (Baker et al., 1973a). In the experiments described the authors did not manage to measure directly calcium current associated with the slow phase of calcium entry into the fibre. However, Tasaki and colleagues observed prolonged action potentials in giant axons in conditions when only divalent cations (Ca^{2+}, Sr^{2+} or Ba^{2+}) were present in the external solution and monovalent cations inside the fibre, which appeared to be caused by both fast and slow calcium entry into the axon (Watanabe et al., 1967; Tasaki et al., 1969). These authors tried to explain their data from a special viewpoint which was not confirmed experimentally. According to Tasaki, electrical excitation is provided by conformational changes of macromolecules in the membrane which results in an increased diffusion of cations across the membrane;

the existence of specialized ion-conducting structures is denied (Tasaki, 1968).

A modern conception of electrical excitation gives the following explanation of biphasic entry of Ca^{2+} ions in the nerve fibre during excitation. Fast calcium current is carried through sodium channels due to the imperfect selectivity of the latter. The slow phase of calcium is due to the existence of proper calcium channels in the axonal membrane which possess special pharmacological properties resembling those in the membrane of muscle fibres. However, unlike muscle membrane, the density of such channels in the axonal membrane is exclusively low so that it becomes impossible to measure electrically the current carried through these channels. Evidently this reflects the functional specialization of the nerve fibre in transmitting a propagating signal that can be effectively secured by a combination of sodium and potassium membrane currents; however one might suppose that in other parts of the neuronal membrane having more complex functions (the somatic membrane and the membrane of nerve terminals) the ionic mechanisms of excitation would have different characteristics.

In fact, Oomura et al. (1961) found that the neuronal soma of the mollusc Onchidium can generate full-scale "calcium" action potentials in Na-free solution. Later this property was observed by Gerasimov et al. (1964) in the neurons of Helix pomatia and by Koketsu and Nishi (1969) in the toad sympathetic ganglion. These data have been confirmed by Meves (1968), Geduldig and Junge (1968) and by other authors. The use of the voltage clamp method (by the introduction of two microelectrodes into the cell) enabled inward calcium currents underlying these action potentials to be recorded (Doroshenko et al., 1973, Kostyuk et al., 1974a). Eckert and Lux (1975) have described a calcium inward current in an autorhythmic neuron of Helix pomatia. The somatic membrane of the nerve cell thus actually revealed essential differences in the ionic mechanisms of electrical excitability compared with the membrane of a nerve fibre.

Katz and Miledi (1967a, b, c) studied ion-dependent acetylcholine release from terminals in the nerve–muscle junction and concluded that a specific inward calcium current develops which links excitation of the surface membrane of the terminal to transmitter release. They subsequently confirmed this experimentally (see Katz and Miledi, 1970).

The presence of a special system of calcium conductance is thus peculiar to those parts of the neuronal membrane which have a more complex function than the nerve fibre and where coupling between the superficial membrane and the intracellular processes takes place.

Many data indicating that the mechanism of neuronal electrical excitability includes a component evoked by the activation of a specific calcium conductance were later obtained. In particular, the presence of a calcium component in the spike has been shown in motoneurons of insects (Pitman, 1975), in Retzius cells of the leech (Kleinhaus, 1976), in neurons of dorsal root ganglia of the mouse (Matsuda *et al.*, 1976; 1978; Ransom and Holz, 1977; Yoshida *et al.*, 1978), in dendrites of cerebellar Purkinje cells (Llinas and Hess, 1976; Llinas *et al.*, 1977), and in hippocampal neurons of the rabbit (Schwartzkroin and Slawsky, 1977).

Because electrically operated calcium conductance occurred so widely and was so important in intercellular integration, ways of distinguishing calcium currents from other transmembrane currents and of studying the properties of such currents were sought. These attempts were successful only in the somatic membrane of nerve cells, and the information about the structure and function of electrically operated calcium channels obtained contributed to the understanding of the peculiarities of calcium conductance in the membrane of other cellular structures.

PROPERTIES OF OPEN CALCIUM CHANNELS

In most cases calcium and sodium currents can be separated by applying specific blockers of sodium channels, such as tetrodotoxin. This method has been used in motor nerve endings by Katz and Miledi (1967b), in *Aplysia* neurons by Geduldig and Gruener (1970) and in squid axons by Baker and colleagues (Baker *et al.*, 1970). However, in some tissues sodium channels are either not very sensitive to this toxin (neurons of the Gastropoda) or they possess, with TTX-sensitive sodium channels, TTX-resistant ones differing from each other in kinetics (some mammalian spinal ganglia neurons). In such cases it is possible to eliminate sodium inward current by substituting external Na^+ ions for impermeable ions; the current–voltage relation of the remaining calcium current substantially differs from that of the sodium one (Kostyuk *et al.*, 1974b; Standen, 1975a, b; Eckert and Lux, 1976).

However, due to their slow development, potassium currents are always superimposed on calcium currents; reliable separations cannot be achieved by pharmacological drugs because even the most potent (tetraethylammonium, aminopyridine) do not suppress them completely. Evidently, removal of carriers of these currents by washing them out from inside the cell would be the most successful way. This was achieved reliably in nerve cells by a modified method of intracellular perfusion (Krishtal and Pidoplichko, 1975) in which openings are made in the membrane of the isolated nerve cell through which a solution of desirable ionic composition flows inside the cell. These openings are fixed in a microscopic pore made in a plastic membrane to prevent resealing. This method is quite ubiquitous. It was originally applied to giant neurons of molluscs (Kostyuk *et al.*, 1975), but was also successful in other excitable cells: e.g. isolated neurons of amphibian dorsal root ganglia (Veselovsky *et al.*, 1977a), mammalian spinal ganglia neurons (Veselovsky *et al.*, 1979), neuroblastoma cells (Veselovsky *et al.*, 1977b; Kostyuk *et al.*, 1978). Modifications of this method have been proposed by Lee *et al.* (1978) and Takahashi and Yoshii (1978). The latter modification has been used for the perfusion of oocytes and separation of ionic currents in their surface membrane.

Intracellular potassium can be replaced during intracellular perfusion by $Tris^+$ ions. Both fast and delayed potassium currents are completely eliminated and it is possible to record undistorted calcium currents over a wide range of membrane potential shifts. Nevertheless, in these conditions, at membrane potentials more positive than $+ 40$ mV, a residual potential- and time-dependent outward current still remains, due to the presence of a proportion of low selective ionic channels which are capable of passing even $Tris^+$ ions (Doroshenko *et al.*, 1978). Pharmacological blockers for such channels have not yet been found; they are not affected by TEA, in particular. The presence of this current makes it difficult to study the current–voltage relations of calcium currents at high positive potentials when it can mimic, for example, a reversal of their direction.

This problem can be partly solved by lowering the pH of the external solution. Such lowering leads to protonation of charged groups at the membrane surface which create the surface potential of the membrane, as well as to protonation of charged groups inside the ionic channels themselves which are essential for the maintenance of conductance (see below). The character of charged groups may be different in various ionic channels and the pH at which they can be titrated will be correspondingly different. It has been shown experimentally that the channels of nonspecific outward current are extremely sensitive to the lowering of pH which produces both a substantial shift of the current–voltage characteristic of this current to more positive membrane potentials and a decrease of its maximum value (Doroshenko *et al.*, 1978). Calcium channels are much less sensitive to decrease in pH. Therefore, uncomplicated inward calcium currents can be recorded at high positive testing potentials by lowering the pH of the external solution.

Calcium channels pass both Ca^{2+} ions and other divalent cations (Ba^{2+}, Sr^{2+}, Mg^{2+}). The selectivity of these channels can be estimated from the maximum inward current produced in solution when Ca^{2+} is replaced by an equivalent amount of other divalent ions; the relative values are: 2.8 : 2.6 : 1.0 : 0.2 (Doroshenko *et al.*, 1978). In normal conditions no measurable conductance of calcium channels for monovalent cations has been found.

The instantaneous current–voltage characteristics of currents through calcium channels are largely nonlinear: they reach exponentially a zero value at high positive testing shifts of the membrane potential. When testing potential is higher than the corresponding equilibrium potential sodium and potassium currents undergo reversal of direction. No such reversal is found for calcium currents.

Such potential-dependence of calcium currents would be expected if the intracellular concentration of free Ca^{2+} ions is very low; in the simplest case when the intramembrane electric field is constant

$$I_{Ca} = P_{Ca} \frac{4F^2 V}{RT} \cdot \frac{[Ca]_o - [Ca]_i \exp(-2EF/RT)}{1 - \exp(-2EF/RT)} \tag{10}$$

where $[Ca]_i \to 0$.

Indeed, under natural conditions, their concentration in the axoplasm of giant axon is about 5×10^{-8} M (Baker, 1976): probably, this value is close to that in the cytoplasm of the nerve cell. So, the equilibrium potential for the calcium currents should be about 200 mV, which is practically unattainable for experimental membrane potential shifts.

Calcium current reversal might be more easily achieved by artificially increasing the intracellular Ca^{2+} content during intracellular perfusion. This is carried out by introducing into the cell buffer solutions such as Ca-EGTA buffer which maintain the Ca^{2+} ion concentration at any desired level. However, tests have shown that increase of internal free calcium ions to a level as low as 10^{-7} M has a reversible blocking effect on calcium channels (Kostyuk and Krishtal, 1977b), an effect which is obviously similar to that observed by Hagiwara and Naka (1964) during injection of these ions into *Balanus* muscle fibres.

The reason for the blocking effect of increased internal free calcium is not yet clear. It can be explained by highly specific binding of Ca^{2+} ions with certain groups located at the inner mouth of a calcium channel, as was suggested by Doroshenko and Tsyndrenko (1978) on the basis of the concentration dependence of the blocking effect. More complex effects of excess internal Ca^{2+} ions may occur, through effects on the metabolic systems which may sustain the conformation of calcium channels in a "working" state (see below).

According to the extensive studies of Hagiwara and Naka on the concentration dependence of calcium action potentials in the muscle membrane, ionic currents through calcium channels of the somatic membrane show a clear saturation with increase of concentration of carrier ions in the external solution. This undoubtedly indicates the presence of a binding site at the outer mouth of a calcium channel. Akaike et al. (1978) and Valeyev (1979) estimated on the nerve cell membrane the dissociation constants for such a binding of divalent cations, using the equation (6) proposed by Hagiwara, which are, correspondingly, $K_{Ca} = 5.4$ mM, $K_{Sr} = 10$ mM and $K_{Ba} = 15$ mM.

Due to the presence of this binding site, cations having higher binding capacity block the calcium channel conductance for less effective ions. These blockers are Mg^{2+}, Ni^{2+}, Co^{2+} and Mn^{2+}, whose depressive effect on the inward current can be successfully described by the Langmuir isotherm; the dissociation constants for the ion channel complexes, estimated in the presence of 4 mM Sr^{2+}, are as follows: $K_{Mg} = 18.2$ mM, $K_{Ni} = K_{Co} = 0.74$ mM and $K_{Mn} = 0.36$ mM. The study of their blocking effects at varying concentrations of carrier ions confirms the suggestion about its competitive nature (Ponomaryov et al., 1980). The most effective blocker of calcium channel is Cd^{2+}; in the presence of 4–30 mM Ca^{2+} in the external solution, $K_{Cd} = 0.07$ mM (Krishtal, 1976).

Data on the effect of internal fluoride on calcium channels are of great interest. Introduction of these anions into the cell in concentrations of the order of several tens of millimolarity produces an irreversible block of calcium currents, though other ionic currents (sodium, potassium) are not affected and the cell membrane remains even more stable than after the introduction of other anions (Kostyuk et al., 1975; Kostyuk and Krishtal, 1977a). Fluoride ions also affect the displacement of gating charges related to the activation of calcium channels (see below). A similar effect of fluoride anions has been described in the oocyte membrane (Takahashi and Yoshii, 1978). A peculiar feature of this effect is its strong dependence on temperature: at low temperatures the blocking effect of fluoride on calcium conductance develops much more slowly, and at about 7° C even

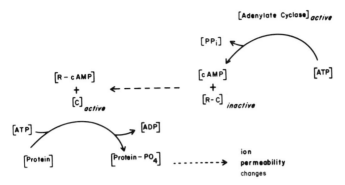

Fig. 1. *Schematic diagram illustrating the possible participation of enzymatic processes in the regulation of membrane permiability (according to Greengard).*

perfusion for one hour does not produce such a blockade. This observation supports the suggestion that the effect of fluoride anions is not due to their physico-chemical interaction with binding sites in the calcium channel, but to more complex conformational changes in the channel structure, which may result from disturbances of cellular biochemical processes that normally maintain calcium channels in their functional state. This is an alternative explanation of the exclusively high sensitivity of calcium channels to increase in internal concentration of calcium ions.

There are ample indications that intracellular Ca^{2+} ions are closely related to the regulation of some enzymatic processes, which, in turn, are involved in the maintenance of ion conductance of the membrane (for details see, for instance Rasmussen and Goodman, 1977). The sequence of reactions which may influence the functional state of ionic channels in the membrane surface by phosphorylation of their constituent proteins has been generally recognized (Fig. 1).

It is known that Ca^{2+} ions may play an essential role in the course of each of these reactions and may either activate or suppress them, depending on concentration. Fluoride anions may also have an effect on the course of this cycle by influencing the adenylatecyclase link. One may think that calcium channels are especially dependent on phosphorylation of their internal functional groups and that disturbance of these processes inactivate them. Recent experiments support this suggestion.

A peculiar feature of calcium channels, unlike sodium or potassium channels, is their fast inactivation during cell perfusion. As observed in early experiments with cell perfusion, calcium currents decreased rapidly during the experiment, while sodium and potassium currents in the same cell remained stable. Obviously, this difference may be due to the higher sensitivity of calcium channels to the disturbance of the normal cource of intracellular biochemical reactions that is inevitable during cell perfusion. Veselovsky and Fedulova (1980) have shown, in the spinal ganglion neurons of the rat, that introduction of cyclic adenosine-3, 5-monophosphate (cAMP), ATP and Mg^{2+} ions into the cell prevents the development of calcium channel inactivation and restores the diminished calcium conductance to a considerable extent. Separate introduction of ATP or Mg^{2+} ions into the cell also partially restored calcium conductance, but it was considerably less stable. Sodium and potassium currents have not been influenced by ATP or Mg^{2+}. It would be interesting to conduct a more detailed study of specific influences on different stages of intracellular cyclic nucleotide metabolism which may affect the functional state of calcium channels.

Existing data on the properties of open calcium channels can be used to model the energy profile in these channels using Eyring's theory of absolute reaction rates for the description of ion penetration through the channel as a series of its successive binding with functional groups of the channel separated by energy barriers. Such an approach had been employed earlier by Chizmadjev and colleagues (Chizmadjev et al., 1974; Chizmadjev and Aityan, 1977) and Hille (Hille, 1975; Hille and Schwartz, 1978) to predict the energy profiles in sodium and potassium channels. Using an analogous approach, the process of ion transport in calcium channels was first theoretically considered by Naruševičius and Rapoport (1979) and Yasui and colleagues (Yasui et al., 1979). A more detailed analysis has since been performed by Kostyuk et al. (1980). These authors suggested the presence of two main binding sites (external and internal) in the calcium channel and of three corresponding main energy barriers. It was also assumed that only one permeable ion may exist at any given moment in the channel; this suggestion is in accordance with data on the potential-dependence of the calcium channel conductance. The calculations of steady-state current through the channel have been made using the diagram technique (see Markin and Chizmadjev, 1974). The diagram (Fig. 2) presents the full set of possible states of the system ion-ionic channel and of the transition rates between them. The resulting ionic flow through the channel can be calculated by computer minimization of the mean square deviation of the calculated current–voltage characteristics from the experimental ones, varying the heights of these barriers and the depth of potential wells which, in turn, determine the transition rate constants.

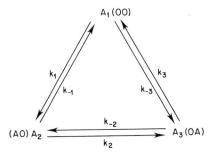

Fig. 2. *Schematic diagram of possible states of the system ion–ionic channel for the calcium channel. O, A: occupation numbers of the outer and inner binding sites; K_i: transition rate constants.*

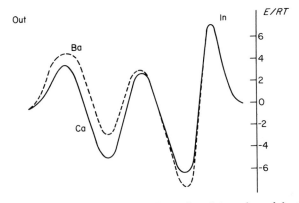

Fig. 3. *Schematic diagram of the energy profile in the calcium channel for Ca^{2+} and Ba^{2+} ions.*

Figure 3 presents the results of such calculations for calcium and barium currents through a calcium channel. A good correlation was obtained between theoretical and experimental current–voltage characteristics for different concentrations of carrier ions. Only for large values of membrane depolarization did the calculated amplitude of the inward current systematically exceed the experimental one. Such a discrepancy is likely to be due to the activation of the above mentioned outward current which appears at large depolarization shifts and is carried by $Tris^+$ ions used for replacing intracellular cations. The standard deviation of the model parameter from their mean values obtained by measurements on several nerve cells did not exceed 10%. The model obtained has been used to predict the current–voltage curves through calcium channels for various volume concentration of penetrating ions; the prediction turned out to be quite satisfactory.

P. G. KOSTYUK

TABLE I. The pK values for H^+, Ca^{2+} and Ba^{2+} ion complexes with different anion residues of proteins in aqueous solutions (from Martell and Smith, 1977) and with functional groups of calcium channel.

Ion species	$-O^-$	$-S^-$	(pyridine)	$-COO^-$	$(-COO^-)_2$	$(-COO^-)_3$	pK_o	pK_i
H^+	9.3	8.4	6.0	2.4	4.3	6.3	5.8	—
Ca^{2+}	1.4	—	—	1.4	3.0	4.0	2.0	2.6
Ba^{2+}	0.8	—	—	0.8	2.4	2.8	1.0	2.8

The model described above has also been used for the description of experimental results of the effect of extracellular pH changes on the calcium currents. As with other transmembrane current, the calcium currents are blocked by lowering the pH of the external solution; this is probably due to protonation of the corresponding functional groups of the channel. The pK of dissociation of protons from the outer binding site was found to be 5.8, which corresponding to a dissociation constant of the order of 0.0016 mM.

Table I presents the pK of dissociation of Ca^{2+}, Ba^{2+} and H^+ ion complexes with binding sites at the outer and inner mouth of the calcium channel compared with those of different anions in groups which may be present in the protein macromolecules forming the ionic channel (according to Martell and Smith, 1977). This comparison points to the conclusion that there is only one carboxylic group both in the outer and inner mouth of the calcium channel responsible for the binding of the penetrating ions. The pK_H for other groups is over 6, and the pK for complexes of Ca^{2+} and Ba^{2+} ions with two or more carboxylic groups exceeds the pK values obtained in experiments on calcium channels. This is confirmed by the order of binding of different ions by carboxylic groups in aqueous solutions, which is as follows: $Ni^{2+} > Co^{2+}$ $La^{3+} > Ce^{2+} > Mn^{2+} >$ $Ca^{2+} > Ba^{2+}$ Sr^{2+} (according to Martell and Smith, 1977), i.e. similar to the order obtained for the calcium channel of the neuronal membrane.

The possible structure of the calcium channel can be compared with that of other electrically operated channels on the basis of the results obtained. It follows from the data of Hille (1975) that the active centre in the outer mouth of the sodium channel of the axonal membrane also includes a carboxylic group. The values of $pK_H = 5.3$ (Hille, 1975) and $pK_{Ca} = 1.7$ (Woodhull, 1973) for the sodium channel are close to the values given above for the calcium channel. The drugs which are specific blockers of these channels are chemically similar. So, tetrodotoxin (blocker of sodium channels) and niphedipine (blocker of calcium channels), according to Fleckenstein et al. (1975), have an $=N-H$ group in their structure which is bound covalently with a large carbohydrate residue. This $=N-H$ group is likely to be the active centre of those toxins which interact with the outer mouth of the channels. The derivatives of verapamil, which contain a tertiary nitrogen atom ($\equiv N$), also have a certain structural similarity.

At the same time there is a considerable difference between the structures of sodium and calcium channels. If the amplitude of transferred current in the sodium channel is determined mainly by the height of the potential barrier inside the channel corresponding to the "selectivity filter" (Hille, 1975), in the calcium channel it depends on the depth of the potential well (the effectiveness of ion binding) at the outer mouth. Less stable binding of Ba^{2+} and Sr^{2+} ions in comparison with Ca^{2+} ions produces larger barium and strontium currents compared with calcium ones. Na^+ ions bind much more weakly to carboxylic groups than do divalent ions and therefore need to produce much stronger currents. The fact that these ions cannot penetrate the channel indicates the presence of a "selectivity filter" in the calcium channel in front of the site of primary binding of the carrier ion; this filter somehow eliminates monovalent cations, preventing them from penetrating inside the channel.

Data on the possible modification of calcium channels into sodium ones by chemical effects are in favour of this suggestion. The introduction of Ca-chelating agents (EGTA or EDTA) in a calcium-free external medium induces in the somatic membrane an intense sodium inward current which has, however, the kinetic characteristics of the calcium current (Kostyuk and Krishtal, 1977b; Krishtal, 1978). The current thus induced can be immediately blocked by the introduction of Ca^{2+} ions in the solution, as well as by those pharmacologically active drugs which block the calcium channels (verapamil, D-600). There are reasons to believe that these phenomena are associated with a modification of the calcium channels which destroys the selectivity filter located at its input, so that Na^+ ions can effectively pass into them. This modification is probably due to the interaction of the chelating agent with those functional groups of the channel which usually interact with Ca^{2+} ions.

It should be noted that a satisfactory description of the experimental results, using a three-barrier model, could be obtained only at considerably smaller values of the binding constants for ions in the inner potential well than those which correspond to the data for dependence of calcium conductance on the increase in internal concentration of free Ca^{2+} ions. If the conductance of a single calcium channel is estimated according to this model, assuming the depth of the potential well at the inner mouth of the calcium channel is $14\,RT$ (according to the data for a complete block of calcium conductance at internal concentration of free Ca^{2+} ions of 10^{-7}M), then the value of ionic current through a single channel would be only 10^{-15}–10^{-16} A. However, the experimental value obtained from the measurements of membrane noise (see below) is 2–3 orders higher. This favours the idea that blocking of calcium channels at very low concentrations of Ca^{2+} ions inside the cell is due to biochemically mediated conformational changes in the structure of the channel, rather than to the filling of the internal binding site in the channel by these ions. Such changes, resulting in inactivation of calcium conductance, develop at those values of $[Ca^{2+}]_i$ where the occupation number of the internal well of the calcium channel is still much less than unity.

ACTIVATION AND INACTIVATION KINETICS OF CALCIUM CHANNELS

A prominent feature of calcium currents is the delayed rise with time at corresponding depolarizing displacement of the membrane potential (compared with sodium current). Therefore such currents could be designated as "slow".

A quantitative study of the activation kinetics of the calcium currents in the membrane of *Helix* neurons (Kostyuk and Krishtal, 1977a; Kostyuk *et al.*, 1979) and of spinal ganglia neurons (Veselovsky and Fedulova, 1980) has shown that they can be satisfactorily described by the Hodgkin-Huxley model, using the square power of their *m* variable. Similar results have been obtained on *Aplysia* neurons but in this case the approximation was not so good (Adams and Gage, 1979b).

Attempts to approximate calcium currents by the Hodgkin-Huxley model have been undertaken for other excitable membranes, such as the presynaptic membrane of the squid giant synapse (Llinas *et al.*, 1972) and the crayfish muscle membrane (Henček and Zachar, 1977). In these cases satisfactory approximation was possible only by using higher powers of the *m* variable (the fifth or sixth). These findings may indicate differences in the "gating" mechanism of calcium channels in these membranes; however, more complex activation kinetics for the calcium currents may be apparent and may be due to residual potassium currents which could not be eliminated without intercellular perfusion.

Another distinct feature of calcium currents is a slow inactivation during sustained depolarization of the cell membrane, with a complex time course. In perfused mollusc neurons two exponential components in the inactivation of calcium current could be separated — the fast (with a time constant of some dozens of milliseconds) and the slow (with a time constant of several hundredths of milliseconds) one. The initial fast decline of calcium current can be erroneous due to activation of the nonspecific outward current mentioned above. In favour of this suggestion are the data obtained by lowering of pH in external medium (which to a considerable extent slows down the activation of nonspecific outward current and shifts its current–voltage characteristic to more positive potentials): this results in a considerable increase of the time constant of the fast component, while that of the slow component remains unaffected (Doroshenko *et al.*, 1978). True inactivation of calcium current is believed to be a very slow process described by the second exponential component.

Views differ on the nature of such inactivation. On the one hand, inactivation of calcium channels, as of sodium of potassium ones, may result from a special gating mechanism which prevents penetration of ion across the channel. On the other hand, according to the above data on the powerful blocking effect of increased concentration of internal Ca^{2+} ions on calcium conductance, Ca^{2+} ions penetrating into the cell may themselves exert an autoblocking effect on calcium channels. Data on the dependence of inactivation of calcium currents on the amount of Ca^{2+} ions brought into the cell during previous depolarization of the membrane are in favour of the latter viewpoint. In experiments when the test shift of the membrane potential was kept constant and the level of preceding depolarizing shift was varied, inactivation of calcium current increased with increase of the latter shift; the potential-dependence of inactivation coinsided with the potential-dependence of the current–voltage characteristics of calcium current. In other words, the degree of inactivation depended not on the membrane potential but on the quantity of Ca^{2+} ions which had previously penetrated into the cell. Sr^{2+} or Ba^{2+} ions produced much less pronounced inactivation of calcium current, despite the fact that the ionic current carried by these ions exceeded twice the calcium current (Brehm and Eckert, 1978; Tillotson, 1979).

Changes of calcium currents observed under steady-state conditions at sustained shifts of the holding potential favour potential-dependent inactivation of calcium channels. As with other ionic channels, an S-shaped dependence of calcium current on the membrane potential ($h_\infty(V)$ function) can be seen; however, the curve is displaced to more positive potentials compared with those for sodium currents in the same membrane; complete inactivation of calcium currents does not occur even at high and prolonged positive shifts of the membrane potential.

This may indicate heterogenity of the population of calcium channels in the neuronal membrane and the possible existence of special "slow" calcium channels which are inactivated especially slowly.

Data indicating the existence of slowly developing calcium currents have been obtained on "bursting" neurons. Such currents have been recorded by depolarizing shifts which were subthreshold to activate faster inward currents. At high depolarizations, fast currents masked the slow calcium currents, but at "off" depolarizing shifts, the latter were revealed by very long current "tails" (Eckert and Lux, 1975; Thompson and Smith, 1976). This component of calcium current is sometimes designated I_B.

The elucidation of membrane mechanisms underlying the kinetic characteristics of calcium currents has made substantial progress since the discovery of the intramembrane charge displacement that accompanies activation of the ionic conductance of the membrane (the so-called "gating currents"). Successful separation and analysis of gating currents related to the activation of sodium conductance (Armstrong and Bezanilla, 1973; Keynes and Rojas, 1973) led to hopes for similar success regarding calcium channels. In fact, Adams and Gage (1976), by analysing charge displacements in the membrane of *Aplysia* neurons (under conditions of a block of ionic transmembrane currents), have found a component which, because of its slow time development, could correspond to the activation of calcium channels. Kostyuk *et al.* (1977) discovered an important property of this component; similar to calcium conductance, it could be abolished by internal perfusion of the cell with fluoride-containing solutions. Thus, the total displacement current may be reliably separated into components which reflect the function of different types of ionic channels.

A detailed analysis of displacement currents in the membrane of perfused mollusc neurons (Kostyuk *et al.*, 1979) gave the following results. The fluoride-sensitive component of this current corresponds well in its potential and kinetic characteristic to the presumed "gating current" of calcium channels. The amount of charge displaced at the beginning of the depolarizing shift (on-response) corresponds to that displaced in the opposite direction at the end of depolarization (off-response) as has to be the case if such displacement occurs inside the membrane. With increasing membrane depolarization, the amount of displaced charge approaches saturation at the same potential value, which produces saturation of the calcium current; the maximum amount of displaced charge is of the order of 1500–2000 electron charges transferred per square micrometre of the membrane (through its whole thickness). The voltage dependence of the amount of transferred charges during on-gating current and the voltage dependence of the m variable of the calcium conductance (calculated as the square root from the normalized calcium conductance according to the m^2 model) coincide satisfactorily. Calculation of the maximum logarithmic slopes for the $Q(V)$ and $G(V)$ functions gave values of 8 mV and 4 mV respectively for an e-fold shift of the displaced charge. If one assumes that such charge displacement represents the Boltzmann distribution of charged particles between two steady-state positions, then the effective valency for displaced charge can be determined and is equal to 3. Thus, the data obtained are in good agreement with the model, according to which two "gating" particles (each having an effective valency of 3) should be simultaneously displaced to open the calcium channel.

Correlation is also observed between the kinetic characteristics of intramembranal charge displacement and the activation of calcium conductance. The time constant of the on-displacement current decline (τ^{as}) coincides with the time constant of ionic current activation on (τ_m) over a wide range of membrane potential shifts; discrepancies are observed only at small depolarizations. Kinetic studies of the off-displacement current have shown that the ratio between its time constant and the time constant of relaxation of the calcium current is close to 2. This accords with the suggestion that two gating particles are displaced in order to open the channel; a reverse movement of only one of them closes the channel.

Finally, stationary decrease of the holding potential produces a decrease of the gating current which occurs in parallel with the inactivation of calcium conductance, as it results from the idea about the "immobilization" of gating particles as the base for inactivation of ionic channels.

Thus, the gating mechanism of calcium channels is, in principle, similar to the gating mechanism of sodium channels, differing only in certain quantitative characteristics (i.e. the number of suggested gating particles). This conclusion should be considered as a first approximation to an understanding of the mechanism of activation of calcium channels. This mechanism may involve one or more intermediate stages whose transition is not associated with appreciable intramembranal charge displacements. A short rising phase in displacement current and certain other observations are complementary lines of evidence in favour of intermediate stages.

STUDIES OF SINGLE CALCIUM CHANNELS

Recently, special attention has been focused on current noise in electrically excitable membranes as a tool for analysing the behaviour of single ionic channel. Such noise represents the discrete nature of the molecular structures which create transmembrane ionic currents. It can be discovered if the number of structures participating in the total reaction is considerably decreased. The efficiency of noise analysis is determined mainly by the value of the single channel conductance. At high enough values of this conductance current fluctuations produced by the opening or closing of single channels may predominate over other kinds of noise (thermal noise, amplifier noise, etc.). For this reason, the analysis of noise produced by activation of chemically operated ionic channels has been most effective; so, the conductance of acetylcholine-operated ionic channel in the muscle fibre end-plate is about 23 pS and the current fluctuations produced by the opening or closing of a single ionic channel can easily be recorded. However, the analysis of noise produced by sodium channels in electrically excitable membrane has shown that the unitary conductance of these channels is one order less than that of the acetylcholine operated channel. It was thus difficult to analyse current fluctuations produced by the opening of single ionic channels; their behaviour can be determined only by statistical analysis of current noise (Conti et al., 1976; Sigworth, 1977).

Similar statistical analysis of the calcium noise of the neuronal membrane has been carried out by Kostyuk and his colleagues (Kostyuk et al., 1978, 1980) and by Akaike and others (Akaike et al., 1978). In the former case fluctuations of transmembrane current were recorded from a micropatch of the membrane sucked into a micropore of the plastic membrane. To increase the contribution of currents in single channels to the total current, the recording was made when unitary current was maximal. For this purpose, external Ca^{2+} ions were replaced by Ba^{2+} ions, which are more permeable;

also the Ba^{2+} ions were introduced in isotonic or even hypertonic concentration. It was then possible to record fluctuations of the order of 10^{-11} A, i.e. sufficient for statistical analysis. In the experiments of Akaike and colleagues the currents were recorded over a large area of the cell membrane.

Statistical analyses of fluctuations have shown that their mean square deviation at first increases with the increasing membrane depolarization, reaching a maximum at a half-value of the recorded current, and then decreasing with further depolarization. Such dependence is to be expected if the current fluctuations reflect random opening or closing of the channels. Therefore, a simple evaluation of a single channel conductance could be carried out which, in the case of a binominal distribution, is as follows:

$$i = \frac{\sigma^2}{\langle I \rangle (1 - \langle 1 \rangle / I_{max})} \tag{11}$$

where $\langle I \rangle$ = mean value of the inward current, σ^2 = mean square deviation of fluctuations, I_{max} = maximal inward current. With 130 mM Ba^{2+} in external solution the current through the single calcium channel was 0.20 ± 0.02 pA (at $22°$ C). If the ratio of permeabilities for Ba^{2+} and Ca^{2+} ions in the single channel is the same as that measured for the total current, the calcium current through the same channel will be about 0.1 pA and the conductance of the channel, 0.5 pS.

The conductance of a single calcium channel can also be evaluated in another way. Since the charge transferred by the gating mechanism of calcium channel is known (see above), the current through the single channel can be determined by the value of displaced charge and the value of calcium current through a certain membrane area is as follows:

$$i_{Ca} = \frac{2 z' e I_{Ca}^{max}}{Q_{max}} \tag{12}$$

where I_{Ca}^{max} = maximum calcium current, Q_{max} = maximum value of displaced charge, e = the value of elementary charge and z' = the effective valency of the gating particle. By this method, the current through a single channel was estimated as 0.02 ± 0.01 pA (Kostyuk et al., 1980) and the density of a calcium channel was $250-330/\mu m^2$. Since in this case the current was measured at a normal concentration of external Ca^{2+} ions, the value thus obtained was close to that obtained from noise analysis.

From these values, the single calcium channel can pass at most 6×10^5 barium or 3×10^5 calcium ions per second.

To obtain time characteristics for the activity of single calcium channels, the spectral analysis of the corresponding membrane noise was obtained and then subjected to Fourier transformation. Several spectra were recorded during depolarization, averaged and then subtracted from those recorded at the holding potential level (to eliminate noise from nonrelaxation origin).

To approximate to stationary conditions, measurements were carried out some time after the onset of depolarization when, due to exponential development of inactivation, the current level already changed slowly. The spectra thus obtained were approximated by the Lorentz function characteristic of the processes with a single relaxation time constant:

$$S(f) = \frac{S(0)}{1 + (f/f_{1/2})^2} \tag{13}$$

where $S(0)$ = spectrum density at zero frequency, $f_{1/2}$ = cut-off frequency at which $S(f_{1/2}) = 1/2 S(0)$. The mean value of the relaxation time constant for the channel determined as $\tau_i = 1/2 \pi f_{1/2}$ is equal to 0.7 ± 0.2 ms for the calcium channel of the membrane of the *Helix* neuron (at 22° C) (Kostyuk *et al.*, 1980).

The relaxation time of the calcium channel determined from the spectral characteristics of the current fluctuations is only slightly dependent on the testing potential. This does not conform with the predictions of the model that displacement of the gating particle directly transfers the channel from a nonconducting state to a conducting one. Such a discrepancy obviously indicates the existence of intermediate stages during activation of ionic conductance. The displacement of charged gating particles is a necessary but insufficient condition in this process; it only creates the conditions for realization of the next stage — transition from opening to closing which is independent of the potential. Such a scheme would account for the change in the ratio between the ionic and the gating currents of calcium channels during prolonged perfusion of the cell. During such perfusion the latter step appears to be selectively blocked without disturbing the movement of gating particles. Therefore, the correlation between calcium conductance and the amount of transferred gating charges changes in favour of the latter (Kostyuk *et al.*, 1980).

CALCIUM CHANNELS IN EMBRYONIC CELLS

Calcium conductance is a primary property of the cell membrane which can readily be seen in oocytes. Microelectrode studies have shown oocytes to possess the mechanism of electrical excitability; the action potential which appears can be preserved in sodium-free medium and is at least partly of calcium origin (Miyazaki et al., 1974, 1975). The peculiarities of ionic dependence of the action potentials have been studied in parthenogenetically differentiated egg cells of an annelid, *Chaetopterus* (Hagiwara and Miyazaki, 1977). The calcium-dependent action potential in the untreated egg cell becomes Ca and Na dependent in the ciliated embryo stage. In dissociated tissue cultures of the nerve lamina of *Xenopus* tadpole also the action potentials are calcium dependent, then become Ca–Na dependent and, finally, become Na dependent only (Spitzer and Lamborghini, 1976). Okamoto et al. (1976a, b, 1977), using the voltage clamp technique on the egg membrane of tunicates, sea urchins and rats, have shown the presence of fast and slow inward currents, the fast being carried by Na^+ ions, and the slow one by Ca^{2+} ions. The difference between these currents in egg cells of these animals is quantitative but not qualitative. The application of a modified method of intracellular dialysis (Takahashi and Yoshii, 1978) enabled both currents in the tunicate egg cells to be analysed in detail. They were similar to sodium and calcium currents of differentiated nerve cells in all their properties. In particular, the calcium current was characterized by the absence of a clear-cut reversal potential and was blocked by the introduction of fluoride anions into the egg cell. This current could be obtained in a pure form by lowering the holding potential to -34 mV, when the sodium current became completely inactivated.

Data on the calcium conductance of nerve cells during differentiation were obtained in rat and mouse dorsal root ganglia neurons isolated immediately after birth and at different periods of postnatal development. Matsuda et al. (1978) and Yoshida and Matsuda (1980) made action potential recordings on these neurons in sodium-free solution and came to the conclusion that only some neurons in newborn animals are capable of generating "calcium" action potentials. In adult animals such a recording could not be obtained. This may indicate that calcium conductance is characteristic of the membrane of undifferentiated nerve cells only. However, Veselovsky and Fedulova (1980) succeeded in detecting the calcium in *all* the neurons investigated by direct measurement of ionic currents in perfused neurons of rat dorsal root ganglia, though their maximal amplitude varied considerably from cell to cell. Apparently, in many neurons, these currents are not sufficient to produce regenerative response of the membrane, and their presence cannot be observed by recording action potentials. Moreover, it is possible to record calcium currents in the somatic membrane of both newborn and adult animals. Thus, in mammals the calcium conductance cannot be regarded as occurring only in undifferentiated nerve cells; such a conductance is always a component of the functional mechanism of nerve cell soma, though it is possible that its expression may vary at different stages of cell differentiation. In this connection, recent studies of Fukada and Kameyama (1979, 1980) are of great interest. These authors investigated the ion dependence of action potentials in mammalian neurons cultivated in artificial medial for a long time. They found that the percentage of cells which show a calcium component in the action potential is constant after 17 days of cultivation. At the same time, a change in the rate of rise of calcium spikes was observed, in parallel with the intensity of the growth of cell processes. On the 5th–6th day of cultivation, when this intensity reached its climax, the rate of rise became maximal; then it declined to the initial level by the 30th day of cultivation. No changes in sodium spikes were seen in such cases. Therefore, the authors came to the conclusion that the growth of processes in the nerve cell is accompanied by an increase in the density of calcium channels in the surface membrane or by conformational changes in their structure which enhance the conductance.

CALCIUM CURRENTS IN THE POSTSYNAPTIC MEMBRANE

Though electrically operated calcium channels are the main route of calcium entry into the cell, inward current of these ions mediated by chemically operated ionic channels may also occur in postsynaptic membranes. Studies of ionic currents underlying the changes in postsynaptic potentials caused by transmitter substances have shown that such currents are much less ion-specific than the currents of electrically excitable membrane; therefore there are reasons for believing the currents can be transferred by both monovalent and divalent cations.

The calcium transmembrane current through the postsynaptic membrane in the end-plate of frog striated muscle fibre resulting from acetylcholine activation has been measured by use of the Ca-sensitive reagent arsenazo III. When this reagent is introduced into the region of the end-plate increase of internal Ca^{2+} content during acetylcholine application is detected. Penetration of the Ca^{2+} ions was nonlinear with membrane hyperpolarization, and the introduction of Co^{2+} ions into the external medium caused a decrease in calcium flow, which also became more linear (Miledi *et al.,* 1980).

A more detailed analysis of the penetration of Ca^{2+} ions through acetylcholine-operated ionic channels has recently been undertaken by Adams *et al.* (1980). These authors introduced ion species of differing size into the external solution, measured the postsynaptic currents and concluded that these ionic channels represent a pore with a cross section of about 0.65 nm X 0.65 nm. The penetration of ions through such a pore is less dependent on specific chemical factors than is the case in electrically operated channels. Accordingly, these channels show a weak selectivity for divalent cations and pass them in the preferred sequence $Mg^{2+} > Ca^{2+} > Ba^{2+} > Sr^{2+}$. Moreover, Co^{2+}, Ni^{2+}, Zn^{2+} and Cd^{2+} ions, which are effective competitive calcium current blockers, penetrate chemically operated calcium channels. This circumstance, as well as the finding that relative permeability of acetylcholine-operated channels for divalent cations is in reverse order to that for calcium channels, undoubtedly show a different way of penetration of Ca^{2+} ions in principle. Evidently there are no specific binding sites for cations in the acetylcholine operated channel and cation penetration is determined mainly by cation diffusion in an aqueous medium.

In normal postsynaptic membrane the inward calcium current must be relatively small compared with sodium and potassium currents due to the much lower concentration of the corresponding carrier ions. However, the presence of fixed negative charges on the outer surface membrane may enhance the contribution of Ca^{2+} ions due to preferential accumulation of these ions near the membrane by fixed charges: therefore their effective concentration at the channel mouth may be increased relative to the concentration of Na^+ ions.

No detailed information is available about possible calcium currents through ionic channels activated by other transmitters.

As mentioned above, catecholamines increase the permeability of heart muscle fibre membrane to Ca^{2+} ions. This has been shown both by measuring penetration of labelled ions into muscle fibre and by recording changes in the shape of action potentials. An additional calcium inward current initiated by catecholamines has been found to be associated with activation of potential- and time-dependent conduction, i.e. apparently with additional activation of electrically operated calcium channels. This was confirmed by detailed study of the effect of adrenaline on heart muscle carried out under voltage clamp conditions (Reuter and Scholz, 1977). It has been shown that adrenaline does not affect the kinetic characteristics of calcium but increases considerably the limiting calcium conductance. The reversal potential of the slow inward current is not changed. All these data indicate that the effect of adrenaline does not result in changes of selective or kinetic properties of the calcium channels and it can best be explained by assuming that additional channels become operative at the expense of some kind of reserve.

There are several hypotheses for this effect. It may be thought that the calcium channels themselves have β-adrenoreceptive groups, whose interaction with adrenaline is a necessary link in the process of potential-dependent activation of the calcium current. However, activation of a large proportion of calcium channels is possible without the presence of catecholamines, as well as in the presence of β-receptor blockers; therefore this explanation requires that different populations of catecholamine-sensitive and catecholamine-insensitive channels be present.

An hypothesis of the indirect influence of catecholamines on the calcium channels mediated by an intracellular system of cyclic nucleotides is now widely used. It was shown in heart muscle membrane that all effects which cause an increase in cAMP content in the muscle also increase the inward calcium current (Tsien et al., 1972; Reuter, 1974). Since catecholamines are known to stimulate the activity of the membrane adenylatecyclase (Sutherland and Robinson, 1966), it is reasonable to suggest that additional calcium channels become activated due to the activation of cAMP synthesis which, in turn, facilitates the protein kinase mediated phosphorylation of the channel-forming membrane proteins. The calcium-binding proteins may also play some part in the calcium-dependent regulation of adenylatecyclase activity (Brostrom et al., 1975).

Extensive experimental results have been obtained concerning the influence of synaptic transmitters on the cyclic nucleotide content of nerve cells. Kakiuchi and Rall (1968) were first to obtain these data, which have been repeatedly confirmed in various systems (see reviews of Rall and Sattin, 1970; Daly, 1973; Perkins 1975). It was found that the effects of catecholamines can be due to both β- and α-adrenoreceptors, as well as to adenosine receptors. It was suggested that membrane adenylatecyclase may be related to a receptor group which causes the activation of enzyme only when it binds simultaneously two different agonistic molecules (Sattin et al., 1975).

By analogy with heart muscle, these data present a complementary evidence in favour of the idea that the changes in the permeability of the nerve cell membrane, caused by the catecholamines or possibly by other transmitters, are mediated by the cyclic nucleotide metabolism system. However, all these suggestions lack precise data on the changes in membrane conductance which might develop parallel to the changes of cyclic nucleotide levels in the cell or to the changes in activity of membrane protein kinases. Therefore, the results given above, which indicate that controlled introduction of cAMP into the cell performed on isolated perfused neurons during direct measurements of specific transmembrane currents prevents inactivation of the calcium channels occurring in the course of perfusion and contributes to recovery of their functional state (Veselovsky and Fedulova, 1980) are of large importance in this respect.

The possible presence of an adrenoreceptor group in the structure of the calcium ionic channel itself should not be ruled out, however, especially in connection with new data indicating the possibility of both β- and α-adrenoreceptors effects on calcium conductance. Thus, it has been shown in sympathetic ganglion neurons of the rat that noradrenaline may block a calcium-dependent component of the action potential (Horn and McAfee, 1979). A similar effect of noradrenaline has been found in the spinal ganglion neurons of the chick embryo (Dunlop and Fischbuck, 1978). These authors consider the presence of α-receptor groups in the structure of potential-dependent ionic channels to be the most probable explanation.

The effect of catecholamines on membrane calcium channels can also be seen in presynaptic terminals. It is known that catecholamines can produce an increase in the frequency of miniature endplate potentials and an intensification of quantal transmitter release during evoked end-plate potentials (Breckenridge et al., 1967). Such potentiation of transmitter release seems to be due to an increase in Ca^{2+} entry into the terminal across the presynaptic membrane; a parallel increase of cAMP content takes place in the latter (Singer and Goldberg, 1970; Miyamoto and Breckenridge, 1974). However, analogous studies by other authors (Wilson, 1974) led them to conclude that the increase in cAMP content is not related to the activation of calcium conductance in this case but reflects the processes of synthesis, storage and mobilization of synaptic transmitter (acetylcholine). The precise role of cyclic nucleotides and the nature of interaction between adrenoreceptive structures and the structural units of the membrane ionic channels apparently needs further clarification.

It is important to note that a serotonin-induced potential-dependent calcium conductance has been recently described in *Aplysia* neurons. This phenomenon can also be interpreted as due to possible incorporation of chemoreceptive groups in the structure of the calcium ionic channel (Pellmar and Carpenter, 1979).

FUNCTIONAL SIGNIFICANCE OF TRANSMEMBRANE CALCIUM CURRENTS

It can be suggested that calcium currents which appear in the surface membrane of excitable cells might play at least two roles: first, in the generation of active membrane reactions; second, in coupling between membrane and cytoplasmic processes.

The data presented above show that when the density of specific calcium channels in the membrane is high enough, they are capable of creating a strong inward current sufficient to recharge the membrane capacity and to cause a regenerative reaction in the form of a propagating action potential. Moreover, the calcium component of the inward current can exceed considerably the value of the sodium component and the action potential may therefore be "calcium" by origin even at the usual concentration of electrolytes in the external medium. This peculiarity is characteristic of many invertebrate neurons and may be related to the ecological conditions of their existence. So in comparison with the freshwater molluscs *Planorbis* and *Limnea* (Gerasimov *et al.,* 1964), the neurons of the terrestrial snail *Helix* are more prone to generate "calcium" action potentials, since they have more pronounced calcium currents (Kostyuk and Krishtal, 1977a). It should be emphasized that in the cell soma of these animals the generation of a propagating impulse is based mainly on the mechanism of calcium conductance, while in the axon the mechanism of sodium conductance is used (Wald, 1972). The reason for an effective use of calcium channels for generation of a propagating impulse in the cell soma in the absence of such a mechanism in the axon is not clear: probably, it is associated with peculiarities of the synthesis and intracellular transport of corresponding channel-forming proteins to the sites of their reconstruction in the membrane. However, there are data indicating the possibility of evoking propagating "calcium" action potentials in the axon (one of *Aplysia* neurons, Horn, 1978).

Nevertheless, in most excitable membranes the inward currents produced by the system of sodium channels are high enough to generate the active membrane response, and participation of the calcium conductance causes only a slight modification of its course. This undoubtedly indicates a predominant role for the membrane calcium conductance in the maintenance of some other cytoplasmic processes. Ca^{2+} ions are more favoured than Na^+ ions in this respect, because their high coordination number and irregular coordination geometry considerably potentiate the possibilities of their binding to biological molecules of irregular structure.

In the present chapter we shall discuss only some possible direct influences of transmembrane calcium currents on cellular physiological processes; the problems of intracellular calcium homoeostasis and its role in regulation of enzymatic processes are considered in other chapters.

Triggering of Muscle Contraction

In striated muscle fibres Ca^{2+} ions may participate in the triggering of muscle contraction due to the processes of their release and the binding by an intracellular T-system which is not considered in this review. However, in more primitive muscle cells, for example in smooth muscle cells, this system is absent and a triggering role may be exerted by those calcium ions which enter the cell directly through the ionic channels of the membrane surface.

Ca^{2+} ions can penetrate into smooth muscle fibres in two ways: through electrically operated calcium ionic channels during the action potential; and through chemically operated ionic channels opening under the influence of synaptic transmitters. Since these processes differ in their time course, two phases of Ca^{2+} flow into the muscle cells can be distinguished — fast and slow, with correspondingly two types of contractile effects. A prolonged increase in calcium conductance under the influence of chemical factors is regarded as the basis for direct "pharmacomechanical coupling" (Somlyo and Somlyo, 1968).

Calculations show that the amount of calcium ions which penetrate through the surface membrane during one action potential is sufficient to activate the contractile process in a smooth muscle fibre (Bolton, 1979). If the membrane capacity is assumed to be $1 \ mF/cm^2$, then about $3 \times 10^{-13} \ mol/cm^2 \ Ca^{2+}$ ions have to pass across the muscle fibre membrane during generation of the calcium action potential of the value of 60 mV. About $0.7 \times 10^{-4} \ cm^3$ of the cell volume corresponds to each square centimetre of membrane area. Accordingly, each action potential may increase the internal Ca^{2+} concentration by 4×10^{-6} M — quite enough to induce a considerable (though not maximum) contraction, because the threshold concentration sufficient to activate the contractile apparatus is equal to $10^{-7}-10^{-6}$ M (Endo *et al.*, 1977).

These calculations may be largely corrected if one takes into account the fact that a certain portion of Ca^{2+} ions passing across the membrane can immediately be bound to intracellular calcium binding systems. Increase of the intracellular calcium content is known to be the factor which induces release of these ions from corresponding intracellular depots.

Regulation of Intracellular Transport

In a nerve cell the transport of substances necessary for the funtioning of all its parts is closely connected to the neurotubular system and takes place in both directions at a constant rate and temperature for a given cell. At the same time, the volume of transported substances may change substantially during cell activity. The study of labelled amino acid transport from the soma of dorsal root ganglion neurons along their axons has shown that the volume of this transport decreased by 60% when the ganglion was placed in calcium-free solution; no appreciable effect of this solution was observed on the axon (Hammerschlag *et al.*, 1975; Ochs *et al.*, 1978). In parallel with changes in amino acid transport, the axonal transport of labelled Ca^{2+} ions was also influenced in the same way. It is possible that the changes described reflect the role of the calcium inward current in the regulation of cytoplasmic transport, in particular correlating the volume of transported substances with the level of cell activity (such a correlation was described by Lux *et al.*, 1970). Further studies have shown that the transfer of glycoproteins from the Golgi apparatus to the tubular transport system may be the calcium-sensitive link in this process (Hammerschlag and Lavoie, 1979). The increased activity of tyrosine hydroxylase which occurs in adrenergic nerve terminals during stimulation of nerve fibres results directly from increase in the internal concentration of Ca^{2+} ions, since this enzyme can be activated *in vitro* by Ca^{2+} ions and the activation in intact tissue is prevented by the removal of Ca^{2+} ions from the external solution (Morgenroth *et al.*, 1974).

Neurotransmitter Release

The release of neurotransmitters from nerve terminals is the final step in axonal transport which is especially closely related to the transmembrane calcium current. The most detailed quantitative study on this topic has been made on the squid giant synapse where two intracellular electrodes could be introduced due to the large size of pre- and postsynaptic elements in order to clamp the pre- and postsynaptic membrane potentials (Llinas *et al.*, 1972). Injection of aequorin allowed calcium influx in the presynaptic terminal to be detected. Depolarization of the presynaptic membrane during blockade of sodium and potassium currents by TTX and 4-amino-pyridine induced a pure inward calcium current in this case. Linear dependence between this current and the postsynaptic response (Llinas *et al.*, 1976) was observed. This observation enabled the quantitative description of the events which led to transmitter release in response to nerve fibre stimulation, though the molecular mechanism of the regulation of this process by Ca^{2+} ions still remains unclear. It has been suggested that Ca^{2+} ions have an effect on the surface of the presynaptic membranes that might facilitate their interaction and subsequent release of the transmitter into the synaptic cleft (Blioch *et al.*, 1968). It is also possible that Ca^{2+} may interact with contractile proteins in the presynaptic terminals which participate in this process (Puszkin *et al.*, 1972; Blitz and Fine. 1974).

Experiments in replacement of Ca^{2+} ions by other cations in the course of transmitter release are of undoubted interest for the solution of this important question. Calcium channels in the membrane of the presynaptic fibre, and in the channels of somatic membrane have the same relative permeability for different divalent cations; the calcium current in the former membrane is subjected to competitive blockade by other divalent cations and by H^+ ions as well. This was specifically shown by measuring the uptake of labelled Ca^{2+} ions by isolated synaptosomes (Nachshen and Blaustein, 1979). No quantitative difference has been found between Ca^{2+} and Sr^{2+} ions in their ability to produce transmitter release upon depolarization of nerve terminals (Mellow, 1979). This favours the physico-chemical rather than the conformational character of ion interaction inside the terminal.

Regulation of the Surface Membrane Conductance

Recurrent influence of calcium ions on the ionic permeability of the cell membrane may be regarded as a specific physiological function of Ca^{2+} ions passing through the surface membrane. As has been discussed above, Ca^{2+} ions have an autoblocking effect on membrane calcium conductance which is likely to be a kind of intracellular negative feedback mechanism contributing to the maintenance of intracellular calcium homoeostasis. At the same time, increased intracellular Ca^{2+} content results in a change of the surface membrane permeability to other ions. In 1974 Meech (1974) found that intracellular Ca^{2+} injection increases potassium conductance in *Helix* neurons. A detailed study of this phenomenon (Meech and Standen, 1975) led to the idea that the inward calcium current produced by excitation of the somatic membrane may be the direct factor which activates potassium ionic channels in the surface membrane; the result of such an activation of potassium channel is a bend in the current–voltage curve of the outward current (the so-called N-shaped characteristic) that occurs in the region of activation of the inward calcium current. It was suggested that specific Ca^{2+}-activated potassium channels distinct from the usual potential-dependent channels may exist in the membrane. The effect of Ca^{2+} ions was highly specific and could not be reproduced in intracellular injection of other divalent cations (Hermann and Gorman, 1978).

Similar effects have been observed in other excitable tissues, in particular in heart and smooth muscle cells; in this case they were also specific for Ca^{2+} ions (Weigel *et al.*, 1979; Siegelbaum and Tsien, 1980). It was suggested that Ca^{2+}-dependent increase of potassium conductance may be the mechanism which breaks the slow inward current, thus preventing possible disturbances of the autogenic rhythmicity of the cell. The direct mechanism of this phenomenon is still far from being clear. Heyer and Lux (1976a, b), in their studies on the effect of Ca^{2+} injection in *Helix* neurons, failed to observe the potentiation of potassium conductance; on the contrary, suppression of potassium conductance was found. Therefore, the authors supposed that the appearance of the N-shaped characteristic of the outward current is associated with some intramembrane effect of the inward calcium current on potassium channels. A further development of this idea is the suggestion that the calcium channels can be modified into potassium ones under the influence of Ca^{2+} ions; a number of similar features in reactions of both types of channels to various extra- and intracellular influences was considered to be in favour of this suggestion (Hofmeier and Lux, 1978). However, contradictory results have been obtained by the experiments of Eckert and Tillotson (1978), who used intracellular injection of aequorin to control the change in intracellular Ca^{2+} concentration. They found good correlation between the increasing Ca^{2+} content inside the neuron and the enhancement of potassium conductance. These authors observed a block of potassium conductance only in some neurons and considered it as a secondary effect, similar to desensiti-zation. The development of the method of intracellular perfusion of isolated neurons made it possible to measure more precisely the effect of internal Ca^{2+} ions on potassium currents by injection of Ca-EGTA buffered solutions that are able to maintain a given level of free calcium ions within the cell. It has been confirmed that an increase of free calcium even to as low as 10^{-7} M is accompanied by a distinct potentiation of outward currents in the membrane of *Helix* neurons. Only those potassium channels which are TEA resistant are involved in this effect and have special kinetic properties (absence of inactivation). They show also a lowered selectivity (Doroshenko *et al.*, 1979). Intracellular Ca^{2+} ions do not affect the activation of ionic channels which produce the fast and delayed potential-dependent potassium currents; moreover, these ions depress the corresponding components of potassium conductance (Kostyuk and Krishtal, 1977b). The effect of Ca^{2+} ions is highly specific; it cannot be repro-duced by Sr^{2+} ions in much higher concentrations. It should be specially noted that the additional potassium conductance activated by internal free Ca^{2+} ions has the same potential-dependent charac-teristics which are peculiar to the background potassium conductance of the membrane in the absence of Ca^{2+} ions inside the cell. It is more likely that the corresponding potassium channels represent electrically operated ionic channels which seem to be transferred from a nonactive state into a state ready for activation in the pres-ence of Ca^{2+} ions rather than a special kind of chemically operated structures.

REFERENCES

Adams, D. J., Dwyer, T. M. and Hille, B. (1980). *J. gen. Physiol.* **75**, 493–510.

Adams, D. J. and Gage, P. W. (1976). *Science, N. Y.* **192**, 783–784.

Adams, D. J. and Gage, P. W. (1979a). *J. Physiol., Lond.* **289**, 115–141.

Adams, D. J. and Gage, P. W. (1979b). *J. Physiol., Lond.* **289**, 143–161.

Akaike, N., Lee. K. S. and Brown, A. M. (1978). *J. gen. Physiol.* **71**, 509–531.

Anderson, N. S., Ramon, F. and Snyder, A. (1971). *J. gen. Physiol.* **58**, 322–339.

Armstrong, C. M. and Bezanilla, F. (1973). *Nature, Lond.* **242**, 459–461.

Baker, P. F. (1976). *Fedn. Am. Socs. exp. Biol. Proc.* **35**, 2589–2595.

Baker, P. F., Hodgkin, A. L. and Ridgway, E. B. (1970). *J. Physiol., Lond.* **208**, 80P–82P.

Baker, P. F., Hodgkin, A. L. and Ridgway, E. B. (1971a). *J. Physiol., Lond.* **214**, 33P–34P.

Baker, P. F., Hodgkin, A. L. and Ridgway, E. B. (1971b). *J. Physiol., Lond.* **218**, 709–755.

Baker, P. F., Meves, H. and Ridgway, E. B. (1971c). *J. Physiol., Lond.* **216,** 70P–71P.
Baker, P. F., Meves, H. and Ridgway, E. B. (1973a). *J. Physiol., Lond.* **231,** 511–526.
Baker, P. F., Meves, H. and Ridgway, E. B. (1973b). *J. Physiol., Lond.* **231,** 527–548.
Bassingthwaighte, J. B. and Reuter, H. (1972). *Biophys. J.* **12,** 214a.
Blioch, Zh. L., Glagoleva, J. M., Liberman, E. A. and Nenashev, V. A. (1968). *J. Physiol., Lond.* **199,** 11–36.
Blitz, A. L. and Fine, R. E. (1974). *Proc. natn. Acad. Sci. U.S.A.* **71,** 4472–4476.
Bolton, T. B. (1979). *Physiol. Rev.* **59,** 606–718.
Breckenridge, B. M., Burn, J. H. and Matschinsky, F. M. (1967). *Proc. natn. Acad. Sci. U.S.A.* **57,** 1893–1897.
Brehm, P. and Eckert, R. (1978). *Science, N. Y.* **202,** 1203–1206.
Brostrom, C. O., Huang, Y-C., Breckenridge, B. M. and Wolff, D. J. (1975). *Proc. natn. Acad. Sci. U.S.A.* **72,** 64–68.
Bülbring, E. and Tomita, T. (1969). *Proc. R. Soc.* B **172,** 121.
Bury, V. A. and Shuba, M. F. (1976). *In* "Physiology of Smooth Muscle" (Eds E. Bülbring and M. F. Shuba), 65–75. Raven Press, New York.
Chizmadjev, Yu. A. and Aityan, S. Kh. (1977). *J. theor. Biol.* **64,** 429–453.
Chizmadjev, Yu. A., Khodorov, B. I. and Aityan, S. Kh. (1974). *Bioelectrochem. Bioenerg.* **1,** 301–312.
Conti, F., Hille, B., Neumcke, B., Nonner, W. and Stämpfli, R. (1976).*J. Physiol., Lond.* **262,** 699–727.
Coraboeuf, E. and Otsuka, M. (1956). *C. r. hebd. Séanc. Paris* **243,** 441–444.
Daly, J. W. (1973). *In* "Frontiers in Catecholamine Research" (Eds E. Usdin and S. Snyder), 301–306. Pergamon Press, New York.
Doroshenko, P. A. and Tsyndrenko, A. Ya. (1978). *Neurophysiology, Kiev* **10,** 203–205.
Doroshenko, P. A., Kostyuk, P. G. and Krishtal, O. A. (1973). *Neurophysiology, Kiev* **5,** 621–627.
Doroshenko, P. A., Kostyuk, P. G. and Tsyndrenko, A. Ya. (1978). *Neurophysiology, Kiev* **10,** 645–653.
Doroshenko, P. A., Kostyuk, P. G. and Tsyndrenko, A. Ya. (1979). *Neurophysiology, Kiev* **11,** 460–468.
Dunlop, K. and Fischbuch, G. D. (1978). *Nature, Lond.* **276,** 837–839.
Eckert, R. and Lux, H. D. (1975). *Brain Res.* **83,** 486–489.
Eckert, R. and Lux, H. D. (1976). *J. Physiol., Lond.* **254,** 129–151.
Eckert, R. and Tillotson, D. (1978). *Biophys. J.* **21,** 178a.
Endo, M., Kitazawa, T., Yagi, S., Iino, M. and Kakuta, Y. (1977).*In* "Excitation-Contraction Coupling in Smooth Muscle" (Eds R. Casteels, T. Godfraind and J. Rüegg), 199–209. Elsevier, Amsterdam.
Fatt, P. and Ginsborg, B. L. (1958). *J. Physiol., Lond.* **142,** 516–543.
Fatt, P. and Katz, B. (1953). *J. Physiol., Lond.* **120,** 171–204.
Fleckenstein, A., Nakayama, K., Fleckenstein-Grün, G. and Byon, Y. K. (1975). *In* "Calcium Transport in Contraction and Secretion" (Eds E. Carafoli *et al.*) 555–566. North-Holland, Amsterdam.
Fukuda, J. and Kameyama, M. (1979). *Nature, Lond.* **279,** 546–548.
Fukuda, J. and Kameyama, M. (1980). *Brain Res.* **182,** 191–197.

Geduldig, D. and Gruener. R. (1970). *J. Physiol., Lond.* **211**, 217–244.

Geduldig, D. and Junge, D. (1968). *J. Physiol., Lond.* **199**, 347–365.

Gerasimov, V. D., Kustyuk, P. G. and Maisky, V. A. (1964). *Bull. exp. Biol. Med., Moscow* **58**, n.9, 3–7.

Giles, W. and Tsien, R. W. (1975). *J. Physiol., Lond.* **246**, 64P–66P.

Hagiwara, S. (1973). *Adv. Biophys.* **4**, 71–102.

Hagiwara, S. and Miyazaki, S. (1977). *J. Physiol., Lond.* **272**, 197–216.

Hagiwara, S. and Naka, K. I. (1964). *J. gen. Physiol.* **48**, 141–162.

Hagiwara, S. and Nakajima, S. (1966). *J. gen. Physiol.* **49**, 793–806.

Hagiwara, S. and Takahashi, K. (1967). *J. gen. Physiol.* **50**, 583–601.

Hagiwara, S., Chichifu, S. and Naka, K. I. (1964). *J. gen. Physiol.* **48**, 165–179.

Hagiwara, S., Fukuda, J. and Eaton, D. C. (1974). *J. gen. Physiol.* **63**, 564–578.

Hammerschlag, R. and Lavoie, P. -A. (1979). *Neuroscience* **4**, 1195–1201.

Hammerschlag, R., Dravid, A. R. and Chiu, A. Y. (1975). *Science, N.Y.* **188**, 273–275.

Henček, M. and Zachar, J. (1977). *J. Physiol., Lond.* **268**, 51–71.

Hermann, A. and Gorman. A. L. F. (1978). *Biophys. J.* **21**, 178a.

Heyer, B. and Lux, H. D. (1976a). *J. Physiol., Lond.* **262**, 319–348.

Heyer, B. and Lux, H. D. (1976b). *J. Physiol., Lond.* **262**, 349–382.

Hille, B. (1975). *J. gen. Physiol.* **66**, 535–560.

Hille, B. and Schwartz, W. (1978). *J. gen. Physiol.* **72**, 409–442.

Hodgkin, A. L. and Huxley, A. F. (1952). *J. Physiol., Lond.* **117**, 500–544.

Hodgkin, A. L. and Keynes, R. D. (1957). *J. Physiol., Lond.* **138**, 253–281.

Hofmeier, G. and Lux, H. D., (1978). *J. Physiol., Lond.* **287**, 28P–29P.

Holman, M. E. (1958). *J. Physiol., Lond.* **141**, 464–488.

Horn, J. P. and McAfee, D. A. (1979). *Science, N.Y.* **204**, 1233–1235.

Horn. R. (1978). *J. Physiol., Lond.* **281**, 513–534.

Inomata, H. and Kao, C. Y. (1976). *J. Physiol., Lond.* **255**, 347–378.

Kakiuchi, S. and Rall, T. W. (1968). *Molec. Pharmac.* **4**, 367–378.

Kao, C. Y. and McCullough, J. R. (1975). *J. Physiol., Lond.* **246**, 1–36.

Katz, B. and Miledi, R. (1967a). *J. Physiol., Lond.* **189**, 535–544.

Katz, B. and Miledi, R. (1967b). *J. Physiol., Lond.* **192**, 407–436.

Katz, B. and Miledi, R. (1967c). *Proc. R. Soc.* B **167**, 8–22.

Katz, B. and Miledi, R. (1970). *J. Physiol., Lond.* **207**, 789–802.

Keynes, R. D. and Rojas, E. (1973). *J. Physiol., Lond.* **233**, 28P–30P.

Keynes, R. D., Rojas, E., Taylor, R. E. and Vergara, J. (1973). *J. Physiol., Lond.* **229**, 409–455.

Kleinhaus, A. L. (1976). *Pflügers Arch. ges. Physiol.* **363**, 97–104.

Kohlhardt, M., Baner, B., Krause, H. and Fleckenstein, A. (1972). *Pflügers Arch. ges. Physiol.* **335**, 309–322.

Kohlhardt, M., Haastert, H. P. and Krause, H. (1973). *Pflügers Arch. ges. Physiol.* **342**, 125–136.

Koketsu, K. and Nishi, S. (1969). *J. gen. Physiol.* **53**, 508–623.

Kostyuk, P. G. and Krishtal, O. A. (1977a). *J. Physiol., Lond.* **270**, 545–568.

Kostyuk, P. G. and Krishtal, O. A. (1977b). *J. Physiol., Lond.* **270**, 569–580.

Kostyuk, P. G., Krishtal, O. A. and Doroshenko, P. A. (1974a). *Pflügers Arch. ges. Physiol.* **348**, 83–93.

Kostyuk, P. G., Krishtal, O. A. and Doroshenko, P. A. (1974b). *Pflügers Arch. ges. Physiol.* **348**, 95–104.

Kostyuk, P. G., Kristal, O. A. and Pidoplichko, V. I. (1975). *Nature, Lond.* **257**, 691–693.

Kostyuk, P. G., Krishtal, O. A. and Pidoplichko, V. I. (1977). *Nature, Lond.* **267**, 70–72.

Kostyuk, P. G., Krishtal, O. A. and Pidoplichko, V. I. (1978a). *Dokl. Akad. Nauk SSSR* **238**, 478–481.

Kostyuk, P. G., Krishtal, O. A., Pidoplichko, V. I. and Veselovsky, N. S. (1978b). *Neuroscience* **3**, 327–332.

Kostyuk, P. G., Krishtal, O. A., Pidoplichko, V. I. and Shakhovalov, Yu., A. (1979). *Dokl. Akad. Nauk SSSR* **249**, 1470–1473.

Kostyuk, P. G., Krishtal, O. A., Pidoplichko, V. I. and Shakhovalov, Yu., A. (1980). *Dokl. Akad. Nauk SSSR* **250**, 219–222.

Kostyuk, P. G., Mironov, S. L. and Doroshenko, P. A. (1980). *Dokl. Akad. Nauk SSSR*, **253**, 978–981.

Krishtal, O. A. (1976). *Dokl. Akad. Nauk SSSR* **231**, 1003–1005.

Krishtal, O. A. (1978). *Dokl. Akad. Nauk SSSR* **238**, 482–485.

Krishtal, O. A. and Pidoplichko, V. I. (1975). *Neurophysiology, Kiev* **7**, 327–329.

Kumamoto, M. and Horn, L. (1970). *Microvascular Res.* **2**, 188–201.

Lee, K. S., Akaike, N. and Brown, A. M. (1978). *J. gen. Physiol.* **71**, 489–507.

Llinas, R. and Hess, R. (1976). *Proc. natn. Acad. Sci. U.S.A.* **73**, 2520–2523.

Llinas, R., Blinks, J. R. and Nicholson, C. (1972). *Science, N.Y.* **176**, 1127–1129.

Llinas, R., Steinberg, I. Z. and Walton, K. (1976). *Proc. natn. Acad. Sci. U.S.A.* **73**, 2918–2922.

Llinas, R., Sugimori, M. and Walton, K. (1977). *Soc. Neurosci.* Abstr. **3**, 58.

Lux, H. D., Schubert, P., Kreutzberg, G. W. and Globus, A. (1970). *Expl. brain Res.* **10**, 197–204.

Markin, V. S. and Chizmadjev, Yu. A. (1974). "The Induced Ion Transport." Nauka, Moscow.

Martell, A. E. and Smith, R. M. (1977). "Critical Stability Constants." Plenum Press, New York.

Matsuda, Y., Yoshida, S. and Yonezawa, T. (1976). *Brain Res.* **115**, 334–338.

Matsuda, Y., Yoshida, S. and Yonezawa, T. (1978). *Brain Res.* **154**, 69–82.

Meech, R. W. (1974). *J. Physiol., Lond.* **237**, 259–277.

Meech, R. W. and Standen, N. B. (1975). *J. Physiol., Lond.* **249**, 211–239.

Mellow, A. M. (1979). *Nature, Lond.* **282**, 84–85.

Meves, H. (1968). *Pflügers Arch. ges. Physiol.* **304**, 215–241.

Meves, H. and Vogel, W. (1973). *J. Physiol., Lond.* **235**, 225–265.

Miledi, R., Parker, I. and Schalow, G. (1980). *J. Physiol., Lond.* **300**, 197–212.

Mironneau, J. (1974). *Pflügers Arch. ges. Physiol.* **352**, 197–210.

Miyamoto, M. D. and Breckenridge, B. (1974). *J. gen. Physiol.* **63**, 609–624.

Miyazaki, S., Takahashi, K. and Tsuda, K. (1974). *J. Physiol., Lond.* **238**, 37–54.

Miyazaki, S., Ohmori, H. and Sasaki, S. (1975). *J. Physiol., Lond.* **246**, 37–54.

Morgenroth, V. H. III, Boadle-Bider, M. and Roth, R. J. (1974). *Proc. natn. Acad. Sci. U.S.A.* **71**, 4283–4287.

Nachshen, D. A. and Blaustein, M. P. (1979). *Biophys. J.* **26**, 329–334.

Naruševičius, E. V. and Rapoport, M. Sh. (1979). *Dokl. Akad. Nauk SSSR* **246**, 217–219.

Niedergerke, R. and Orkand, R. K. (1966). *J. Physiol. Lond.* **184**, 312–334.

Ochs, S., Chan, S. Y., Worth, R. M. and Jersild, R. (1978). *Biophys. J.* **21**, 187a.

Okamoto, H. Takahashi, K. and Yoshii, M. (1976a). *J. Physiol., Lond.* **254**, 607–638.

Okamoto, H., Takahashi, K. and Yoshii, M. (1976b). *J. Physiol., Lond.* **255**, 527–561.

Okamoto, H. Takahashi, K. and Yamashita, N. (1977). *J. Physiol., Lond.* **267**, 465–495.

Oomura, Y., Ozaki, S. and Maeno, T. (1961). *Nature, Lond.* **191**, 1265–1267.

Pellmar, T. C. and Carpenter, D. O. (1979). *Nature, Lond.* **277**, 483–484.

Perkins, J. P. (1975). *In* "The Nervous System" (Ed. D. B. Tower) Vol. 1, 381–394. Raven Press, New York.

Pitman, R. M. (1975). *J. Physiol., Lond.* **251**, 62P–63P.

Ponomaryov, V. N., Naruševičius, E. and Chemeris, N. K. (1980). *Neurophysiology, Kiev* **12**, 221–223.

Puszkin, S., Niklas, W. J. and Berl, S. (1972). *J. Neurochem.* **19**, 1319–1333.

Rall, T. W. and Sattin, A. (1970). *In* "Advances in Biochemical Psychopharmacology" (Eds E. Costa and P. Greenhard) Vol. 3, 113–133, Raven Press, New York.

Ransom, B. R. and Holz, R. W. (1977). *Brain Res.* **136**, 445–453.

Rasmussen, H. and Goodmen, D. B. P. (1977). *Physiol. Rev.* **57**, 421–509.

Reuter, H. (1965). *Naumyn-Schmiedebergs Arch. exp. Path. Pharmak,* **251**, 401–412.

Reuter, H. (1967). *J. Physiol., Lond.* **192**, 479–492.

Reuter, H. (1974). *J. Physiol., Lond.* **242**, 429–451.

Reuter, H. and Scholz, H. (1977). *J. Physiol., Lond.* **264**, 49–62.

Sattin, A., Rall, T. W. and Zanella, J. (1975). *J. Pharmac. exp. Ther.* **192**, 22–32.

Schwartzkroin, P. A. and Slawsky, M. (1977). *Brain Res.* **135**, 157–161.

Shimomura, O., Johnson, F. H. and Saiga, Y. (1962). *J. cell. comp. Physiol.* **59**, 223–239.

Siegelbaum, S. A. and Tsien, R. W. (1980). *J. Physiol., Lond.* **299**, 485–506.

Sigworth, F. J. (1977). *Nature, Lond.* **270**, 265–267.

Singer, J. J. and Goldberg, A. L. (1970). *In* "Advances in Biochemical Psychopharmacology" (Eds E. Costa and P. Greenhard) Vol. 3, 335–348. Raven Press, New York.

Somlyo, A. V. and Somlyo, A. P. (1968). *J. Pharmac. exp. Ther.* **159**, 129–145.

Spitzer, N. C. and Lamborghini, J. (1976). *Proc. natn. Acad. Sci. U.S.A.* **73**, 1641–1645.

Standen, N. B. (1975a). *J. Physiol., Lond.* **249**, 241–252.

Standen, N. B. (1975b). *J. Physiol., Lond.* **249**, 253–268.

Stanley, E. J. and Reuter, H. (1965). *Naunyn-Schmiedebergs Arch. exp. Path. Pharmak.* **252**, 159–172.

Sutherland, E. W. and Robinson, G. A. (1966). *Pharmac. Rev.* **18**, 145–161.

Takahashi, K. and Yoshii, M. (1978). *J. Physiol., Lond.* **279**, 519–549.

Tasaki, I. (1968). "Nerve Excitation: A Macromolecular Approach." Charles C. Thomas, Springfield, Ill.

Tasaki, I., Lerman, L. and Watanabe, A. (1969). *Am. J. Physiol.* **216**. 130–138.

Thompson, S. H. and Smith. S. J. (1976). *J. Neurophysiol.* **39**, 153–161.

Tillotson, D. (1979). *Proc. natn. Acad. Sci. U.S.A.* **76**, 1497–1500.

Tsien, R. W., Giles, W. and Greengard, P. (1972). *Nature, New Biol.* **240**, 181–183.

Valeyev, A. E. (1979). *Neurophysiology, Kiev* **11**, 371–374.

Veselovsky, N. S. and Fedulova, S. A. (1980). *Dokl. Akad. Nauk SSSR* **253**, 1493–1495.

Veselovsky, N. S., Kostyuk, P. G., Krishtal, O. A. and Pidoplichko, V. I. (1977a). *Neurophysiology, Kiev* **9**, 638–640.

Veselovsky, N. S., Kostyuk, P. G., Krishtal, O. A., Naumov, A. P. and Pidoplichko, V. I. (1977b). *Neurophysiology, Kiev* **9**, 641–643.

Veselovsky, N. S., Kostyuk, P. G. and Tsyndrenko, A. Ya. (1979). *Dokl. Akad. Nauk SSSR* **249**, 1466–1469.

Vitek, M. and Trautwein, W. (1971). *Pflügers Arch. ges. Physiol.* **323**, 204–218.

Wald, F. (1972). *J. Physiol., Lond.* **220**, 267–281.

Watanabe, A., Tasaki, I. and Lerman, L. (1967). *Proc. natn. Acad. Sci. U.S.A.* **58**, 2246–2252.

Weigel, R. J., Connor, J. A. and Prosser, C. L. (1979). *Am. J. Physiol.* **237**, C247–C256.

Wilson, D. F. (1974). *J. Pharmac. exp. Ther.* **188**, 447–452.

Woodhull, A. M. (1973). *J. gen. Physiol.* **61**, 687–708.

Yasui, S., Brown, A. M. Akaike, N. and Lee, K. S. (1979). *Biophys. J.* **25**, M-MP-C10, Abstr.

Yoshida, S. and Matusuda, Y. (1980). *Brain Res.* **188**, 593–597.

Yoshida, S., Matsuda, Y. and Samejima, A. (1978). *J. Neurophysiol.* **41**, 1096–1106.

2

The Plasma Membrane Calcium Pump of Erythrocytes and Other Animal Cells

H. J. SCHATZMANN

Department of Veterinary Pharmacology, University of Bern, Switzerland

INTRODUCTION

As a consequence of the great abundance of calcium-containing minerals in the earth's crust and the fair solubility of calcium salts, calcium ions are present in rather high concentrations in fresh water, sea water and all feedstuffs of animals. It stands to reason that arising life had to adjust to the omnipresence of this metal ion. One way of achieving this would have been to organize functional macromolecules such as to make them insensitive towards all possible Ca^{2+} concentrations. However, at least in the Animal Kingdom, the unavoidable – or, reliably present – calcium has been harnessed to assume, as it does, the role of a transmitter between signals at the plasma membrane and the intracellular machinery. To the best of our knowledge *all* animal cells keep the ionic calcium concentration in the cytosol far below that in the external medium. A host of cellular mechanisms perform differently at the steady-state intracellular calcium concentration and at even slightly elevated intracellular calcium concentration. In response to electrical or the most varied chemical stimuli at the plasma membrane, calcium enters the cytosol from the outside across the plasma membrane, from calcium storage organelles or from both, down its chemical gradient, thus altering at short notice the functions of very many cellular processes. A rise in internal Ca^{2+} concentration leads to muscle contraction, disaggregation of microtubuli and exocytosis in exocrine and endocrine glands and nerve terminals. In Na-transporting epithelia it may reduce net Na movement by lowering Na permeability at the luminal membrane (Taylor and Windhager, 1979). In epithelia it also blocks permeability between adjacent cells (Loewenstein and Rose, 1978). Stimulatory effects on DNA replication and/or protein synthesis have been claimed (Gallien-Lartigue, 1976; Andersson and Norrby, 1977; Whitfield et al., 1979, 1980; Boynton et al., 1980) and questioned (Rink et al., 1980). Ca^{2+} entry into the egg cell, seconds after attachment of the spermatozoon, seems to be the decisive step for the spectacular events in the ensuing egg development (Epel, 1980; Chambers, 1980). Uncontrolled Ca^{2+} entry possibly is the common denominator in the action of many lethal poisons (Schanne et al., 1979). In most instances, the trains of events set in motion by incoming calcium are far from being completely understood but it is obvious that they belong to the very essence of life. Maintenance of a large calcium gradient across the plasma membrane is the prerequisite of their susceptibility to regulation by changes in passive calcium flux. It seems appropriate, therefore, to study the processes maintaining this calcium gradient.

It was recognized rather early that the low intracellular calcium concentration cannot be accounted for by some sort of an equilibrium distribution and that therefore processes must be involved which are thermodynamically uphill transport phenomena, that is, must be driven in the last analysis by metabolic energy. At the level of the plasma membrane there are two active transport mechanisms: (1) *Heteroexchange* means calcium extrusion in exchange for inward moving sodium, a process which exploits the energy originally invested in the Na gradient created by outward sodium pumping. (2) By *Ca^{2+} pump* we understand outward calcium movement by a system directly dependent on ATP hydrolysis as energy supply. As we shall see, in some cells (the squid axon, cardiac muscle, renal tubular cells) both processes may be operating in parallel and in others the second process may be present alone (red blood cells, L cells) and it will emerge that they subserve slightly different purposes in the economy of different cell types.

In the classical calcium storage organelle, the sarcoplasmic reticulum of muscle, only the second process has been detected, which makes sense because there the sodium pump seems to be lacking too (Jones et al., 1979). Little is known about the mechanism which accumulates calcium in the so-called dense bodies of blood platelets. The uptake of calcium into mitochondria is a special case. Here the driving force for calcium movement seems to be the membrane potential created by the extrusion of protons which acts on calcium attached to a passive carrier in the membrane (Carafoli, 1974; Bygrave, 1978; Carafoli and Crompton, 1978).

It must be emphasized that Ca^{2+} storage organelles are suitable to remove rapidly small amounts of Ca^{2+} from the cytosol but are not able to keep the cytosolic Ca^{2+} concentration low indefinitely. For this an extrusion mechanism at the plasma membrane is the only expedient.

In the following we shall be concerned with the ATP-driven system (2) — to be denoted by "Ca pump" for brevity's sake — at the level of the plasma membrane, which has been studied to some length in the red cell membrane but indubitably exists in many other animal cells as well.

THE RED CELL

The Calcium Pump in Human Red Cells: Basic Facts

The topic has been reviewed before (Schatzmann, 1975; Vincenzi and Hinds, 1976; Sarkadi and Tosteson, 1979; Roufogalis, 1979; Schatzmann and Bürgin, 1978; Sarkadi, 1980; Vincenzi and Hinds, 1980), but it may be as well to summarize the established facts before accosting more recent advances and to discuss controversial points and unsolved problems.

The great advantage of red cells is their simplicity. There are no internal transporting membranes and no clear-cut evidence for internal comparting of calcium exists.

The passive permeability of the plasma membrane for Ca^{2+} is exceedingly low and can be neglected in most studies on active Ca^{2+} transport.

The mechanisms involved are interesting in that saturable specific carriers are involved (Porzig, 1970; Ferreira and Lew, 1977; Lew and Ferreira, 1978), decrease of Ca^{2+} permeability was observed during maturation (Wiley and Shaller, 1977) and abnormal Ca^{2+} leaks have been found in different diseased states (sickle cell anaemia (Eaton et al., 1973, 1978), microcytic (Wiley and Gill, 1976) and nonspherocytic haemolytic anaemias (Bucher et al., 1978)).

Intracellular protein binding of Ca^{2+} is slight and constant over a wide range of concentration and has been measured to be 50–80% (Schatzmann, 1973; Ferreira and Lew, 1976). At total Ca concentrations of $\sim 10^{-6}$ M high affinity Ca buffering by calmodulin, whose concentration in red cells has been estimated to be more than 10^{-6} M (Jarrett and Penniston, 1978; Jung, quoted by Vincenzi and Hinds, 1980), may become prominent.

Red cells lend themselves to manipulations facilitating the study of Ca movements: they can be loaded with Ca^{2+}, ATP and Ca buffers by reversal of haemolysis, made reversibly leaky to Ca^{2+} by PCMBS (p-chloromercuribenzene sulphonate; Schatzmann, 1973),

Na-salicylate (Bürgin and Schatzmann, 1979) or specific ionophores (e.g. A 23187; Ferreira and Lew, 1976; Sarkadi *et al.*, 1976) and finally they can be transformed into inside-out vesicles, representing a Ca^{2+}-tight preparation in which the concentration dependence of the transport process at the exposed inner surface can easily be studied (Weiner and Lee, 1972; MacIntyre and Green, 1976, 1978; Sarkadi *et al.*, 1978, 1980b; Hinds *et al.*, 1978; Quist and Roufogalis, 1977).

With a rate of Ca^{2+} leak of $1-10\,\mu$mol/litre cells. h (Lew and Ferreira, 1976; Ferreira and Lew, 1977) the human red cell would equilibrate within its lifespan of 100 d to a total internal Ca concentration of several millimoles per litre cells (owing to the membrane potential of 10 mV, inside negative, and the internal Ca binding to proteins, which amounts to 2–5 times the free Ca^{2+} concentration between 0 and 10 mM (Schatzmann, 1973; Ferreira and Lew, 1976). It is a fact, however, that the total Ca concentration is about 0.01 mmol/litre cells, and there is agreement that the ionized Ca^{2+} concentration in the cell water is less than 10^{-6} M, most of the total Ca being associated with the membrane. If the steady state situation is characterized by a transmembrane Ca^{2+} gradient of 1000–10 000 a Ca extrusion mechanism must exist. Making use of ATP depleted and ATP replenished red cells it could be shown that this extrusion mechanism is directly dependent on ATP and that it has the property of a $(Ca^{2+} + Mg^{2+})$— activated ATPase. The requirement for Mg^{2+} is essential. This was already clear when the red cell Ca pump was discovered (Schatzmann, 1966, 1969) and was recently confirmed by Sarkadi *et al.* (1978). In other tissues it was found whenever a Ca^{2+}-sensitive ATPase was unequivocally demonstrated to be a Ca pump. Mg is not or not strictly necessary for ATP binding to the system (or in other words Mg-ATP is not the obligatory substrate), but is required in later steps of the reaction (this may be different in the sarcoplasmic reticulum, Makinose and Boll, 1979). Klinger *et al.* (1980) recently pointed out that Mg^{2+} competes with Ca^{2+} above ~ 1 mM free Mg^{2+} concentration, shifting the Ca^{2+} concentration for half maximal activation of the ATPase to higher values. Mg^{2+} is definitely not translocated across the membrane in or against the direction of the Ca^{2+} movement (Schatzmann, 1975). It has to be presented to the internal membrane side as, of course, also goes for ATP. To complete this it may be recalled that the ultimate liberation of inorganic phosphate takes place at the internal surface, too (Schatzmann and Roelofsen, 1977).

The affinity for Ca^{2+} at the internal surface is high ($K_{Ca} \sim 1\,\mu$M) in the presence of Ca-EGTA buffer; certain ways of preparing membranes even yield values below $1\,\mu\text{M}$ (Downes and Michell, 1981). K_m (for free ATP) is also in the micromolar range. In the presence of 2 mM ATP and almost saturating Ca^{2+} concentration K_{Mg} for free Mg^{2+} was found to be about $13.8 \pm 3.2\,\mu\text{M}$ (SEM, five experiments; $CaCl_2$ and EGTA concentration was approximately 1 mM. In the required range of Mg^{2+} concentration the Ca^{2+} concentration was not kept constant but stayed above 5×10^{-6} M Schatzmann, unpublished)). High Ca^{2+} concentration (above $1 \times 10^{-4} - 5 \times 10^{-4}$ M) is inhibitory, probably by competition with Mg^{2+} at the Mg-specific site. Specificity for Ca^{2+} in transport is nearly exclusive; with the exception of Sr^{2+} no other divalent metal has been found to be transported although some stimulate (less well than Ca^{2+} or Sr^{2+}) the ATPase activity (Mn, Zn, Ni, Co, Cu) (Pfleger and Wolf, 1975), in the presence of Mg^{2+}.

Soon after the early reports on the (Ca + Mg)-ATPase (first described by Dunham and Glynn, 1961; Wins and Schoffeniels, 1966) being identical with the Ca pump (Schatzmann, 1966; Schatzmann and Vincenzi, 1969; Lee and Shin, 1969; Olson and Cazort, 1969; Romero and Whittam, 1971) a controversy arose over the question whether two (Ca + Mg)-ATPases, differing in affinity for Ca^{2+}, exist. Some authors found activity versus Ca^{2+} concentration curves for the ATPase that were not (more or less sigmoidally biased) rectangular hyperbolas but were composed of a steep part at low and a more gradual part at higher Ca^{2+} concentrations. It was soon shown that this depends on the way membranes are prepared, the presence of Ca chelators over long periods leading to the kinked curve (Scharff, 1972, 1976, 1978; Schatzmann, 1973). It is the merit of Scharff and Foder (1978) to have shown very convincingly that the appearance of a low affinity part is due to partial removal of calmodulin from the membranes when prepared in the total absence of Ca^{2+}. Since it is clear today that calmodulin can shift the affinity for Ca^{2+} by more than one order of magnitude, it has become pointless to discuss any further the question whether the transporting system has high or low Ca^{2+} affinity. The only question which remains is whether the Ca pump in the physiological state in intact red cells is fully activated by calmodulin or not (see below).

Conceptually it is unavoidable to postulate that the Ca^{2+} transport site must alternately face the internal and external membrane side and that during or after the outward journey the affinity must drop in order to allow the system to transport Ca^{2+} uphill. The ratio of the Ca^{2+} dissociation constants in the out-position (K_{Ca}^o) and the in-position (K_{Ca}^i) must be at least as large as the ratio of the external and internal Ca^{2+} concentration:

$$\frac{K_{Ca}^o}{K_{Ca}^i} \geqslant \frac{[Ca^{2+}]_o}{[Ca^{2+}]_i} \qquad (1)$$

It was in fact found that the external Ca^{2+} concentration which inhibits pump activity was larger than 10^{-3} M in experiments where K_{Ca}^i was 4×10^{-6} M (Schatzmann, 1973). Since the rate of Ca^{2+} leakage at the physiological Ca^{2+} gradient is $5-10\,\mu mol/litre$ cells. h (Ferreira and Lew, 1977; Lew and Ferreira, 1978), whereas the maximal rate of the Ca pump is $5-10\,mmol/litre$ cells. h (at saturating $[Ca^{2+}]_i$), the expected concentration ratio $[Ca^{2+}]_o/[Ca^{2+}]_i >$ 1000 thus seems possible (this makes of course the assumption that the leak is not through the pump but in channels parallel to it).

The temperature dependence of Ca^{2+} pumping has been investigated and values for "activation energy" were found to be 56.8 kJ/mol (Lee and Shin, 1969), 80.2 kJ/mol (Sarkadi et al., 1980b) and 104.6 kJ/mol (Schatzmann and Vincenzi, 1969) which corresponds to a Q_{10} of 2.1, 2.8 or 3.8. The large differences may be due to the fact that the temperature dependence is steeper below 27°C than above (Mollman and Pleasure, 1980).

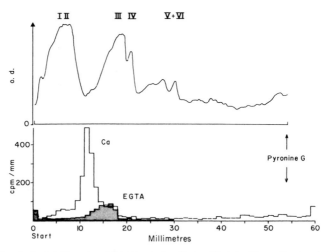

Fig. 1. *SDS gel electrophoresis of human red cell membranes. Top: Coomassie blue stain giving protein scan (arbitrary scale); bottom: radioactivity per millimetre of gel length after phosphorylation of membranes for 30 s at 0–2°C. Medium (mM): $CaCl_2$ 0.05 (white columns), Tris-Cl 150 (pH 7.4), $[\gamma\text{-}^{32}P]\text{-}ATP$ 0.00057. Control (shaded) 0.5 mM EGTA instead of $CaCl_2$. Reaction stopped with 6% trichloracetic acid, 50 mM H_3PO_4, 0.5 mM ATP. (Schatzmann and Bürgin, 1978.)*

The Red Cell Calcium Pump: New Aspects

Molecular constitution
The main protein. The search for the pump protein has been greatly facilitated by the fact that it forms a phosphorylated intermediate, as was detected by Knauf *et al.* (1972, 1974) and Katz and Blostein (1975). When ATP, labelled with ^{32}P in the γ position, is added to isolated red cell membranes in the presence of Ca^{2+} and the total absence of Mg^{2+} for a few seconds at 0°C a subsequent SDS gel electrophoresis shows a sharp single radioactivity peak at an apparent molecular weight of 140 000 (see Fig. 1). Obviously, protein phosphorylation by kinases which are undoubtedly present in red cell membranes is much too slow to interfere with this technique. The Ca^{2+}-induced phosphorylated protein can easily be distinguished from the sodium-induced phosphoprotein signalling the existence of the Na-K pump α-subunit of a molecular weight of 100 000 (see Fig. 2).

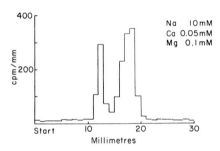

Fig. 2. *SDS gel electrophoresis of human red cell membranes as in Fig. 1 but with 0.05 mM $CaCl_2$, 10 mM NaCl, 0.1 mM $MgCl_2$. Na (+ low Mg^{2+} concn) induces phosphate incorporation into a protein of mol. wt ~ 100 000 in addition to the Ca^{2+}-evoked peak (at 140 000). With EGTA instead of $CaCl_2$ the left hand peak is absent (not shown). (Schatzmann and Bürgin, 1978.)*

Attempts have been made to isolate the protein from the membrane. These meet with two difficulties: (1) the protein is not functioning and unstable without its natural surrounding of phospholipids or some detergents. This is similar in the sarcoplasmic reticulum Ca pump (Dean and Tanford, 1978). (2) The amount of the Ca pump protein is exceedingly low, namely about 0.2% or less of the total membrane proteins.

Wolf and coworkers (Wolf and Knipser, 1975a, b; Wolf and Dieckvoss, 1976; Lichtner and Wolf, 1977; Wolf *et al.*, 1977) were the first to succeed in considerably purifying the Ca pump protein by way of gel chromatography. Their method was briefly as follows:

they partially solubilized red cell membranes in 0.2% Triton-X-100 and applied the supernatant to a Sepharose CL 6 B column equilibrated with a sonicated suspension of lecithin and Tween 20. Upon elution with the same lipid–detergent mixture they succeeded in retaining proteins in the gel and recovered the (Ca + Mg)-activated ATPase in a single peak, apparently attached to mixed micelles of lecithin and Tween. This material showed three protein bands in SDS gel electrophoresis with the respective molecular weights of 145 000, 115 000 and 105 000. Only the 145 000 molecular weight protein showed the characteristic Ca^{2+}-dependent rapid phosphorylation. The properties of the (Ca + Mg)-activated ATPase were in very satisfactory agreement with those of the membrane-bound enzyme. K_{Ca} was 1.9 μM and K_m (for total ATP) was similar to that of the membrane-bound enzyme. The purified material was further activated by Na, K or NH_4, but not Li. Co^{2+}, Zn^{2+} and Mn^{2+} could partly replace Ca^{2+}. In the purified material the decomposition of the phosphorylated intermediate could be accelerated by the addition of Mg^{2+} (Lichtner and Wolf, 1980a). The phosphorylated intermediate was acid stable and sensitive to hydroxylamine and molybdate (Lichtner and Wolf, 1980b). The authors found it essential to have protease inhibitors present throughout the preparation procedure beginning with the membrane isolation.

Recently a rather ingenious idea was applied to the problem of isolating the Ca pump protein. Calmodulin binds to the Ca pump at (low) Ca^{2+} concentration and the complex dissociates upon rigorous removal of Ca^{2+} by EGTA or the like. The dodge was to couple calmodulin to Sepharose and use this material for affinity chromatography of solubilized red cell membrane proteins. Niggli *et al.* (1979) were first to use a calmodulin affinity column with success. They prepared calmodulin-depleted red cell members by extensive washing in 1 mM EDTA-containing 10 mM Tris-Cl solution, pH 7.4. Bovine brain calmodulin was attached to 4 B CNBr-activated Sepharose. The Triton-X-100 solubilized membrane material was mixed with a phosphatidylserine suspension and applied to the column, equilibrated with a solution containing Triton-X-100 (0.5%), phosphatidylserine (0.05%) and $CaCl_2$ (0.1 mM). The column was perfused with the same solution until all protein escape had ceased. Upon switching to 5 mM EDTA instead of Ca^{2+} in the perfusing solution, a minor peak of protein appeared which displayed (Ca + Mg)-activated ATPase. The material contained in addition 5% Mg-ATPase and could *not* be stimulated by calmodulin. This is at first sight puzzling, because what comes from the column must be the Ca pump protein devoid of calmodulin, provided the principle works as expected. However, it has long been known that Triton-X-100 activates the protein and Gietzen *et al.* (1980c) demonstrated that Triton-activation and calmodulin-activation are not additive but mutually exclusive. We found that taking the protein into phosphatidylserine (PS) instead of phosphatidylcholine (PC) gives increased specific activity (equal to that in PC plus calmodulin), but loss of calmodulin sensitivity (Stieger and Luterbacher, 1981). The same discovery was made by Niggli *et al.* (1981). Either of these reasons may explain the inefficacy of calmodulin here.

In the SDS gel electrophoresis Niggli's product shows two proteins of molecular weight of 125 000 and 205 000 respectively, both of which display Ca^{2+}-induced phosphorylation. The relative amounts seem to have varied considerably from experiment to experiment. The authors' suggestion is quite plausible that they are dealing with a monomeric and a dimeric form of the pump protein.

Gietzen *et al.* (1980c) used the same principle of affinity chromatography with calmodulin to purify the main protein. They haemolysed cells by freezing–thawing in the presence of 5% Tween 20, washed the membranes in EDTA containing solutions and solubilized the membranes in deoxycholate (DOC) with a DOC/protein ratio of 0.05–0.1 (g/g) in a medium containing 0.2 mg phosphatidylcholine/ml at pH 7.0. The solubilized material was applied to the affinity column in a buffer with phosphatidylcholine, Tween 20, Triton-X-100 and $100 \mu M$ $CaCl_2$, the column was washed until no more protein was eluted and then the perfusing fluid was replaced by a similar solution with 1 mM EDTA instead of Ca^{2+}, which brought forth a fraction containing the (Ca + Mg)-ATPase. The material gave a single band in SDS gel electrophoresis corresponding to a molecular weight of 135 000–150 000 and its ATPase activity was stimulated nine-fold by calmodulin. The authors showed that the calmodulin sensitivity is lost if the Triton concentration is too high. Their procedure is also carried out under constant protection by protease inhibitors (TLCK = N-α-p-tosyl-l-lysine chlormethyl-ketone and trasylol). The feasibility of Gietzen's method was confirmed in our laboratory (Stieger and Luterbacher, 1981). The protein appears as a single peak in SDS gel electrophoresis (see Fig. 3). However, as in Niggli's experiments, occasionally two peaks of variable size were present. The second peak has a molecular weight of very nearly twice that of the lighter protein. So it might well be that a dimeric species can form which does not dissociate in SDS.

In the presence of calmodulin, purified (Ca + Mg)-ATPase has a

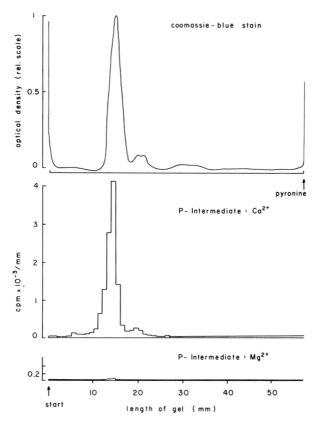

Fig. 3. *SDS gel electrophoresis of pump protein purified according to Gietzen* et al. *(1980c) and collected into lecithin suspension (compare with Fig. 1). Top: protein scan (arbitrary scale); middle: radioactivity per millimetre gel length of the same material after phosphorylation for 30 s at 0–2° C. Medium: EGTA 0.575 mM, CaCl$_2$ 0.62 mM (Ca^{2+} ~ 50 μM), MOPS 25 mM (pH 7.0 at 0°C), KCl 120 mM, [γ-^{32}P]-ATP 0.5 μM, calmodulin from bovine brain 2 μg/ml. Reaction stopped with 10% trichloroacetic acid, 10 mM H$_3$PO$_4$, 2 mM ATP. Bottom: the same, but instead of CaCl$_2$ 2 mM MgCl$_2$. (Stieger and Luterbacher, 1981.)*

K_{Ca} of 3.3 μM and a K_{Mg} of 35 μM (free Mg^{2+}). Two K_m (for total ATP) of 3.5 μM and 120 μM are detectable (Stieger and Luterbacher, 1981). The activity decays with a half-life of 1.5 days at 0°C but is quite stable (half-life ~ 7 d at 0°C) in the presence of calmodulin and Ca^{2+} (Luterbacher and Stieger, unpublished). Haaker and Racker (1979) found a similar protective action of Ca^{2+} and calmodulin on Triton-X-100-solubilized pig red cell membranes.

These recent experiments seem to rule out the participation in the ATPase function of further proteins of a molecular weight of around

100 000 postulated by Wolf *et al.* (1977). They make it highly unlikely that the system is composed of a protein kinase, a substrate protein and a phosphohydrolase.

With the successful purification of a functional protein, the way seems paved for a detailed analysis of the chemical characteristics of the main protein of the Ca pump.

The role of calmodulin. Bond and Clough (1973) were first in showing that the red cell cytosol contains a soluble proteinacious activator of the $(Ca + Mg)$-ATPase. Two years later Luthra *et al.* (1976, 1977) were able to purify this activator protein considerably. Jarrett and Penniston (1977, 1978) described a reliable method of purifying the activator protein to apparent homogeneity and indicated that it belongs to a family of Ca-binding "modulator" proteins found in many living cells (brain and heart cells, sperm, ova, unicellular organisms) where these proteins, among other functions, are activators for adenylate cyclase and cAMP-phosphodiesterase. The adenylate cyclase activator has been baptized "calmodulin" (for reviews see Cheung, 1980; Means and Dedman, 1980), and we shall see that it is warranted to attribute this name also to the Ca pump activator of red cells. Gopinath and Vincenzi (1977), Larsen and Vincenzi (1979) and Vincenzi and Hinds (1980) demonstrated in elegant experiments that bovine brain calmodulin can replace the red cell activator protein in the ATPase function and in transport of Ca^{2+} in the red cell (Fig. 4), but that troponin C cannot. Vincenzi and Larsen (1980) reported that for half maximal effect 2.5 nM calmodulin is required in $(Ca + Mg)$-ATPase and 4.4 nM in transport (see Fig. 4).

It appears that the red cell factor is functionally and very probably chemically identical with calmodulin from other sources. Its monomeric molecular weight is 16 700 (Jarrett and Penniston, 1978). It has its isoelectric point at a pH of 3.5–4.5 and is resistant to heating to 95°C for a few minutes. Its amino acid composition is very similar to that of calmodulin from bovine brain (Jarrett and Penniston, 1978), and after trypsin digestion the peptide mapping gave the same pattern for erythrocyte and bovine brain calmodulin (Jarrett and Kyte, 1979). Another method for its isolation has been proposed recently by Mualem and Karlish (1979b).

The literature is cluttered up with reports about different Ca^{2+} affinity and complex dependence of ATPase activity on Ca^{2+} concentration. In 1972 it was surmised that use of Ca^{2+} chelators during preparation of membranes caused the low affinity state (Scharff,

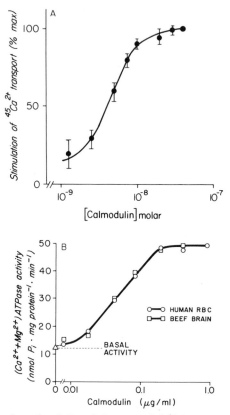

Fig. 4. *A, stimulation by calmodulin of the rate of Ca^{2+} transport in inside-out vesicles prepared from human red cells according to Sarkadi et al. (1977). Maximal rate was 14.4 nmol/mg of protein. min. Mean ± SD from 3 expts. Apparent K_D for calmodulin 4.4 nM. (Vincenzi and Larsen, 1980.) B, (Ca + Mg)-ATPase activity of fragmented human red cell membranes as a function of calmodulin concentration (from red cells or beef brain). Apparent K_D for calmodulin 2.5 nM (Vincenzi and Larsen, 1980). The same authors also showed that calmodulin from human red cells and beef brain are equipotent on red cell Ca^{2+} transport (Larsen and Vincenzi, 1979).*

1972; Schatzmann, 1973). Scharff and coworkers have painstakingly analysed the reaons for the apparently very unpredictable behaviour of the enzyme (Scharff, 1972, 1976, 1978, 1980; Scharff and Foder, 1978), showing that in fact omission of all calcium during preparation of isolated membranes yields low affinity for Ca^{2+} (and low V_{max}). In his terminology such membranes are said to be in the A-state. On the other hand low Ca^{2+} concentrations (such as may be present as contaminant in chemicals used to prepare the washing solutions) are able to bring about (or to preserve) the behaviour characterized by a K_{Ca}

of 1 μM, high V_{max} and slight positive cooperativity with respect to Ca^{2+}. These are B-state membranes according to Scharff. Farrance and Vincenzi (1977a, b) and Scharff and Foder (1978) clearly demonstrated that Ca^{2+}-free solutions remove the activator and that in Ca^{2+}-containing solutions the activator is again attached to the enzyme, the rate of the process being dependent on the Ca^{2+} concentration. Scharff and Foder did not use a purified calmodulin but employed EGTA extracts (1.2 mM EGTA in 70 mM Tris-Cl, pH 7.4, 37°C, 5 min) of B-membranes as such.

Although the simple procedure of diluting calmodulin away from membranes in Ca^{2+}-free medium is quite convincing, Lynch and Cheung (1979) were at pains to show that solubilized red cell (Ca + Mg)-ATPase and brain calmodulin stay together on a Sephacryl S200 column in a calcium medium but appear separately in an EGTA medium. Niggli et al. (1979) essentially confirmed the Ca^{2+} requirement for calmodulin binding also for solubilized (Ca + Mg)-ATPase, but made it a point that low ionic strength is important for detaching calmodulin. This was originally proposed by Farrance and Vincenzi (1977a, b), who observed that hypotonic phosphate or imidazole during cell lysis removes calmodulin, whereas isotonic imidazole (which rapidly penetrates) does not, and that after haemolysis hypotonic media no longer dissociate the complex, but EGTA does.

Removal of activator is sluggish (Wüthrich, unpublished). Reattachment is a slow process unless high calmodulin and Ca^{2+} concentration are employed (Scharff and Foder, 1978; Hinds et al., 1980). Even very prolonged extraction does not bring ATPase activity to nil (Gopinath and Vincenzi, 1977; Scharff, 1980; Wüthrich, unpublished). I am inclined to think that what calmodulin does is to increase the affinity for Ca^{2+} of the Ca^{2+} transport site which has a finite value in the calmodulin-depleted enzyme (Fig. 5). It seems difficult to decide whether calmodulin has in addition an increasing effect on V_{max}, because in calmodulin depleted enzyme the Ca^{2+} activation curves move towards the region of inhibitory Ca^{2+} concentration and the affinity for this Ca^{2+} self-inhibition is clearly not affected by calmodulin (Wüthrich, unpublished) (Fig. 5). An experiment similar to that of Fig. 5 was presented by Klinger et al. (1980). The curves published by Gopinath and Vincenzi (1977) are compatible with the assumption of a single effect on Ca^{2+} affinity only. However, Au (1978b), working with pig red cells and activator prepared according to Luthra et al. (1977) from pig red cells (Chromatography on Sephadex C50, heat treatment, preparative electrophoresis) found only an effect on V_{max} and none on affinity.

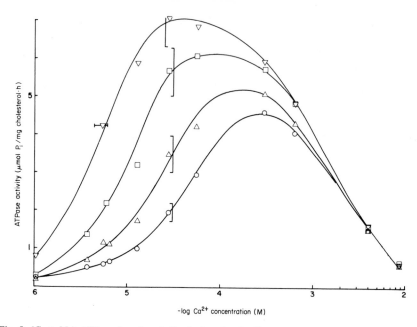

Fig. 5. *(Ca + Mg)-ATPase in calmodulin-depleted red cell membranes as a function of Ca^{2+} concentration at different concentrations of added calmodulin. Membranes were freed from calmodulin by washing in 50-fold volume of (mM): Tris-Cl 15, pH 7.75, EGTA 1; incubation for 1 h at 37°C in this solution and 3 further washings. Assay: Incubation for 90 min at 37°C in (mM): KCl 120, imidazole-Cl 30, $McCl_2$ 4, Na_2ATP 2, ouabain 0.17, $CaCl_2$ to give desired Ca^{2+} concentration. Ca^{2+}-free controls: 0.5 mM EGTA. Up to 12 μM Ca^{2+} solutions contained 0.5 mM EGTA. Calculated Ca^{2+} in EGTA buffers corrected according to measurements with a Ca^{2+}-selective electrode. Over-all Ca chelation in final medium calculated by iterative procedure for EGTA + ATP + Mg + Ca. Red cell calmodulin added (ng/ml): ○ none, △ 6.4, □ 25.6, ▽ 410. Activity referred to cholesterol because protein loss during calmodulin depletion prevents comparison with intact cells on the basis of membrane protein. Mean of 4 expts with 4 different preparations ± SEM (bars). Horizontal bar: largest SEM in points with unequal Ca^{2+} concentration (Wüthrich, unpublished).*

Scharff and Foder (1978) found both an effect on Ca^{2+} affinity (by a factor of 30) and roughly a doubling of V_{max}. Muallem and Karlish (1981) conclude from careful experiments that calmodulin affects both Ca^{2+} affinity and turnover rate and is required for the accelerating effect of ATP at the low affinity, "regulatory" site.

Wolff *et al.* (1977) studied the binding of divalent cations to calmodulin. They came to the conclusion that there are four metal binding sites and that three of them are occupied by Ca^{2+} and one by Mg^{2+} at physiological Mg^{2+} concentration. Under these conditions the dissociation constant for Ca^{2+} was 3×10^{-6} M. They think that Ca^{2+} and Mg^{2+} are competitive at the same sites, whereas Dedman *et al.* (1977) found a $K_D = 2.4 \times 10^{-6}$ M for Ca^{2+} but no competition between Ca^{2+} and Mg^{2+} (see also Seamon, 1980). Jarrett and Kyte (1979) also studied Ca^{2+} binding to erythrocyte calmodulin. By equilibrium dialysis they found a dissociation constant for high affinity sites of 7×10^{-6} M (at $0°C$), and a probable number of gram-atoms of Ca per mole of protein of 4. In addition there were low affinity binding sites present. By measuring change in fluorescence due to Ca^{2+} binding to calmodulin a concentration for half effect of about 10^{-6} M was found by the authors. At any rate both measuring techniques gave nearly identical results for erythrocyte calmodulin and bovine brain calmodulin.

When studying the dependence of rate of ATPase on calmodulin concentration at $30\,\mu M$ Ca^{2+} and by assuming the reaction scheme shown on the left-hand side below

I E + Ca $\rightleftharpoons$ E . Ca   II A + Ca $\rightleftharpoons$ A . Ca
   +   +   +   +
   A   A   E   E
   $\updownarrow$ $\updownarrow$ $\uparrow$ $\updownarrow$
   A.E + Ca $\rightleftharpoons$ A.E.Ca   A.E. + Ca $\rightleftharpoons$ Ca.A.E.

(A = free calmodulin, E = free enzyme, AE concentration of
calmodulin-occupied enzyme sites)

Jarrett and Kyte (1979) found for the apparent dissociation constant for calmodulin binding to whole isolated membranes

$$K_{app} = \frac{([E] + [E.Ca]) . [A]}{[A.E.] + [A.E.Ca]} = 14.5\,nM \qquad (2)$$

They chose a reaction scheme for the interaction between pump protein, calmodulin and Ca^{2+} in which the Ca^{2+}–calmodulin complex does not appear (I). Scheme II shown on the right-hand side above seems to comply better with what little we know: (i) calmodulin binds Ca^{2+}, (ii) Ca^{2+} is required for the association between pump protein and calmodulin. However, since the Ca^{2+} concentration used was saturating calmodulin, this should not matter in the authors' determination of K_{app} for binding of calmodulin to the pump protein (i.e. whole membranes).

Vincenzi and Hinds (1980), by matching the probabilities of different Ca–calmodulin complexes with the activation curve of (Ca + Mg)-ATPase at variable Ca^{2+} concentration, came to the conclusion that the activating species are the calmodulin complexes with two or more Ca^{2+} ions attached. Scharff (1980) proposed three Ca to one Mg in active calmodulin and gave some evidence that 4-Ca-calmodulin might be responsible for Ca self-inhibition.

The reader is warned not to get confused by Ca^{2+} appearing on one hand in calmodulin binding and on the other hand in the Ca^{2+} pumping action of the system. First it is not known whether calmodulin binding is via Ca^{2+} cross-bridges or whether Ca^{2+} favours binding by changing the conformation of calmodulin. Secondly, if Ca^{2+} required in the calmodulin binding were to make direct contact with the main protein, it would be highly unlikely that it is this Ca^{2+} which is transported and activates the ATPase, in view of the demonstrated slowness of dissociation of the complex. If, on the other hand, calmodulin were, so to speak, the ionophoric part of the system and the main protein the energizing part, one cannot readily understand why by partial stepwise removal of calmodulin from the system the affinity for Ca^{2+} in the transport process (and in the ATPase activity) falls accordingly, because this should only reduce the number of active units and thus the V_{max}, unless one makes complicated assumptions such as interactions between several calmodulin molecules per functional unit.

Recently Mauldin and Roufogalis (1980) described a further protein in human red cells which activates (Ca + Mg)-ATPase by a factor of 1.2–2.0, has a molecular weight of 63 000 and was obtained virtually free of calmodulin.

The cytosolic inhibitor(s). Wang and Desai (1977) described the isolating from bovine brain of a protein that antagonizes the activation of cAMP-phosphodiesterase by calmodulin. They also showed that the inhibition was due to combination with calmodulin and that this association required Ca^{2+}. Larsen et al. (1978b) tested a bovine brain inhibitor on the red cell (Ca + Mg)-ATPase and Ca^{2+} transport in inside-out vesicles and found "competitive" behaviour between it and calmodulin from red cells, with the "basal" activity (not sensitive to calmodulin) remaining unaffected by high concentrations of the inhibitor. Au (1978a) described for pig red cells a protein fraction of membrane-free cell lysate (obtained by ion exchange chromatography) containing calmodulin plus an inhibitor protein. He was unable to separate the two properties by heating which, as we have seen, is not detrimental to calmodulin but destroys the inhibitor protein. Sarkadi et al. (1980b) demonstrated that membrane-free haemolysate or a partially purified fraction from it was stimulatory at low and inhibitory at higher concentration (at constant membrane concentration in the assay system). Wüthrich (unpublished) succeeded in obtaining a fraction containing several proteins from membrane free human red cell lysate which was inhibitory at moderate calmodulin concentrations and whose action was overcome by high calmodulin concentration.

Maimon and Pushkin (1978) described a troponin-inhibitor-like protein from red cells. It seems risky to equate this with the other calmodulin-antagonistic inhibitory proteins because troponin C is, if at all, at least 100 times less active than calmodulin on Ca^{2+} transport in red cells (Larsen and Vincenzi, 1979).

It seems probable, therefore, that one or several proteins, able to combine with calmodulin and apt to inhibit thereby its action on the Ca-pump, also exist in red cells. The physiological significance of this is at present obscure, particularly in view of the possibility that the inhibitory association requires Ca^{2+}, or, more cautiosly expressed, the inhibitor binds to the Ca form of calmodulin. On the other hand it may be significant in this context that studies into Ca^{2+} transport in intact cells loaded with Ca^{2+} (with ionophore A 23187 or salicylate) revealed a very much lower apparent affinity for Ca^{2+} at the internal site of the transport system than what is seen in isolated membranes (Ferreira and Lew, 1976; Bürgin and Schatzmann, 1979). This might mean that in fact under normal conditions the intracellular calmodulin is not fully active, perhaps because of combination with other soluble proteins. This will become clearer when the abundance of the inhibitor protein(s) relative to calmodulin and their affinities will be known.

Varghese and Brown-Cunningham (1980) described an inhibition of the (Ca + Mg)-ATPase by cyclic AMP in nanomolar concentrations of the order of 10% which was absent at higher cAMP concentrations. They think that this has something to do with phosphorylation or dephosphorylation of the main protein by a protein kinase–phosphatase system.

Lipid requirement and reconstitution. The main protein is indubitably an integral membrane protein, i.e. it spans the membrane and must be built into the membrane by hydrophobic interaction between its surface and other lipophilic components of the membrane. It seems to be a pertinent question, therefore, whether function and stability depend in any way on the lipid surrounding of the protein. Roelofsen (Roelofsen, 1977; Roelofsen and Schatzmann, 1977) and Ronner *et al.* (1977) embarked on studying this question with respect to (Ca + Mg)-ATPase. Their results diverge vastly. Roelofsen employed the following technique: intact washed red cells or (calmodulin saturated) membranes were used. The outer phospholipids were degraded by successive incubation of cells with pure phospholipase A_2 from *Naja naja* venom and sphingomyelinase C from *Staphylococcus aureus*. The enzyme reactions were stopped by adding EGTA. Attack on the total phospholipids was achieved by treating disrupted membranes with pure phospholipase A_2 from *N. naja* venom or from pancreas (whereby in some experiments lyso-compounds and free fatty acids were removed with bovine serum albumin) or treatment with phospholipase C from *Bacillus cereus* or *Clostridium welchii*. Content in different phospholipids was measured and (Ca + Mg)-ATPase activity determined after such treatment. Restoration of activity was tested after adding sonicated pure phospholipids, fatty acids or lysophosphatidylcholine. Complete hydrolysis of glycerophospholipids and sphingomyelin of the outer membrane leaflet did not affect the (Ca + Mg)-ATPase nor did complete hydrolysis of all the sphingomyelin. Phosphlipase A_2 attack from both sides, carried essentially to completion, inhibited by 80–85% and the residual 15–20% activity disappeared when lysocompounds (and fatty acids) were removed. The degree of inactivation by phospholipase C (from both sides) was directly proportional to the fraction of total glycerophospholipids removed. After essentially complete inactivation such membranes could be reactivated by any of the glycerophospholipids but not by sphingomyelin, free fatty acids or Triton-X-100. The effect of removing phospholipids clearly bears on V_{max} and not on affinity for Ca^{2+} (Roelofsen and Schatzmann, 1977). The conclusion is that any of the glycerophospholipids can sustain the activity of (Ca + Mg)-ATPase and that their presence is essential only in the internal membrane leaflet. This is in marked contrast to the behaviour of the (Na + K)-ATPase where Roelofsen and van Deenen (1973) and Roelofsen (1977) showed that negatively charged glycerophospholipids, such as phosphatidylserine and (for rabbit kidney) also phosphatidylinositol (Mandersloot *et al.*, 1978) are essential for maintaining functional integrity. Ronner *et al.* (1977), who applied similar techniques to (Ca + Mg)-ATPase of red cells, came to different conclusions. In their experiments a marked predilection for phosphatidylserine existed and free fatty acids were also effective in preserving activity. Niggli (Niggli *et al.*, 1979; Carafoli and Niggli, 1981) confirmed this claim to superiority of phosphatidylserine (PS) over phosphatidylcholine (PC) in maintaining activity in her purified (Ca + Mg)-ATPase, whereas Stieger and Luterbacher (1981) convinced themselves that the enzyme purified according to Gietzen *et al.* (1980c) performed perfectly well in pure phosphatidylcholine.

Two recent experiments might explain the disagreement. Niggli *et al.* (1981) prepared purified pump protein in Triton-X-100 and PC and reconstituted this material with either PC or PS. Stieger and Luterbacher (1981) prepared purified pump protein according to Gietzen *et al.* (1980c) with either PC or PS. Both experiments gave the result that the PC system behaves as the native membrane, i.e. is calmodulin dependent, whereas the PS system is fully activated in the absence of calmodulin and unreactive to calmodulin. Thus the better reactivation by PS (Ronner *et al.*, 1977) in (enzyme degraded) membranes may have been due to calmodulin depletion. The experiments with purified (Ca + Mg)-ATPase are somewhat difficult to interpret, because in the micelles obtained some detergent is always included.

Nearly complete removal of phospholipids by enzymatic attack leading to total abolition of (Ca + Mg)-ATPase function does not destroy the enzyme, because it was clearly shown that it can be reactivated by a later addition of phospholipids. On the other hand it is common knowledge that solubilized (Ca + Mg)-ATPase does not survive for any length of time as soon as phospholipids and the used detergent are removed. It is puzzling that the protein in the isolated state is extremely unstable without phospholipids, but that the functionally pertinent phospholipids obviously can be removed from the membrane structure without irreversible damage being done to the protein.

Richards *et al.* (1977b) also degraded red cell membranes by phospholipase A_2 (from pancreas) and phospholipase C (from *Bacillus cereus*). On top of finding reduction of the (Ca + Mg)-ATPase activity not unlike that described by Roelofsen and Schatzmann (1977) they noticed that phospholipase C reduced the steady-state level of Ca^{2+}-dependent phosphorylated intermediate drastically (to a few per cent of the original value). Since in their experiments there was hardly any effect of the phospholipase C on (Ca + K)-activated phosphatase activity and an unequivocal stimulation of the Ca^{2+}-activated phosphatase, they concluded that phospholipids are essential only in the phosphorylation step. This conclusion depends on the view, advocated strongly by these authors (Pouchan *et al.*, 1969; Rega *et al.*, 1972, 1973, 1974), that phosphatase substrates (for instance *p*-nitrophenylphosphate) can replace the phosphate group in the phosphorylated intermediate in the normal sequence of events of the (Ca + Mg)-ATPase and that phosphatase activity thus reflects the performance of the second half of the ATPase cycle. Incidentally, the authors found a difference between red cells and sarcoplasmic reticulum: resistance to phospholipase C of (Ca^{2+}-stimulated) phosphatase activity could not be demonstrated in sarcoplasmic reticulum.

By *reconstitution* of a transport system is meant the reassembly of the functional membrane protein with its own lipid surroundings or with pure lipids possibly in a way which, somehow, allows testing of sidedness or, with some luck, of transport. The most popular way is to suspend lipid vesicles (liposomes) together with a solubilized protein and to remove the added detergent (for instance by dialysis for deoxycholate or on Biobeads SM-2 for Triton). If the detergent is successfully removed the system will no longer form lipid–detergent–protein micelles but the lipids will arrange themselves in protein-containing bilayers which have the tendency to form closed spherical vesicles. The proteins will in all probability be inserted into the vesicle walls randomly, i.e. 50% head-on and 50% tail-on. However, in an ATP-energized system 50% of the protein assuming the correct orientation will be activated by externally added ATP and the other 50% will be silent such that the whole system will transport Ca^{2+} (in the present case) from the outside to the vesicular lumen. Other, experimentally more exacting, procedures exist, such as inserting the protein into a planar lipid bilayer expanded as a diaphragm in a small aperture between two solvent compartments (black films). The purification experiments described on p. 48–49 are essentially reconstitution experiments, except for the removal of detergent and thus probably always yielding micelle structure rather than laminar bilayers.

Peterson *et al.* (1978) solubilized red cell membranes with Triton-X-100 prior to or after an attempt at purifying the proteins by electrofocusing, added phospholipids, sonicated the mixture and removed the Triton-X-100 on a Biobead column. Finally, vesicles of 0.05–1 μm diameter were harvested free of soluble proteins from a Sepharose 4B column. The ATPase activity in this reconstituted system was tested, whereas no attempt was made to ascertain whether the system did transport Ca^{2+}. The best yield in specific ATPase activity (calculated per milligram of protein) was obtained when the added phospholipid was phosphatidylserine and was 14-fold higher than the specific activity of the original red cell membranes. Results seem to have been similar regardless of whether the crude solubilizate or proteins collected between pH 3 and pH 7 in the electrofocusing procedure were used. The electrofocusing procedure somewhat lowered the total protein content of the final vesicles and considerably shifted the relative composition in different proteins.

Gietzen *et al.* (1980b) published a solubilization–reconstitution using deoxycholate as detergent applied to erythrocyte membranes treated with Tween 20 which seems to "stabilize" the system. The degree of solubilization was defined by the fact that the material which was further processed remained in the supernatant during 60 min centrifugation at 140 000 g. Sonicated lecithin was added to the supernatant and deoxycholate removed by dialysis at 18° C The authors collected unilamellar vesicles of 0.2–0.8 μm diameter in which the protein–lipid ratio was similar to that in the original red cell membranes and in which a definite gain in specific activity of (Ca + Mg)-ATPase (10–15 times compared to red cell ghosts) was achieved. At high external Ca^{2+} concentration (0.1 mM) and oxalate concentration (5 mM) the vesicles accumulated Ca^{2+} when ATP was added to the medium. The ratio of Ca^{2+} transport rate to ATP-splitting rate was at best 0.3. The Ca^{2+} uptake was drastically reduced by ionophore A 23187; however, accumulated Ca was not released by the ionophore, possibly because of formation of intravesicular Ca–oxalate crystals. Increasing the lecithin supplement increased both the rate of Ca uptake and the steady-state vesicular Ca concentration.

Haaker and Racker (1979) solubilized and purified (Ca + Mg)-ATPase from pig red cells essentially according to Wolf *et al.* (1977) and incorporated it into liposomes made from crude soybean phospholipids by freeze–thaw sonication. They obtained ATP-dependent [45]Ca uptake, which could be stimulated considerably by added calmodulin. A 23187 abolished the Ca uptake and increased the ATP splitting rate. Passing the reconstituted vesicles over a phenyl Sepharose 4B column improved the Ca to ATP ratio in the transport experiment from 0.25 to 0.4.

Carafoli and Niggli (1981) incorporated Niggli's purified protein (Niggli *et al.*, 1979a) into liposomes made of total soybean lipids, removing Triton-X-100 by treatment with Biobeads. The resulting product showed high affinity for Ca^{2+} ($K_{Ca} \sim 0.3$ μM), was not stimulatable by calmodulin and did accumulate Ca^{2+} when ATP was present in a medium which did not contain oxalate (Fig. 6). The accumulation ratio was not measured but accumulated Ca was very probably present in the intravesicular space because it could be released by A 23187 (see Fig. 6). Ca uptake reached equilibrium in about 3 min at 37° C. Comparison of initial rate of Ca uptake with initial ATP hydrolysis gave a Ca/ATP ratio of 1.1. This does of course not mean that 1 : 1 stoichiometry for the pump has been proved, because there may have been protein not incorporated into tight vesicles, contributing

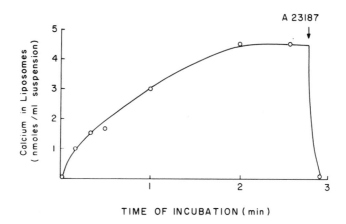

Fig. 6. *Ca²⁺ transport by isolated (Ca + Mg)-ATPase, reconstituted into soybean phospholipid (asolectin) liposomes. Medium (mM): KCl 120, Hepes 30 (pH 7.4), MgCl₂ 0.5, ⁴⁵CaCl₂ 0.02 (no oxalate); temp. 37°C. Ordinate: intravesicular calcium, expressed per ml of suspension. A 23187 added to give 1 μM at arrow. Ionophore addition reveals a considerable concentration gradient after 3 min between vesicular interior and medium. (Redrawn from Carafoli and Niggli, 1981, with permission.)*

to ATPase but not to Ca accumulation. What these experiments do prove, however, is that the isolated high affinity (Ca²⁺ + Mg²⁺)-ATPase is the Ca pump. This identity is thus no longer an inference but a demonstrated fact.

Summary. The Ca pump protein (or the (Ca + Mg)-activated ATPase) is an integral protein of molecular weight ~ 140 000 requiring phospholipids to be active, which can form dimers and may be present as a dimer in the membrane. It has high affinity for Ca²⁺ (K_{Ca} ~ 1 μM) when saturated with the activator calmodulin. Upon removing calmodulin a K_{Ca} as high as 80 μM has been observed. Calmodulin found in the cytosol of red cells is indistinguishable from that stemming from other cells. The attachment of calmodulin to the main protein requires in its turn intracellular Ca²⁺. Calmodulin may be bound to intracellular proteins (in the presence of Ca²⁺) which abolishes its stimulating action. In the presence of calmodulin any glycerophospholipid may sustain the functional state; in its absence phosphatidylserine gives full activation. The purification and reconstitution studies establish beyond reasonable doubt that the (Ca + Mg)-ATPase and the Ca pump are identical.

The transport cycle
Reaction sequence. Knauf *et al.* (1972, 1974) and Katz and Blostein (1973, 1975) demonstrated that in the presence of Ca^{2+} the terminal phosphate group of ATP is transferred to a membrane protein appearing in SDS gels at a molecular weight of 140 000. The experiment is very simple: isolated disrupted red cell membranes are exposed to [γ-^{32}P]-labelled ATP at 0°C for a few seconds, the proteins are precipitated and the precipitate washed extensively with trichloroacetic acid containing unlabelled ATP and inorganic phosphate. The precipitate is redissolved and used for SDS gel electrophoresis or counted directly for radioactivity. The Ca^{2+}-induced phosphorylation is distinguished from that of other membrane proteins by its rapidity: at 0°C the time for reaching half maximum value is in the order of 8–15 s (Rega and Garrahan, 1975; Garrahan and Rega, 1978; Schatzmann and Bürgin, 1978). From Fig. 1 it may be seen that up to 30 s there are no other proteins phosphorylated. Knauf *et al.* (1974) also made it clear that the Ca^{2+} induced protein phosphorylation cannot be confounded with the equally rapid phosphorylation elicited by $Na^+(+ Mg^{2+})$ in the protein of the (Na + K)-ATPase, because the molecular weights of the labelled proteins obtained from SDS gel electrophoresis coincide with the monomeric molecular weights (determined independently) which are 140 000 for the Ca system and 100 000 for the α-subunit of the Na–K system (Fig. 2; see also Drickamer, 1975). Lichtner and Wolf (1980a, b) have shown that in their purified material which consists of three major proteins electrophoretically, only the one of 140 000 molecular weight was susceptible to Ca^{2+}-induced phosphorylation. Niggli *et al.* (1979a) who obtained with their purification procedure essentially only two bands in SDS gel electrophoresis, corresponding to the pump protein and presumably its dimer, were able to demonstrate that both these peaks could be phosphorylated in a Ca^{2+}-dependent fashion. Strictly Ca^{2+}-dependent phosphorylation of the protein isolated according to Gietzen *et al.* (1980c) is shown in Fig. 3. Lichtner and Wolf (1980b) pointed out that the phosphorylation is completely independent of cyclic AMP. The observations on the nearly pure pump protein seem to exclude the possibility that the phosphorylation is the result of the action of a protein kinase and make it very likely that the phosphoprotein is the intermediate reaction product of the (Ca + Mg)-ATPase, as was strongly suggested by Knauf *et al.* (1974). A detailed picture of the sequence of events taking place in the cyclic phosphorylation–dephosphorylation of the Ca pump is available mainly thanks to a

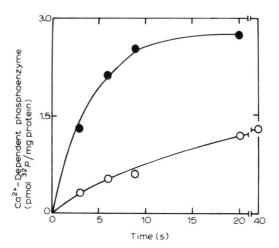

Fig. 7. *Fragmented membranes from human red cells. Time course of the formation of Ca²⁺-
dependent phosphoenzyme in the presence (•) and absence (○) of 0.5 mM MgCl₂;
medium (mM): Tris-Cl 150 (pH 7.4), [γ-³²P]-ATP 0.03, EGTA 0.1, CaCl₂ 0.15;
temp. 0–3°C. (Garrahan and Rega, 1978.)*

series of publications of high acumen by Rega, Garrahan and coworkers who studied the requirements of the reactions taking place.
The facts that have been emerging successively are summarized in
the following and it will appear that they can be arranged to give a
coherent view of the sequential reaction steps.

(i) Phosphorylation occurs in the presence of Ca^{2+} without any
Mg^{2+} (Rega and Garrahan, 1975) (see Figs 1 and 3). Affinity for
Ca^{2+} is high even in the absence of ATP (Richards *et al.*, 1977a)
and Ca^{2+} does not influence the affinity for ATP (Schatzmann,
1977). The phosphorylation experiments were performed in 0.5 mM
EGTA and 0.1 mM $CaCl_2$ and Mg^{2+}-free samples contained 0.1 mM
EDTA, so that possible Mg^{2+} contamination was probably markedly
reduced. Mg^{2+} has no effect by itself but accelerates phosphorylation
in the presence of Ca^{2+} (Rega and Garrahan, 1975; Garrahan and
Rega, 1978) and causes the steady-state amount of phosphoprotein
(maximum reached after 20 s) to increase (Rega and Garrhan, 1975;
Bürgin and Schatzmann, 1978) (see Fig. 7). Since Mg^{2+} is not
required, it is clear that Mg-ATP is not the substrate. Graf and
Penniston (1981) presented evidence for Ca-ATP to serve as substrate
with a $K_m \sim 0.01\ \mu M$. In the absence of Mg^{2+}, the ATP-concentration
for half-maximal phosphorylation was found to be 1.6 μM. Adding
0.012 mM $MgCl_2$ increased this value to 6.5 μM (Rega and Garrahan,
1975). Ca^{2+} concentration for half-maximal phosphorylation was

$7 \mu M$ in their trials (Rega and Garrahan, 1975), but $0.2 \mu M$ in the experiments by Szasz *et al.* (1978a).

(ii) When a chase with high concentration of unlabelled ATP is applied, the phosphorylated intermediate made in the presence of Ca^{2+} decays rapidly upon adding Mg^{2+} (Rega and Garrahan, 1975). The phosphorylated intermediate made in the presence of both Ca^{2+} and Mg^{2+} decays rapidly in the absence and equally rapidly in the presence of Mg^{2+} (Rega and Garrahan, 1975; Schatzmann and Bürgin, 1978). This suggests that Mg^{2+} is not necessary for hydrolysis itself, but for a step prior to hydrolysis, i.e. that it brings the phosphoprotein into a form which can react with water. The K_{Mg} (for free Mg^{2+}) in this reaction step was found to be $80 \mu M$ (Garrahan and Rega, 1978), which is not in too good agreement with $14 \mu M$ for the ATPase reaction (Schatzmann, unpublished).

(iii) When using high (30 mM) EGTA concentration instead of adding cold ATP to stop further incorporation of phosphate (reducing the Ca^{2+} concentration to 4×10^{-10} M) Garrahan and Rega (1978) demonstrated clearly that 0.15–0.2 mM Mg^{2+} did not affect at all the dephosphorylation of the phosphorylated intermediate (formed in the presence of Ca^{2+}) unless high concentrations of ATP (0.155 mM free or 1 mM total) were present (Fig. 8A). The free ATP and metal concentrations are not given in the paper, but were calculated from the composition of the media for a two-metal double-buffer system. In addition the authors showed that after Mg^{2+} had been acting on the system it could be removed (by chelation with CDTA) without abolishing the subsequent stimulation of decay by ATP (see Fig. 8B). This clearly indicates that there must be a receptor of low affinity for free ATP (rather than Mg-ATP) whose occupancy makes the phosphoryl bond prone to hydrolysis. From calculations of ATP, Mg-ATP and Ca-ATP in their media Muallem and Karlish (1981) reach the opposite conclusion, namely that Mg-ATP is the active agent which accelerates at the regulatory site (and that Ca-ATP is a competitive inhibitor).

There is good evidence from three independent studies on the ATPase (Richards *et al.*, 1978; Mualem and Karlish, 1979a, 1980; Stieger and Luterbacher, 1981) and on Ca^{2+} transport (Mualem and Karlish, 1979a; Mollman and Pleasure, 1980) for two ATP sites of different affinity, based on biphasic activation curves (see Fig. 9). Stieger and Luterbacher (1981) demonstrated that this also holds for the purified protein. Biphasic activation curves can have three different meanings: (1) negative cooperativity at the substrate site, (2) mixture of two enzyme populations or (3) activation at the

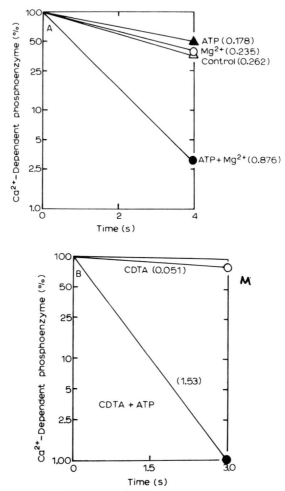

Fig. 8. A, fragmented membranes from human red cells. Dephosphorylation of Ca^{2+}-dependent phosphoenzyme made in the absence of $MgCl_2$ during 20 s; temp. 0–3°C. Conditions as in Fig. 7. △ control, ▲ medium with 1 mM ATP, ○ with 0.5 mM $MgCl_2$, ● with both (1 mM ATP + 1.5 mM $MgCl_2$). All media contained during dephosphorylation 30 mM EGTA to stop phosphorylation. B, effect of removal of Mg^{2+} on dephosphorylation in media with and without added ATP. Phosphoenzyme was made in the absence of Mg^{2+}. The concentration of $MgCl_2$ was raised to 0.5 mM 5 s before phosphorylation was terminated. Dephosphorylation was initiated by addition of 20 mM CDTA (trans-1,2-diaminocyclohexanetetraacetic acid) or 20 mM CDTA plus 1 mM ATP. Figures in brackets are rate constants (per second) calculated assuming first order kinetics. The experiments show that Mg^{2+} and ATP are necessary for dephosphorylation but that Mg^{2+} is required at a step preceding that accelerated by ATP. (Garrahan and Rega, 1978.)

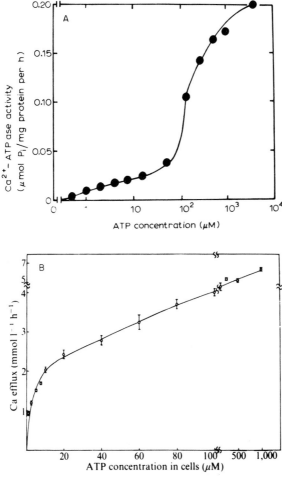

Fig. 9. A, (Ca + Mg)-ATPase activity of fragmented human red cell membranes as a function of ATP concentration. Medium (mM): Tris-Cl 150, MgCl₂ 1, EGTA 0.5, CaCl₂ 0.7; pH 7.8, ATP above 0.1 mM added with equimolar amount of $MgCl_2$. Below 0.1 mM ATP determination of phosphate liberated by use of $[\gamma\text{-}^{32}P]$ ATP. K_m = 2.46 μM and 143 μM respectively. (Richards et al., 1978.) B, ATP dependence of active Ca^{2+} efflux from resealed human red cell ghosts. Lysis of starved cells in: ATP as indicated, creatine phosphate 10 mM, creatine kinase 25 U, $^{45}CaCl_2$ 2.5 mM, HEDTA 1 mM, $MgCl_2$ 2.5 mM, Hepes–Tris 10 mM (pH 7.4), inosine 5 mM, ouabain 0.1 mM. Isotonicity restored by NaCl. Next cells were washed in 130 mM NaCl, 10 mM KCl, 2.5 mM $MgCl_2$, 10 mM Hepes-Tris (pH 7.4), the last wash solution containing 0.5 mM EGTA. Incubation in wash solution. K_m ~ 2 μM and 190 μM respectively. (Mualem and Karlish, 1979a.)

substrate site plus acceleration at a second site with lower affinity. Possibility (3) is the interpretation required here. It is made probable by the further observation that the high affinity site has ATPase activity even in the absence of Mg^{2+} (Richards *et al.*, 1978, see below). K_m for one site was found to be 2.5 μM (Richards *et al.*, 1978) and 1–2 μM (Mualem and Karlish, 1979a) and for the other site 145 μM (Richards *et al.*, 1978) and 180 μM (Mualem and Karlish, 1979a). The high affinity is attributed to the catalytic site proper and the low affinity to the hydrolysis stimulating site. K_m values refer to total ATP and it is likely that both sites operate with free ATP. Inserting free ATP as variable in the results of Richards' experiments shifts the low affinity site K_m to the micromolar range. This does not alter the fundamental finding of two ATP sites. It must be mentioned that Mualem and Karlish used whole resealed cells in order to show the two-site behaviour also on Ca^{2+} transport. However, since the system displayed high Ca^{2+} affinity, its state with regard to calmodulin must have been optimal. Here again the free ATP concentration for half saturation of either site may have been considerably lower than the figures given, since 2.5 mM $MgCl_2$ was present in the lysing medium and thus the intracellular Mg^{2+} concentration was high. At very high Ca^{2+} concentrations (> 3 mM) the biphasic curve merges into a simple hyperbola giving an intermediate K_m of 5–10 μM (Mualem and Karlish, 1979a). Mualem and Karlish (1980) observed a hyperbolic curve in calmodulin-stripped membranes ($K_m \sim 7\,\mu$M total ATP) but a biphasic curve in the presence of calmodulin ($K_{m_1} \sim 1.4\,\mu$M and $K_{m_2} \sim 330\,\mu$M). The absence of any sign of such a biphasic behaviour in our experiments (Schatzmann, 1977) cannot be attributed to either high Ca^{2+} concentration or absence of calmodulin, because there the Ca^{2+} concentration was 0.04 mM and Ca^{2+} affinity was high. It might mean that under certain conditions of preparing membranes the K_m of the low affinity ATP site may differ less from that of the high affinity site than described in the experiments of the two groups.

Richards *et al.* (1978) showed that with only the high affinity site occupied a small Ca^{2+}-stimulated ATPase activity remains in the absence of Mg^{2+} and that Mg^{2+} enhances the maximum stimulatory effect of Ca^{2+} and has no effect on affinity for Ca^{2+} or ATP. This Mg^{2+} action obeyed simple saturation kinetics with a Mg^{2+} concentration of 0.33 mM for half-maximal activation. In contrast, the rate of rapid dephosphorylation (in the presence of high ATP concentration – (1 mM)) also increases with rising Mg^{2+} concentration along a rectangular hyperbola (from a finite value in the absence of

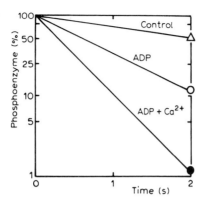

Fig. 10. *The effect of ADP and ADP + Ca²⁺ on dephosphorylation of the Ca²⁺-dependent phosphoenzyme, made in the presence of 0.5 mM MgCl₂. [γ-³²P]-ATP 1 μM. Conditions otherwise as in Fig. 8. After 20 s (zero time in fig.) chase solution was added to final concentrations (mM): ATP 0.03 μM, CaCl₂ 0.1, EGTA 0.067, MgCl₂ 0.34, Tris-Cl 150 (pH 7.4); temp. 0°C. ○ 1 mM ADP + 20 mM EGTA, ● 1 mM ADP. (Rega and Garrahan, 1978.)*

Mg^{2+}) but with a free Mg^{2+} concentration for half-maximal effect of 80 μM (see above; Garrahan and Rega, 1978).

(iv) Both the phosphorylated intermediate formed in the presence of Ca^{2+} alone and that resulting from the combined action of Ca^{2+} and Mg^{2+} can rapidly be decomposed by adding high ADP concentration (Rega and Garrahan, 1978; Schatzmann and Bürgin, 1978). The most probable interpretation for this is that it reflects the backward reaction, forming ATP. Another possibility is that high ADP mimics the action of ATP in facilitating the dephosphorylation in the forward direction. Since decomposition of phosphorylated intermediate made in the presence of $Ca^{2+} + Mg^{2+}$ was rapid in the presence of ADP without Ca^{2+}, but was further accelerated by addition of Ca^{2+} (see Fig. 10), whereas with the phosphorylated intermediate made in the presence of Ca^{2+} only, the ADP effect was strictly Ca^{2+} dependent, both pathways are involved in the former and the back reaction alone is operating in the latter case (Rega and Garrahan, 1978). Interestingly, the La^{3+}-induced phosphorylated intermediate (Schatzmann and Bürgin, 1978; Szasz et al., 1978a) can also be back-reacted with ADP, but cannot decay in the (normal) forward direction.

(v) The phosphorylated intermediate(s) is (are) acid stable and can be decomposed by hydroxylamine (Rega and Garrahan, 1975; Lichtner and Wolf, 1980b). This is usually taken to indicate that the bond in question is an acylphosphate.

(vi) Rega and Garrahan (1975) have calculated from the maximal amount of phosphoenzyme that 700 copies of the pump may be present per cell. From the amount of (Ca + Mg)- ATPase protein retrieved from affinity columns we arrive at 1000–2000 molecules per cell. Thus, with an average pumping rate at saturating Ca^{2+} concentration of 10 mmol/litre of cells. h (assuming one Ca^{2+} ion per cycle), a turnover rate of 4000–8000 cycles/min results (Luterbacher $et\ al.$, unpublished).

The existing knowledge can be summarized as shown in Table I. In view of the normal ATP concentration in human cells being 1.5 mM and the free Mg^{2+} concentration 0.3–0.4 mM (Flatman and Lew, 1977) all the activating steps will be switched on in the physiological state as soon as Ca^{2+} is present inside the cells.

What we believe then to be happening under physiological conditions in intact cells can be depicted as (for the role of Na and K see below):

$$ATP + E_1 \rightleftharpoons E_1 - ATP \underset{k_2}{\overset{k_1}{\rightleftharpoons}} E_1 \sim P \underset{k_4}{\overset{k_3}{\rightleftharpoons}} E_2 \sim P \underset{k_6}{\overset{k_5}{\rightleftharpoons}} E_2$$

$$k_7\ \mathrm{IV}$$

The Mg^{2+} requirement in reactions Ia and Ib calls for a comment. It is based on the finding that with both reaction I and II turned on (with $Ca^{2+} + Mg^{2+}$ but low ATP) the steady state level of $[E_1 \sim P + E_2 \sim P]^{I,II}$ may be larger than the equilibrium level $[E_1 \sim P]^I$ (with reaction II blocked by lack of Mg^{2+}), i.e.

$$\frac{[E_1 \sim P + E_2 \sim P]^{I,II}}{[E_{tot}]} > \frac{[E_1 \sim P]^I}{[E_{tot}]} \tag{3}$$

Assuming for the argument's sake, that Mg^{2+} acts only at reaction II and if in the steady state $E_1 \neq 0$ (ATP concentration limiting) and $E_2 \approx 0$ ($k_7 \gg K_5$), it can be shown that the conditions for (3) to be true are

$$\frac{k_5}{[ADP] \cdot k_2} + \frac{k_4}{k_3} < 1 \tag{4}$$

or

$$[ADP] \cdot k_2 \gg k_5 \quad \text{and} \quad k_3 \gg k_4 \tag{5}$$

This means that the reaction II must be poised to the right and the

Table I. Partial reactions of the Ca pump according to Garrahan and Rega (1978). Effect of Mg^{2+} and ATP on the partial reactions and on the activity of the Ca-ATPase.

ATP (μM)	Mg^{2+} (μM)	Partial reaction	Rate	EP level	ATPase (% of maximum)
30	0	$E_1 + ATP \xrightarrow{Ca^{2+}} E_1 \sim P + ADP$	low	low	4
		$E_1 \sim P + H_2O \longrightarrow E_1 + P_i$	low		
30	500	$E_1 + ATP \xrightarrow{Ca^{2+} + Mg^{2+}} E_1 \sim P + ADP$	high	high	10
		$E_1 \sim P \xrightarrow{Mg^{2+}} E_2 \sim P$	high		
		$E_2 \sim P \longrightarrow E_2 + P_i$	low		
1000	500	$E_1 + ATP \xrightarrow{Ca^{2+} + Mg^{2+}} E_1 \sim P + ADP$	high	unknown	100
		$E_1 \sim P \xrightarrow{Mg^{2+}} E_2 \sim P$	high		
		$E_2 \sim P \xrightarrow{ATP} E_2 + P_i$	high		

back reaction with ADP must be faster than the forward dephosphorylation. The former may easily be true and the latter is not
unlikely in view of the reported rate constant for the Ca^{2+}-induced
back reaction of $\sim 1.5/s$ (Rega and Garrahan, 1978) and that for
dephosphorylation at low ATP concentration ($30\,\mu M$) of $\sim 0.25/s$
(Garrahan and Rega, 1978). However, it seems safer to assume, as
Garrahan and Rega do, that Mg^{2+} acts in addition before reaction II
in conjunction with Ca^{2+} for which there is an absolute requirement.
The reason is that the affinities for Mg^{2+} in phosphorylation and
dephosphorylation are different (see p. 68). In this interpretation a
higher phosphoprotein level reached more rapidly can result from an
increase of $[E_1\text{-ATP}]$ or a shift of reaction Ib to the right. An
increase of affinity for free ATP by Mg^{2+} has been demonstrated
(Schatzmann, 1977), whereas an increase of k_1/k_2 by Mg^{2+} is
hypothetical.

Rega and Garrahan (1980) measured, under identical conditions,
ATPase activity and the steady-state level of phosphorylated intermediate in calmodulin-depleted membranes (obtained by haemolysis
in 1 mM EDTA, 30 mM Tris-Cl, pH 7.2) with and without added
calmodulin at low ATP ($15\,\mu M$). In the absence of Mg^{2+} the steady
state level of phosphorylated intermediate increased less by the
addition of calmodulin (1.4-fold) than the ATPase (2.7-fold). With
Mg^{2+} the steady-state level of phosphorylated intermediate was
decreased (to 0.7) by addition of calmodulin and ATPase activity
was increased (3.8-fold). The conclusion is, that calmodulin increases
mainly the turnover rate of the phosphoenzyme, i.e. that it accelerates both phosphorylation and dephosphorylation. Mualem and
Karlish (1980) found a 10-fold acceleration of phosphorylation by
calmodulin at $50\,\mu M$ Ca^{2+} (Fig. 11), no increase of steady state
concentration of phosphoenzyme in the presence of Mg^{2+} and an
increase in the absence of Mg^{2+}. They point out that the easiest way
to reconcile this with calmodulin increasing the Ca^{2+} affinity is the
assumption that calmodulin accelerates the $E_2 \rightarrow E_1$ reaction. Luthra
et al. (1980) demonstrated that the calmodulin effect is indeed on
the 140 000 mol. wt protein.

Sarkadi et al. (1980a) and Enyedi et al. (1980a) made the
interesting observation that mild trypsin digestion of the internal
surface, removing a peptide of 30 000 mol. wt, mimics the effect of
calmodulin (Taverna and Hanahan, 1980) and abolishes calmodulin
binding. This effect was accompanied by a reduction of the steady
state phosphorylation (in the absence of Mg^{2+}).

As mentioned above, removal of phospholipids abolishes the

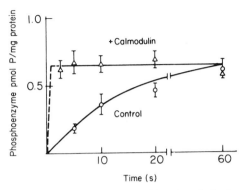

Fig. 11. *Fragmented membranes from human red cells, stripped of calmodulin. Time course of Ca²⁺-dependent phosphorylation at 0°C. Medium (mM): Hepes-Tris 150 (pH 7.4), MgCl₂ 1 and either Tris-EGTA 0.5 or CaCl₂ 0.05, ATP 0.002. Added calmodulin greatly accelerates formation of phosphoenzyme. (Mualem and Karlish, 1980.)*

ability of the system to undergo Ca^{2+}-induced phosphorylation (Richards *et al.*, 1977b).

Szasz *et al.* (1978a) showed that *in intact red cells* a membrane protein of 150 000 mol. wt is phosphorylated in a Ca^{2+}-dependent fashion. They also agree that rapid dephosphorylation of the phosphorylated intermediate requires Mg^{2+} and ATP.

The effects of La^{3+} on the system are of considerable interest. La^{3+} inhibits the pump and the (Ca + Mg)-ATPase at approximately 0.2 mM completely (Quist and Roufogalis, 1975; Sarkadi *et al.*, 1977) (which is also true for other lanthanides as was shown by Schatzmann and Tschabold, 1971). Schatzmann and Roelofsen (1977) demonstrated that Ho^{3+} inhibition is reversible by washing. The inhibition is accompanied by an increase in the amount of phosphorylation in the 150 000 mol. wt protein in disrupted membranes (Schatzmann and Bürgin, 1978; Szasz *et al.*, 1978a). The product can be back-reacted with ADP (Schatzmann and Bürgin, 1978) but the "natural" dephosphorylation by Mg^{2+} and ATP is inhibited (Szasz *et al.*, 1978a; Schatzmann, unpublished). Obviously, inhibition is also possible from the external membrane side since Quist and Roufogalis (1975) and Sarkadi *et al.* (1977) observed inhibition of the system by external La^{3+} in resealed or intact cells under conditions not allowing entry of La^{3+} (Szasz *et al.*, 1978b). However, the inhibition from outside is complete for transport when (Ca + Mg)-ATPase activity is still at half its original value (see p. 75). The stimulation of phosphorylation seems to occur at the internal side at an amazingly high affinity for La^{3+} ($K_{La} = 5$ pM; Szasz *et al.*,

1978a), whereas for inhibition of dephosphorylation in disrupted membranes the required La^{3+} concentration is the same as for inhibition of transport from outside (0.25 mM for 65–75% inhibition; Szasz et al., 1978a). Thus increase of phosphorylated intermediate by La^{3+} is not a consequence of reduced dephosphorylation. Szasz et al. (1978a) observed that La^{3+} has the same effect in inside-out vesicles as in disrupted membranes: at low concentration it enhances the steady-state level of phosphoprotein and at high concentration it is able to prevent dephosphorylation of the phosphoprotein made in the presence of Ca^{2+}. Making the reasonable assumption that La^{3+} replaces Ca^{2+} at the transporting site when having this blocking effect, they concluded that in the phosphorylated state the Ca^{2+} site faces the internal side of the membrane, which means that the phosphorylation step precedes the Ca^{2+} transport step. As they did the phosphorylation in the absence of Mg^{2+} the transport step can be attributed to reaction II or III in the above scheme. Since it is economical to associate the transport step with conformation changes in the protein molecule (reaction II and IV) and since it is clear that the Ca^{2+} site faces the interior in the E_1 conformation the most likely reaction for the transport step is reaction II, which means that in the E_2 conformation the Ca^{2+} site faces the exterior. According to Szasz et al. (1978a) La^{3+} can also block dephosphorylation in the E_2 conformation from the outside but acts there at an even lower affinity (its action can be abolished by externally added ATP) which makes sense if one assumes that it is again the Ca^{2+} transporting site which is the locus of La^{2+} blockade.

A rather interesting suggestion with regard to the role of Ca^{2+} has been made. Sarkadi et al. (1978), using inside-out vesicles, found a K_{Ca} for transport of 40–50 μM in the *absence* of EGTA (which might mean that their vesicles are partly depleted of calmodulin). Szasz et al. (1978a) presented evidence indicating that the apparent Ca^{2+} concentration for half-effect on the phosphorylation is below 1 μM in these same vesicles in the *presence* of EGTA. In order to explain this discrepancy they propose that in addition to the site inducing phosphorylation there must be a second, low affinity Ca^{2+} site accepting also Ca-EGTA, which has to be occupied for Ca^{2+} transport to occur (see p. 79). This idea is not without support from other evidence: (1) activation of (Ca + Mg)-ATPase shows positive cooperativity, with a Hill coefficient of 1.3 (Schatzmann and Roelofsen, 1977). (2) Ferreira and Lew (1976), using an elegant method for studying Ca^{2+} transport by the pump in cells displaying a controlled leak (brought about by the action of graded concentrations of ionophore) found a kinetic behaviour which could best be fitted by assuming two Ca^{2+} sites on the protein. What they neither showed nor claimed is that two Ca^{2+} ions are transported at a time. (3) Upon stepwise removal of calmodulin a behaviour of the ATPase appears that can be described by assuming two Ca^{2+} sites of different affinity, activation by Ca^{2+} following a curve which is steep at low Ca^{2+} concentration and flattens at elevated Ca^{2+} concentrations (Schatzmann, 1973; Scharff, 1976). The second site then could be one whose affinity for Ca^{2+} is increased by calmodulin (but is not the site on calmodulin necessary for assembly of the pump–calmodulin complex). Curiously, when calmodulin is added to calmodulin-depleted membranes this biphasic curve is not recovered but with different calmodulin concentrations a family of smooth curves of different K_{Ca} results.

Utmost care must be used in such experiments in controlling Ca^{2+} concentrations. Published values for the dissociation constants of different Ca-chelating agents used in buffering the Ca^{2+} concentrations vary quite markedly and are considerably changed by temperature and ionic strength. Moreover, other factors may affect them. We have found that 30 mM imidazol, pH 7.0, in combination with 0.5 mM EGTA cause the free Ca^{2+} concentration, measured with a Ca^{2+} electrode calibrated with $CaCl_2$, to be higher than the calculated Ca^{2+} concentration by a factor of 1.7; other pH buffers (Tris–maleate or borate) did not have this effect. Scharff (1979) expressed the suspicion that the Ca^{2+} concentration calculated for EGTA buffers at high Ca/ligand ratio (> 0.95) may be quite wrong.

Stoichiometry. When measuring the number of gram-atoms of calcium transported per number of moles of ATP hydrolysed in Ca^{2+}-tight membrane preparations (correcting for a possible Ca^{2+} leak if necessary, suppressing Na-K-ATPase by ouabain and subtracting Mg-ATPase) one finds a 1:1 ratio (Schatzmann, 1973; Schatzmann and Roelofsen, 1977; Mualem and Karlish, 1979a). In inside-out vesicles Sarkadi *et al.* (1979b; see also Sarkadi, 1980) found the ratio to depend on Ca^{2+} concentration. Below $100\,\mu M$ Ca^{2+} it was less than 1 and steadily approached 2 above $500\,\mu M$ Ca^{2+}, mainly because ATP consumption fell with increasing Ca^{2+} concentration. Quist and Roufogalis (1975, 1977) and Sarkadi *et al.* (1977) found that external La^{3+}, in concentrations which blocked transport completely, reduced (Ca + Mg)-ATPase to only 50%, which leads unevadably to the statement that the La^{3+}-sensitive system has a stoichiometry of Ca:ATP = 2:1, irrespective of whether one refers to the author's measurement or to the previously made overall (Ca + Mg)-ATPase measurement. This finding has been disputed

(Larsen *et al.*, 1978a; Sarkadi *et al.*, 1979c). The question which arises is, whether there are one or several (Ca + Mg)-stimulated ATPases not related at all to the Ca transport system present in the membrane, which make up about 50% of the total and have all the characteristics of the Ca pump's ATPase function (the same affinity for Ca^{2+}, Mg^{2+}, ATP, alkali cations and inhibitors like quercetin; Wüthrich and Schatzmann, 1980). The alternative is that La^{3+} uncouples the pump ATPase partly from Ca^{2+} transport. For the first choice it can be said that it is quite possible, although the mimicry of the Ca-pump seems an amazing coincidence. Several authors claimed to have removed water-soluble Ca^{2+}-activated ATPases from the membrane. (What is recovered in the supernatant of such experiments must not necessarily be the same which is simultaneously lost from the membranes during the mostly rather rough treatment in the solubilization procedure; Rosenthal *et al.*, 1970; Avissar *et al.*, 1975; Weidekamm and Brdiczka, 1975.) Some of these were clearly not similar to the membrane-bound (Ca + Mg)-ATPase (not requiring Mg^{2+} or even being inhibited by it) or were insufficiently characterized. Quist and Roufogalis (1977) added such an extract with putative (Ca + Mg)-ATPase activity to membranes and found an increase in ATPase activity in the mixture (but of course no stimulation of Ca^{2+} transport). A more likely candidate might be a system composed of a kinase (stimulated by Mg^{2+}) and a phosphohydrolase (stimulated by Ca^{2+}) acting either on membrane proteins (Greenquist and Shohet, 1973) or on membrane lipids such as phosphoinositides (Buckley and Hawthorne, 1972; Griffin and Hawthorne, 1978; Kawaguchi and Konishi, 1980), or phosphatidic acid. For the phosphorylation-dephosphorylation system related to phosphoinositides the turnover might be fast enough to come into the order of (apparent) ATPase activity necessary. In favour of the alternative, namely that blocking of the transport site for Ca^{2+} leaves the second Ca^{2+} site active which maintains ATPase activity at half the normal speed, one might advance the fact that a number of bivalent metal ions can stimulate the ATPase activity in the presence of Mg^{2+} at a reduced maximal rate although such metals are not transported (Pfleger and Wolf, 1975) (a mode of action which is obviously obtained also with highly purified pump protein, Wolf *et al.*, 1977). The rationalization of this is that these ions might not fit the transport site, yet activate the ATPase at the second site for Ca^{2+} (Schatzmann and Roelofsen, 1977).

   The controversy cannot be resolved at the present time. The 1:1 stoichiometry would give a comfortable safety margin thermodynamically but (with not too extreme assumptions about the internal steady-state Ca^{2+} concentration in viable red cells) a $2:1$ stoichiometry is thermodynamically not strictly impossible. For the pump to run forwards the work done (W) in the Ca^{2+} movement must be less than the free energy change ($\Delta G_{hydr.}$) of ATP hydrolysis i.e. the overall free energy change must be negative.

$$\Delta G_t(kJ) = \underbrace{-\Delta G^0 - RT \ln \frac{[ATP]}{[ADP] \cdot [PO_4]}}_{\Delta G_{hydrol.}} + \underbrace{n RT \ln \frac{[Ca_0^{2+}]}{[Ca_i^{2+}]}}_{W_{osm.}} + \underbrace{n z F V_m}_{W_{electr.}} \quad (6)$$

With ΔG^0 = standard free energy change of ATP hydrolysis = 30.55 kJ (= 7.3 kcal), [ATP] 1.5 mM, [ADP] 0.3 mM, [PO$_4$] 0.3 mM (for human red cells; Bartlett, 1959; Beutler, 1975), $[Ca_i^{2+}]$ = 10^{-6}–10^{-8} M, $[Ca_0^{2+}]$ = 10^{-3} M, n = number of Ca^{2+} ions per cycle, V_m = membrane potential 9 mV, $R = 8.31$ J, $T = 310$ K, z = charge of Ca^{2+} = 2.

Solving equation (6) for Ca_i^{2+} gives 6×10^{-8} M for the theoretical limit ($\Delta G_t = 0$) of internal Ca^{2+} concentration at a stoichiometry of 2 Ca:1 ATP ($n = 2$). Provided that the coupling of ATP hydrolysis to Ca^{2+} transport is very tight and since the leak is only about 1/1000 of the maximal pump flux it seems not improbable that the pump can cope with the leak not far above this concentration and therefore might maintain a steady-state internal Ca^{2+} concentration corresponding to the lowest estimate ($< 3 \times 10^{-7}$ M) for the intra-cellular Ca^{2+} concentration made so far by actual measurements (Simons, 1976a, b).

Reversal. If the reactions of the scheme shown above (p. 70) are all true equilibria it should be possible to run the pump backwards as a whole, i.e. to synthesize ATP on account of a Ca^{2+} gradient. For this a large inward gradient of Ca^{2+} must be set up and ATP concentration must be made small and phosphate and ADP concentration made as large as possible. A first report on incorporation of $^{32}P_i$ into ATP by the Ca pump under conditions favourable to reversal was promising (Ferreira and Lew, 1975). Rossi et al. (1978) using ATP-depleted, phosphate-enriched intact red cells were able to induce ^{32}P uptake into ATP in a Ca^{2+}-dependent way. The cells were virtually Ca^{2+} free and the external Ca^{2+} concentration used was 2 mM. The proof for the Ca^{2+} gradient to be decisive rather than the presence of Ca^{2+} *per se*, was that the ionophore A 23187 abolished the effect. During the formation of ADP-^{32}P there was a net decrease of total ATP. Although it is difficult to see any flaw in these experiments, we tried to obtain *net* ATP synthesis by the Ca pump (Wüthrich *et al.*, 1979). To this end inside-out vesicles were used which were loaded by pump action with about 20 mM Ca^{2+}. The external Ca^{2+} concentration was $< 10^{-8}$ M (with EGTA), PO_4^{3-} was made 10 mM and ADP 2 mM in the medium. ATP was measured chemically (by the luciferin–luciferase method). There was an ADP-dependent leak of Ca^{2+} from these vesicles after ATP was carefully washed away, and concomitantly a small net ATP increase was noted which was sensitive to 10 μM A 23187.

The role of alkali cations. Schatzmann (1970), Schatzmann and Rossi (1971) and Bond and Green (1971) showed that there is considerable further stimulation of (Ca + Mg)-ATPase in disrupted membranes by Na^+ and K^+, whereby the affinity for K exceeds that for Na ($K_D \approx 6$ mM and 33 mM respectively). High Ca^{2+} concentration depressed the Na effect relative to the K effect (Schatzmann and Rossi, 1971). NH_4^+ also has a stimulatory effect (Wolf, 1973a, b; Wolf *et al.*, 1977), but Li^+ is much less effective. For comparison media with organic cations such as choline or tris were used. Later Sarkadi *et al.* (1978, 1980b) demonstrated that this effect of alkali cations is also present when Ca^{2+} transport is used as the criterion for activity. Enyedi *et al.* (1980b) reported that K accelerates not the phosphorylation but the dephosphorylation reaction which, taken together with the over-all acceleration of ATPase function and Ca^{2+} transport, means that in the absence of K (or Na) but presence of Mg^{2+} dephosphorylation is the rate-limiting step (but see also Larocca *et al.*, 1981).

Interestingly a similar behaviour towards sodium and potassium is also displayed by the Ca^{2+}-sensitive, ATP-requiring nitrophenylphosphatase activity of red cell membranes (Pouchan *et al.*, 1969; Rega *et al.*, 1972, 1973; Richards *et al.*, 1977.)

Whereas Schatzmann and Rossi (1971) found no stimulatory effect of 135 mM external Na + 6 mM K compared to an alkali cation-free medium on Ca^{2+} transport from resealed cells, Romero (1980) recently showed that even in K-rich resealed cells alkali cations in the external medium stimulate Ca^{2+} transport. The affinities for Na and K were very similar to those found on the ATPase and the depression of the Na effect by high (internal) Ca^{2+} was also present. In contrast, Wierichs and Bader (1980) suggested that there is sidedness for the effect in that potassium is somewhat more stimulatory from the inside and Na from the outside of the membrane. It is difficult to say what this means. What it does *not* mean is that there is a functional connexion between $(Ca + Mg)$ and $(Na + K)$-ATPase. There are good reasons to believe that the two systems are separate (see above). There is also no evidence that sodium might be countertransported by the system in exchange for Ca^{2+} because Ca transport proceeds in the total absence of sodium (Schatzmann, 1975).

The problem of calcium affinity. We have seen that in isolated membranes and purified reconstituted pump protein the behaviour is reasonably predictable with respect to Ca^{2+} affinity: when saturated with calmodulin the protein displays high affinity and without calmodulin low affinity for Ca^{2+}. Since calmodulin, relative to pump sites in the membrane, is present in excess in the cytosol (Vincenzi and Larsen, 1980), one would predict that in intact cells or resealed cells after haemolysis in 5–10-fold volume the affinity should be high. However, it is not, or not always. Ferreira and Lew, studying transport in intact cells loaded with $CaCl_2$ by ionophore A 23187 action, found that K_{Ca} instead of 10^{-6} may be as low as 10^{-4}–10^{-3} M, once Ca^{2+} in the interior has exceeded 3–4 μM (pers. comm.; see also Scharff, 1980), due allowance being made for the Ca^{2+} buffering capacity of internal constituents. We could confirm this in intact cells loaded by the salicylate method (Bürgin and Schatzmann, 1979). The same observation was made by Schatzmann (1973) on ATPase activity of resealed, $CaCl_2$-loaded cells. The apparent affinity was at least 100-fold lower when $CaCl_2$ instead of Ca-EGTA buffers was introduced into the cells. A very similar difference between $CaCl_2$ and Ca-EGTA buffers was recently reported by Sarkadi *et al.* (1979a) examining transport in inside-out vesicles. They extended the finding by showing that at 50 μM total Ca in the buffers the affinity for Ca^{2+} was high, but V_{max} was only about half of that in 500 μM Ca buffers of the same Ca^{2+} concentration. In unbuffered media calmodulin had a large effect on affinity for Ca^{2+} and in buffered media its effect was negligible. As proposed by the authors, the enigma can be solved by assuming two Ca^{2+} sites. Site 1 has a high affinity ($K_{Ca} = 0.5$–1.0 μM) and accepts Ca^{2+} only. Site 2 has low affinity and accepts Ca^{2+} an Ca-EGTA. Both sites must be occupied for transport to occur. Interestingly when calmodulin was added the Site 2 affinity increased ($K_{Ca} = 50$ μM $\rightarrow$ 15 μM) whereas the Site 1 affinity stayed unchanged. This indicates that, in the framework of this hypothesis, Site 2 is the locus of action of calmodulin. Calmodulin also increased the observed V_{max} in buffered and unbuffered media (Sarkadi *et al.*, 1979a). This may not be a direct effect but could be due to the fact that Site 2 moves out of the region where Ca^{2+} turns inhibitory (above 10^{-4} M Ca^{2+}) (see Fig. 5). The difference between $CaCl_2$ and Ca-EGTA is not observed in disrupted membranes which normally are freeze–thawed. To reconcile this with the proposal, one has to postulate that the treatment of membranes shifts the Site 2 affinity upwards so that the gap between the K_{Ca} of the two sites (revealed by buffered and unbuffered media) becomes small enough to be swamped by the inaccuracies incurred in calculating free Ca^{2+} concentration. We saw above (p. 74) that two Ca^{2+} sites on the system seem acceptable on different grounds.

Since in isolated, freeze–thawed membranes neither the EGTA effect on Ca^{2+} affinity nor the different behaviour of calmodulin in the presence or absence of EGTA is seen, an alternative explanation seems conceivable. If in the intact system the transport site were screened from the cell interior by a barrier of positive charges the Ca-EGTA complex might equilibrate rapidly across this barrier while Ca^{2+} would not. Since the pump keeps the Ca^{2+} concentration low in the pocket underneath the barrier, this would result in a falsely low Ca^{2+} affinity whose measurement is based on Ca^{2+} concentration in the bulk fluid outside the pocket. Calmodulin might partly neutralize the positive charges and thereby raise the apparent affinity for Ca^{2+} in the absence of the complexing agent (the same might be true for negatively charged phospholipids which abolish the requirement for calmodulin). The assumption must be made that freeze–thawing of the membranes improves accessibility by removal of the positive charges from the vicinity of the Ca^{2+} binding site.

Summary. While Ca^{2+} is translocated across the hydrophobic barrier ATP hydrolysis proceeds in three steps. First phosphorylation of the protein at an acyl group takes place. Next the phosphoprotein undergoes a conformation change and/or (?) transphosphorylation before P_i is liberated. Ca^{2+} is required for phosphorylation. Translation of Ca^{2+} across the hydrophobic barrier most probably occurs at the step of conformational change. Mg^{2+} is necessary for the conformation change and ATP at a nonhydrolysing site with low affinity is required for cleavage of the acyl–phosphate bond in the phosphoprotein. In addition Mg^{2+} is probably needed at the phosphorylation step and a second Ca^{2+} ion might be required somewhere in the cycle at a site dependent in its Ca^{2+} affinity on calmodulin. The question is not settled whether the stoichiometry is $1:1$ or $2:1$ for Ca:ATP. The $2:1$ stoichiometry is compatible with the laws of thermodynamics down to an internal Ca^{2+} concentration of 10^{-7} M approximately. K and/or Na are not an absolute requirement but accelerate the reaction rate somewhat. The system is reversible as a whole, i.e. when Ca^{2+} is forced across the pump from outside to inside, ATP is synthesized from phosphate and ADP.

New findings about inhibitors

A specific inhibitor which neither affects other ATPases nor the Na–K pump has not yet been found. *Lanthanides* and *ruthenium* red are inhibitors probably because they are cations able to combine with the Ca^{2+} site. *Sulphydryl reagents* do block the system (Hg^{2+}: Pfleger and Wolf, 1975; ethacrynic acid: Vincenzi, 1968; NEM: Richards *et al.*, 1977b).

Interestingly *NEM* (*N*-ethylmaleimide) acts differently in the presence of different ligands (ATP, Ca^{2+}, Mg^{2+}), which suggests that binding of such ligands in fact do induce conformational changes in the protein, producing a more hidden or a more exposed state of a sulphydryl group in the active centre (Bond, 1972; Luterbacher and Stieger, unpublished). Whereas, after exposure in the presence of EGTA even high concentrations of NEM (0.5 and 1 mM) inhibit the ATPase only by 70%, in the presence of Ca^{2+} (0.5 mM) the inhibition is 90% at 0.2 mM NEM and complete at 0.5 mM NEM. ATP (2 mM) drastically reduces inhibition (to 30% at 1 mM NEM, 15% at 0.2 mM NEM). Mg^{2+} (4 mM) slightly inhibits the NEM effect in the absence of ATP and further reduces the NEM effect together with ATP (2 mM) (to 0% at 0.2 mM NEM) (Bond, 1972; Luterbacher, unpublished).

Sarkadi *et al.* (1980a) have shown that 4-chloromercuribenzoic acid is (reversibly) inhibitory without affecting affinity for Ca^{2+} or ATP. Their interpretation is that -SH groups are implicated in conformational changes necessary for the completion of the full cycle of the system.

Vanadate (whose prevalent form at physiological pH is probably $[VO_3(OH)]^{2-}$) whose inhibitory action on the Na pump was discovered because it was present as contaminant in some brands of ATP (Cantley *et al.*, 1977, 1978; Beaugé and Glynn, 1978) was reported to inhibit also the Ca pump (Bond and Hudgins, 1978, 1979, 1980). Barrabin *et al.* (1980) gave a rather complete account of the behaviour of vanadate as an inhibitor of the Ca system. Full inhibition can be achieved at 15 μM vanadate at low ATP concentration (with the regulatory ATP site unoccupied). The interaction with ATP at the high affinity (catalytic site) is uncompetitive. The sites for vanadate show positive interaction with sites for K^+ and Mg^{2+} (and Na^+, but not Li^+), i.e. K^+ and Mg^+ enhance the inhibitory effect by increasing the affinity for vanadate (K_I reaching 1.5 μM if K^+ and Mg^{2+} are present at 60 and 10 mM respectively, Bond and Hudgins, 1980). Stieger (unpublished) could confirm this fully (in addition he showed that the very high affinity for vanadate in a high K^+ medium (120 mM) at 2 mM ATP concentration is also present in the purified protein, K_I being 1.2 μM). However, the affinities for Mg^{2+} and K^+ for this enhancing effect are considerably less than those for the activating effect on th unpoisoned ATPase. Vanadate inhibition is independent on the Ca^{2+} concentration up to 50 μM (Barrabin et al., 1980). Rossi et al. (1980) have demonstrated for Ca^{2+} transport in resealed cells that 1 mM external Ca^{2+} relieves the vanadate inhibition by about 50%. It seems possible that vanadate is inhibitory because of its similarity to phosphate.

Quercetin was found to inhibit both Ca^{2+} transport and (Ca + Mg)-ATPase at concentrations of 4–6 μM for half maximal effect (Wüthrich and Schatzmann, 1980). Sarkadi et al. (1980b) reported similar activity for *phloretin* but found other flavonoids ineffective.

Phenothiazine neuroleptics have long been known to be inhibitors of the Ca pump in sarcoplasmic reticulum (Balzer et al., 1968) and red cells (Schatzmann, 1969). It was recently shown by Gietzen et al. (1980a), Raess and Hinds (1979) and Hinds et al. (1980) that these drugs, notably trifluoperazine and the butyrophenone penfluridol, act with a 5–10-fold higher affinity on the calmodulin stimulated activity than on the residual activity. There is an antagonism between the drugs and added calmodulin which exhibits apparent competitive behaviour. This agrees well with the observation that such drugs bind to calmodulin in other tissues (Levin and Weiss, 1968, 1969).

Suramin-Na, known to inhibit the Na–K pump if present on the internal membrane surface with a K_i of ~ 50 μM (Fortes et al., 1973), is effective on the (Ca + Mg)-ATPase of disrupted membranes with a K_i of ~ 40 μM and at slightly higher concentration also on the Mg-ATPase (Schatzmann, unpublished).

Recently, Gietzen and Bader (1980) reported that *vinblastine* inhibits the calmodulin-activated fraction of red cell (Ca + Mg)-ATPase with a K_i of 36 μM, but does not affect the basal activity (remaining in the absence of calmodulin). Another mitosis inhibitor, colchicine, was without effect. Ca^{2+} transport in inside-out red cell vesicles is inhibited in the same way by vinblastine (Gietzen et al., 1981).

Comparison Between the Red Cell Calcium Pump and the Sarcoplasmic Reticulum Calcium Pump

There is no doubt that the red cell membrane and the sarcoplasmic reticulum perform very similarly with respect to active Ca^{2+} transport. Affinity for Ca^{2+} is in the same order of magnitude and specificity for Ca^{2+} (and Sr^{2+}) compared to other divalent cations is high. Both systems require Mg^{2+}. A phosphorylated intermediate is formed in both systems by phosphorylation at an acyl group. The turnover rate of both systems seems to be similar. The stoichiometry (Ca: ATP) is a moot point in the red cell system, but is clearly 2 in the sarcoplasmic reticulum. Sensitivity to Na and K can be demonstrated in both systems (Jones et al., 1977; Shikegawa and Pearl, 1976). In both systems presence of phospholipids is necessary for function. There seems to be a difference in that the detergent Triton-X-100 does not reactivate the red cell system after treatment with phospholipase C (Roelofsen and Schatzmann, 1977), whereas several nonionic detergents including Triton-X-100 could restore activity in a nearly completely delipidated preparation of sarcoplasic reticulum (Dean and Tanford, 1978).

There are however, some differences: the molecular weight in the red cell system is 140 000 and in the sarcoplasmic reticulum 100 000. The substrate in the red cell system is with strong predilection ATP (and ITP or UTP) (Watson et al., 1971; Cha et al., 1971; Sarkadi et al., 1980b), whereas Ca^{2+} transport in sarcoplasmic reticulum is sustained also by p-nitrophenylphosphate or acylphosphates (in the presence of ATP the red cell system does hydrolyse p-nitrophenylphosphate slowly but this enzyme activity is not coupled to Ca^{2+} transport). The red cell system seems to accept free ATP as substrate whereas for the sarcoplasmic reticulum opinions about this point diverge; Garrahan et al. (1976) found that the extent to which the sarcoplasmic reticulum protein was phosphorylated in the steady state was the same in the presence or absence of Mg^{2+} but that Mg^{2+} accelerated the phosphorylation. On the other hand Makinose and Boll (1979) presented evidence that Mg-ATP is required as substrate in sarcoplasmic reticulum. But we do not really know what the role of Mg^{2+} is in increasing the amount of phosphorylated intermediate in the red cell system. A sweeping statement about the role of Mg-ATP seems to be premature for either system. Richards et al. (1977b) found that the red cell system does not lose its phosphatase activity upon treatment with phospholipase C, whereas the sarcoplasmic reticulum did.

Garrahan *et al.* (1976) showed that the Mg site for dephosphory-lation in sarcoplasmic reticulum is occluded after phosphorylation in the mere presence of Ca^{2+} (by which is meant that Mg^{2+} does no longer accelerate dephosphorylation when added after phosphory-lation has taken place, even in the presence of high ATP concen-tration). This seems not to be so with the red cell system, at least under the conditions chosen by the same group of authors (Rega and Garrahan, 1975). Interestingly this occlusion could partially be reversed by treatment with *t*-CDTA (*trans*-1,2-diaminocyclohexane-tetraacetic acid).

Slight stimulation by calmodulin was reported for the sarco-plasmic reticulum of cardiac muscle (Katz and Remtulla, 1978). However, it is not clear whether this has the same meaning as in red cells, because Le Peuch *et al.* (1979) claimed that this effect was due to stimulation of a kinase, phosphorylating phospholamban, which can be distinguished from that elicited by cAMP.

Red Cells from Other Species

Bird red cells
Ting *et al.* (1979), studying vesicles made from pigeon red cells, found an ATP-dependent Ca pump with a very high V_{max} (1.5 mmol/l/min at 27°C), high affinity for Ca^{2+} ($K_{Ca} = 0.18 \mu M$) and a depen-dence of rate on Ca^{2+} concentration which could be described by a straight-line relation between reciprocal rate and $[Ca^{2+}]^2$. The latter finding is strongly supporting the idea of two Ca^{2+} binding sites.

Dog red cells
Brown (1979) and Parker (1979) have presented good evidence for the existence of a Mg^{2+}- and ATP dependent Ca^{2+} extrusion mechan-ism similar to what has been described for human cells. This is important since Parker (Parker *et al.*, 1975; Parker, 1977, 1978) has shown unequivocally that in dog cells the high intracellular Na and low K concentration is due to virtual absence of a Na–K pump, but that those cells display the Na–Ca exchange mechanism which is observed in many other cells but is clearly absent from human red cells. However, contrary to nerve, kidney or heart membranes, where the Na gradient maintained by the Na pump drives Ca^{2+} out of the cell interior (see below), in dog red cells it is the Ca^{2+} gradient which moves some Na out of the cells. The Ca pump thus keeps the cells from reaching electrochemical equilibrium with respect to Na, which would unescapably lead to swelling and eventually to colloid osmotic

haemolysis. The Ca pump therefore is the expedient by which species lacking a Na–K pump (the dog and probably also the cat) maintain the normal volume of their red cells. It remains obscure why Rega *et al*. (1974) found hardly any (Ca + Mg)-ATPase in dog red cells. Parker (1979) proposes as an explanation a lack of calmodulin in their preparation.

Ruminant red cells

Ruminants (sheep, goats, cattle) have red cells which display three special features with respect to cation transport.

(i) Within the same race (breed) there are individuals with low internal K concentration in their red cells (and high Na concentration) and others (a minority) with high internal K concentration (and low internal Na concentration). The difference is explained by the observation that low-K cells have fewer pump sites per cell and, what is more important, that their pump has a higher affinity for K on the internal surface of the membrane, leading to competition of K with Na for the transport site at considerably lower internal K levels than in human red cells (for review see Ellory, 1977).

(ii) Cells in all individuals lack the Ca^{2+}-sensitive K^+ channel (Jenkins and Lew, 1973) after the fetal period (Brown *et al*., 1978).

(iii) The (Ca + Mg)-ATPase activity and the Ca pump activity in the adult are considerably lower than in human red cells (Schatzmann, 1974; Bürgin, unpublished).

The interesting point is that in these species the fetal cells differ from adult cells: they are high in K, regardless of whether the individual will be a high-K or a low-K animal in later life, they do have the Ca^{2+}-sensitive K channel which is lost in the adult (Brown *et al*., 1978) and they have a high (Ca + Mg)-ATPase activity (Schatzmann and Scheidegger, 1975). For cattle we have shown that the latter declines *pari passu* with the fetal haemoglobin. Thus the change is not a matter of young (reticulocytes) or old cells, but depends entirely on the transition from the fetal generation to the adult generation of cells.

Pathological Alterations of the Red Cell Calcium Pump

Abnormalities of red cell cation transport occuring in disease have been reviewed by Parker and Welt (1972), Parker *et al*. (1978), Palek *et al*. (1976), Wiley (1976). Whereas Ca^{2+} leaks have been described for microcytic haemolytic anaemia (Wiley and Gill, 1976), high PC/PE haemolytic anaemia (Bucher *et al*., 1978) and particularly in sickle cell anaemia (Eaton *et al.*, 1973) pump defects in otherwise normal red cell membranes leading to deviating red cell shape or function are unknown. Quite recently, Bookchin and Lew (1980), suggested that in addition to the enormous leak for Ca^{2+} in sickle cells there is also an abnormality in the Ca pump in this disease. Under conditions inducing sickling they found a progressive decline of the Ca pump rate. A reduction of the specific activity of the (Ca + Mg)-ATPase to one third of the normal value (Dixon and Winslow, 1979), probably due to a decreased response to calmodulin (Gopinath and Vincenzi, 1979), has been described in sickle cells.

An extremely interesting observation made by Scharff and Foder (1975) is the decreased (Ca + Mg)-ATPase activity in red cells from polycythaemia vera patients. This finding can be brought into context with claims to the effect that intracellular Ca^{2+} may stimulate reproduction processes in cells which have the potentiality to proliferate. If nucleated precursors of red cells in polycythaemia were unable to pump Ca^{2+} at a normal rate in a given state of maturation it is conceivable that this might be at the root of the overproduction of red cells characteristic for this disease.

Horton *et al.* (1970) described a reduction in (Ca + Mg)-ATPase activity in red cells from children suffering from cystic fibrosis. This claim was recently confirmed by work of Ansah and Katz (1980) (but see also Foder *et al.*, 1980).

The much debated question of whether, in muscular dystrophy of the Aran-Duchenne type, abnormalities in cation transport and especially Ca^{2+} transport of red cells occur and, if so, what they exactly are remains unsettled (Parker *et al.*, 1978).

Significance of the Calcium Pump for the Red Cell

There is no doubt that without the Ca pump red cells would gain intracellular calcium. High internal Ca^{2+} concentration, however, has undesirable effects.

(i) In human red cells (Gardos, 1958a, b; Blum and Hoffman, 1971; Hoffman and Blum, 1977; Lew and Ferreira, 1977) and in dog red cells (Richardt *et al.*, 1979) there is a channel for rapid K passage across the membrane which under physiological conditions is closed and opens under the influence of internal Ca^{2+}. Curiously this mechanism shows the same vacillating behaviour with respect to Ca^{2+} affinity as the Ca pump (Lew and Ferreira, 1976). Affinity was found to be high (measurable effects at $0.3 \mu M$) in Simons' (1976a, b) experiments with resealed cells containing Ca buffers and low ($K_{Ca} = 0.33-1.5$ mM) in intact cells as studied by Lew and

Ferreira (1976). Whatever the true affinity for Ca^{2+} of the K channel under natural conditions may be, cells deprived of their Ca pump will sooner or later lose K and consequently shrink.

(ii) Increase in intracellular Ca^{2+} concentration leads to shape changes unfavourable for cell survival. The membrane starts budding teat-like evaginations (echinocyte formation), the disc-shaped echinocyte is transformed into a spherical echinocyte and finally the evaginations are shed as closed membrane vesicles leaving a spherical smooth cell with reduced surface area behind (Kirkpatrick *et al.*, 1975; Sarkadi *et al.*, 1976). The process, short of the shedding phase, is reversible if the cells are allowed to pump out Ca^{2+}. Szasz *et al.* (1978b) showed that this recovery process can be stopped at will at any phase by La^{3+}, which allows to establish a precise correlation between internal Ca^{2+} concentration and different cell forms. The shape transitions take place between 23 and 94 μmol total (?) Ca/litre cells. It is likely that at least one component of these shape changes is due to the activation by Ca^{2+} of contractile processes in the spectrin–actin system within the cell, leading to shearing off in circumscribed areas of the membrane proper from the underlying cytoskeleton; it seems that these membrane domains loosened from the underlying spectrin layer are able to fuse at their base and detach from the remainder of the cell. It is possible that there are other ways of bringing about these shape changes, but it cannot be questioned that Ca^{2+} entry is one way to trigger them.

(iii) Rather high internal Ca^{2+} concentrations (10^{-4}–10^{-3} M) are inhibitory for the Na–K pump (Dunham and Glynn, 1961; Davis and Vincenzi, 1971).

CELLS OF HIGHER ORGANIZATION

Since in many cells an ATP-driven Ca pump coexists with the Ca extruding, Na_o–Ca_i exchange system, a glance at the latter is necessary for an understanding of those more complicated constellations. Depending on the Na^+ and Ca^{2+} gradients which result from pump actions and leaks the exchange system will cause either an uphill Ca^{2+} or Na^+ efflux from the cells.

The Sodium–Calcium Exchange System

Low intracellular Ca^{2+} concentration seems to be a universal feature of all animal cells. When Baker *et al.* (1969) discovered the existence in the squid axon and Reuter and Seitz (1968) in cardiac muscle of a

Ca extrusion mechanism based on exchange of internal Ca^{2+} for external Na^+ a first answer was available to the question how cells manage to keep internal Ca^{2+} at a low level, namely by using the energy stored in the Na gradient maintained by the Na–K pump. This system is also encountered in the kidney tubule cells (Gmaj et al., 1979; Kinne, 1980). The system is well characterized (Baker, 1972; Glitsch et al., 1970) and may be shortly described as follows: a carrier (protein) can accept either Na^+ or Ca^{2+} at transport sites and the sites change their orientation from inside to outside and vice versa when occupied by ions. It can be shown that the following quantitative relationship holds for such a system (Blaustein, 1977). If the ratio K_{Na}/K_{Ca} is equal inside and outside the equilibrium internal Ca^{2+} concentration is

$$[Ca_i^{2+}] = [Ca_o^{2+}] \frac{[Na_i^+]^n}{[Na_o^+]^n} \cdot \exp\left[(n-2) \cdot \frac{E_m \cdot F}{RT}\right] \tag{7}$$

where subscripts i and o designate concentrations inside and outside the cells and n is the number of Na^+ ions which are transported in exchange for one Ca^{2+} ion. The exponential term in equation (7) is taking into account the membrane potential (E_m, measured inside the cell) which matters in the case that n is not equal to 2. The most natural assumption for n is of course 2, if one supposes that electrostatic attraction is involved in fixating the ions to the protein and provided that it is the same site which alternately binds Na or Ca. However, this may not be so. By arguments similar to those used for the Na–K pump (Garay and Garrahan, 1973) a simultaneous model has been postulated (the protein accepting Na and Ca at different sites simultaneously and motion occuring with both ionic species in their respective places but guided in opposite direction (Blaustein, 1977)). With $n = 2$, $Ca_o^{2+} = 1$ mM, $Na_o = 140$ mM, $Na_i = 5$ mM and $E_m = -80$ mV (for cardiac muscle) and $RT/F \approx 27$ mV (at 37°C) Ca_i^{2+} is 1.9×10^{-6} M at best, provided that any leaks may be disregarded and that the system is truly passive, i.e. that there is no energy input (which for instance changes affinities during the motion of the ion binding site). The Ca^{2+} concentration actually prevailing within resting muscle is clearly below this figure, since actomyosin is activated at physiological pH at Ca^{2+} concentrations far below 10^{-6} (Portzehl et al., 1969). To satisfy the physiological requirement it has been postulated, therefore, that three Na^+ ions migrate in exchange for one Ca^{2+} ion, which ideally results in an internal Ca^{2+} concentration of 3.5×10^{-9} M for cardiac muscle or 0.6×10^{-7}–2×10^{-7} M for the squid axon (Blaustein, 1977). Strong support for a 3 Na^+: 1 Ca^{2+} stoichiometry is provided by the fact that depolarization reduces the Na_o-dependent Ca^{2+} efflux (Brinley and Mullins, 1972; Baker and McNaughton, 1976a, Blaustein, 1977), indicating that there is a net inward movement of positive charge. It was recently shown by other means that the exchange sets up a potential across the membrane (Reeves and Sutko, 1980; but see also Baker and McNaughton, 1976b).

The system apparently works in the total absence of ATP. However, Baker and Glitsch (1973), Jundt and Reuter (1977) and Blaustein (1977) have shown that the presence of ATP increases the rate of Ca^{2+} transport in the range of unsaturation with external Na^+, which means that ATP increases the affinity of the system for Na^+ (and possibly also for Ca^{2+}).

Excitable Cells

Nerve

For the giant axon of a squid (*Dorytheutis plei*) Di Polo (1978) was able to demonstrate that under cyanide poisoning in the absence of a transmembrane concentration gradient for Ca^{2+}, Mg^{2+}, Na^+ and K^+ (Ca^{2+} concentration 0.45 μM) addition of ATP (1 mM) to the internal dialysing fluid increased Ca^{2+} efflux 7.5-fold whereas influx increased insignificantly, the result being, therefore, a net Ca^{2+} efflux. Removal of external Na did not abolish the ATP-dependent net efflux of Ca^{2+}, but reduced it somewhat. It seems, therefore, that a fraction of it is coupled to Na influx, whereas under these conditions the better part of it is not coupled to any cationic movement in the opposite direction. An uncoupled residual Ca^{2+} efflux has been postulated before by Baker and McNaughton (1978).

Di Polo and Beaugé (1979) demonstrated for the giant axon of *Loligo pealei* that the uncoupled, ATP-driven Ca^{2+} efflux is still of considerable magnitude at very low internal Ca^{2+} concentration ($K_{1/2}$ for Ca_i was $\approx 0.18\ \mu M$) and at such concentrations exceeds the Na–Ca exchange flux. The authors conclude that the Ca^{2+} influx through leaks under physiological conditions is mainly compensated by Ca^{2+} extrusion through this ATP-driven pump. They made it clear that the affinity for ATP of this pump is much higher ($K_m = 30\ \mu M$) than that of the Na–Ca exchange system ($K_{1/2}$ for ATP $= 230\ \mu M$), which, as we saw above, is stimulated by the presence of ATP but has not an absolute requirement for it. This suggests strongly that these are separate systems in spite of the fact that both are ATP dependent (Di Polo and Beaugé, 1980). According to the authors the Na–Ca exchange system becomes important only under conditions which markedly increase intra-axonal Ca^{2+} concentration. The capacity of the Na–Ca exchange is much (10 times) higher than that of the ATP-driven pump. At saturating internal Ca^{2+} concentration its rate may be as high as $2 \, pmol/cm^2/s$, whereas Di Polo and Beaugé (1979) found a maximal ATP-dependent efflux of about $150 \, fmol/cm^2/s$. Beaugé et al. (1981) described in another nerve (optic nerve of *Sepiotheuitis sepiodea*) the presence of a membrane-bound (Ca + Mg)-activated ATPase with a high affinity for Ca ($K_{1/2} \approx 0.5 \, \mu M$) and a K_m for total ATP of about $20 \, \mu M$, which could well be the enzymatic basis of the observed Ca^{2+} transport. Di Polo et al. (1979) pointed out that vanadate inhibits the uncoupled Ca^{2+} efflux, but not the Na-Ca exchange. The K_I for this effect was 7×10^{-6} M.

Nerve terminals

A large body of evidence supports the idea that neurotransmitter release (from acetylcholine to endorphines) is dependent on entry of Ca^{2+} across the plasma membrane, triggered by the action potential. It seems certain, therefore, that synaptic nerve terminals must be equipped with a Ca pump. A Ca^{2+} efflux dependent on external Na has been demonstrated in "pinched-off" nerve terminals (synaptosomes) (Blaustein and Oborn, 1975; Blaustein and Ector, 1976; Blaustein et al., 1977). The difficulty in showing an ATP-dependent extrusion of Ca^{2+} is that mitochondria (at a low affinity) and other storage organelles (at a high affinity) accumulate Ca^{2+} by an ATP-dependent mechanism inside synaptosomes (Kendrick et al., 1977; Blaustein et al., 1977). Robinson (1976) found a (Ca + Mg)-stimulated ATPase activity in a microsomal preparation from rat brain which during sucrose density gradient centrifugation distributed with Na-K ATPase and might, therefore, be associated with the plasma membrane. The activity was increased by Na or K and could be inhibited by ruthenium red. The preparation shows rapid Ca^{2+} stimulated phosphorylation by ^{32}P-ATP, the label being found at an acylphosphate group in a protein of 100 000 molecular weight; the system has two K_m values for ATP and phosphorylation requires ATP at the high affinity site (Robinson, 1978).

Rahamimoff and Abramowitz (1978) described Mg- and ATP-dependent Ca^{2+} uptake into synaptosomes from rat brain, but curiously the same preparation showed ATPase activity which was stimulated by either Ca^{2+} *or* Mg^{2+} and the simultaneous presence of both ions resulted in lower maximal activity than that obtained in the presence of one divalent cation alone.

Several authors have succeeded in separating nonmitochondrial transport systems for Ca^{2+} (Rahamimoff and Abramowitz, 1978; Blaustein et al., 1978). Recently Papazian et al. (1979) were able to transfer a $(Ca + Mg)$-ATPase from synaptosomes into soybean phospholipid liposomes. The system accumulated Ca^{2+} together with oxalate in the presence of ATP but not in the presence of the β-γ-imido analogue of ATP. Protein–lipid relation was such that one or a few protein molecules were present in one vesicle. After Ca accumulation these vesicles, due to increased density, could be separated from others by centrifugation and thereby a high purification of the Ca pumping system resulted. Interestingly, two major proteins were identified in this material (by SDS gel electrophoresis) with respective molecular weights of 94 000 and 140 000. These correspond astoundingly closely to those of the monomers of the sarcoplasmic reticulum Ca pump and the red cell Ca pump, and it is tempting to use these similarities in speculating that synaptosomes contain Ca-accumulating structures akin to the sarcoplasmic reticulum and have in addition a plasma membrane pump of the red cell type.

Cardiac muscle

Cardiac muscle is another example of a tissue whose Na_o-Ca_i exchange mechanism is well established and in which recently an ATP-driven Ca^{2+} extrusion mechanism was added to the picture (Trumble et al., 1979). Caroni and Carafoli (1980) applying the isolation technique of Jones et al. (1979) to dog myocardium produced a vesicular preparation enriched in sarcolemma markers. These vesicles showed ATP-dependent accumulation of Ca^{2+} from a K medium containing 75 μM $CaCl_2$ and 5 mM $MgCl_2$, which could be reversed by adding 5 μM A 23187. Removal of Mg^{2+} apparently abolished this uphill Ca^{2+} movement. ^{45}Ca taken up in 4 min could be released by 30 mM Na in the medium nearly completely within 1 min, whereas lower Na concentration (20 and 10 mM) resulted in a near steady-state at vesicular ^{45}Ca concentrations below the one in the Na-free control. This release by Na seems to reflect the presence of the Na–Ca exchange system and is a strong point in maintaining that the vesicles under investigation were mainly made up of plasma membrane material and not of sarcoplasmic reticulum. In the Ca-uptake phase the Na–Ca exchange system cannot have interfered because Na concentration was kept low (20 μM) and ouabain was present, which was intended to prevent a build-up of a Na gradient. The material displays $(Ca + Mg)$-activated ATPase activity and its apparent

dissociation constant for Ca^{2+} is given as 0.2–0.6 μM. Trumble *et al.*
(1980) reported nearly identical results, obtained with similar inside-
out vesicles from dog heart sarcolemma. As a safeguard against build-
up of a Na gradient in nominally Na-free media they used digitoxin,
which is more likely than ouabain to enter the vesicles, and showed
that nigericin, which levels Na gradients by causing an alkali cation
leak, did not affect the uphill Ca^{2+} transport. They made the import-
and observation that, contrary to the case of the sarcoplasmic
reticulum, *p*-nitrophenylphosphate does not replace ATP as substrate
(in accord with the behaviour of the red cell Ca^{2+} pump) and that
oxalate does not stimulate the Ca^{2+} uptake. They also found a K_{Ca}
of less than 1 μM.

It is likely, therefore, that the ATP-dependent Ca pump operates
at very low internal Ca^{2+} concentrations, where the Ca–Na exchange
mechanism becomes very slow or reverses. Morcos and Drummond
(1980) reported the existence of a (Ca + Mg)-ATPase in cardiac
sarcolemma which was activated by 5–10 μM Ca^{2+} and further
stimulated by K or Na.

Smooth muscle
For phasic smooth muscle (e.g. intestinal smooth muscle) there is
evidence for internal calcium stores (Casteels and Raeymakers,
1978, 1979; Bürgin, 1979), notwithstanding the scanty morphological evidence for intracellular tubular or vesicular structures. The
search for a Ca pump in the plasma membrane is fraught with the
difficulty of separating such a putative analogue of the sarcoplasmic
reticulum from the sarcolemmal membrane. Nevertheless, Casteels
et al. (1973) put forward arguments in favour of a Ca pump at the
level of the plasma membrane, likely to be of the ATP-driven type.
In an O_2-free, glucose-free medium total Ca content of taenia coli
more than doubled, while ATP concentration fell to one-tenth. Upon
readmission of O_2 and glucose Ca^{2+} was pumped out of the cells.
In a K-free medium or under ouabain poisoning Ca concentration in
metabolizing cells also rose, but at a much slower rate and with larger
changes in $[Na_i]$ and $[K_i]$. The latter Ca uptake was more slowly
reversed upon normalizing the medium and may be accounted for
by an increase in Ca permeability. All this argues for an ATP-driven
pump and against a Na–Ca exchange mechanism. Recently a (Ca +
Mg)-ATPase present in microsomes from porcine stomach smooth
muscle was partially purified and demonstrated to pump Ca^{2+} after
reconstitution (Wuytack *et al.*, 1981). This might stem from the
plasma membrane.

From porcine coronary arteries Wuytack *et al*. (1978) isolated vesicular membranes showing ATP-dependent Ca accumulation which could be reversed by adding A 23187. These vesicles, however, were admittedly a mixture of internal and plasma membranes. Wuytack and Casteels demonstrated the existence in such membranes of a (Ca + Mg)-activated ATPase and showed that there was concomitant Ca pumping (1980). Later Wuytack *et al*. (1980) added to this finding that the Ca transport was not stimulated by calmodulin, whereas the ATPase activity was slightly enhanced (1.4-fold) by it.

Transepithelial Calcium Transport

The epithelia of the small intestine (particularly the duodenum) and of kidney tubules accomplish a net uphill Ca^{2+} transport from the apical to the basal extracellular space. There is evidence for low, free intracellular Ca^{2+} concentration in these epithelia (Ghijsen and van Os, 1979; Suki, 1979) and without much ado the most likely assumption may be adopted, that Ca^{2+} is pumped out of the cell at the basolateral membrane and enters the cell at the brushborder membrane in a downhill, but not necessarily simple diffusional, fashion.

Intestinal mucosa

Needless to say that in the intestine active Ca^{2+} transport is controlled by homoeostatic mechanisms involving 1,25-dihydroxy-vitamin D (1,25-OH-D). A cytosolic binding protein for the hormone, and a receptor in the cell nucleus exist. There is no doubt that synthesis of the soluble intracellular Ca binding protein (Wassermann *et al*., 1968, 1971) is stimulated by 1,25-OH-D. However, it has become questionable whether the appearance of this intracellular protein has a causal relation to the increase in Ca transport (Spencer *et al*., 1976; Bikle *et al*., 1978; Morrissey *et al*., 1978, 1980). First, appearance of the protein does not precede increased Ca transport, when 1,25-OH-D is added, but outlasts it after removal of 1,25-OH-D. Secondly, block of protein synthesis by cycloheximide prevents the appearance of the Ca binding protein after 1,25-OH-D, but does not abolish stimulation by 1,25-OH-D of Ca transport. This seems to rule out the possibility that the Ca binding protein plays the part of an activator for a Ca pump. Recently evidence was presented to the effect that the protein has the role of buffering intracellular Ca^{2+} concentration (Morrissey *et al*., 1980).

It is not clear whether Ca^{2+} transport is dependent on external Na on the serosal side (Avioli and Birge, 1978), mainly because efforts have been focused on the luminal side. The inefficacy of ouabain to block the Ca^{2+} absorption seems not a strong argument against a heteroexchange of Ca^{2+} for Na^+, because the poison was added to the luminal side (Holdsworth, 1965). A vitamin D-stimulated Ca^{2+}-dependent ATPase in brushborder membranes of the chick was described by Melancom and De Luca (1970). It appeared after vitamin D treatment of rachitic chicks, preceded the appearance of binding protein and required Ca *and* Mg. Mircheff and Wright (1976) neatly separated a Ca-ATPase in the basolateral membranes (accompanying (Na + K)-ATPase) from another one present in brush border membranes. However, this entity was tested in 2 mM $CaCl_2$ (in the presence of 5 mM $MgCl_2$ and 3 mM Na-ATP). It is satisfactory that recently Ghijsen and von Os (1979) demonstrated that these enzymes have K_{Ca} values in the order of 1 μM at least for part of the Ca binding sites (the assay was in the presence of Mg^{2+}, but the Mg requirement was not investigated in detail). Ghijsen *et al.* (1980) were able to show that alkaline phosphatase found in intestinal epithelial cell membranes is not identical with the basolateral (Ca + Mg)-ATPase which seems to be a good candidate for a Ca pumping mechanism driving the transepithelial Ca movement. Recently Hildmann *et al.* (1979) indeed showed that vesicles made from basolateral plasma membranes of rat small intestinal epithelium accumulate Ca^{2+} in an ATP dependent fashion and in addition show a Na–Ca exchange.

The point of attack of 1,25-OH-D in this concept remains unclear. The finding of Borle (1974) that the intracellular calcium pool is depleted in vitamin D deficiency, whereas both "uptake" and efflux of Ca^{2+} are decreased, may give a clue. This might be interpreted by assuming that 1,25-OH-D primarily facilitates Ca^{2+} entry at the brush border membrane and that the rate of pumping rises as a consequence of increased intracellular Ca^{2+} concentration. The role of the brush border Ca-ATPase has not been elucidated.

Kidney tubules

All sections of the kidney tubule (except the medullary part of the thick ascending limb of Henle (Suki and Rouse, 1980)) seem to be able to transport Ca^{2+} actively, but 60% of the filtered Ca^{2+} is removed from the tubular fluid in the proximal convoluted tubule (Suki, 1979). The cytosolic Ca^{2+} concentration is estimated to be low (10^{-7} M) (Suki, 1979) and a Ca–Na exchange mechanism exists (Taylor and Windhager, 1979). Kinne-Saffran and Kinne (1974) isolated from proximal tubule cells of the rat a membrane fraction which was displaying (Na + K)-ATPase activity, but little alkaline phosphatase and little succinic dehydrogenase activity and was therefore fairly pure material from the basolateral plasma membranes. Confirming earlier claims (Parkinson and Radde, 1971; Rorive and Kleinzeller, 1972) they found in these membranes a Ca^{2+}-activated ATPase of very low Ca^{2+} affinity not requiring Mg^{2+} which, for these reasons, is not likely to be a Ca-transporting system operating at the low intracellular Ca^{2+} concentration. But the authors also found a Mg^{2+}-activated ATPase (5 mM Mg) at nominally zero Ca^{2+} concentration. Since they did not control Ca^{2+} concentration by adding EGTA it may well be that this apparent Mg-ATPase was wholly or in part made up of a (Ca + Mg)-ATPase of high Ca^{2+} affinity. Moore *et al.* (1974) described vesicles from kidney plasma membranes which accumulated Ca^{2+} in strict dependence on the presence of ATP and Mg^{2+} in a Na-free medium. Gmaj *et al.* (1979) produced vesicles from basolateral membranes of cortical tubule cells which accumulated Ca^{2+} in the presence of ATP in the medium. The system showed an apparent K_{Ca} of 0.5 μM. The accumulation was abolished by A 23187 and could be reduced to basal value (uptake in the absence of ATP) by increasing the Na concentration (replacing K) in the medium to 62 mM. Since Na had no effect on unidirectional influx of ^{45}Ca, it is very likely that this apparent inhibition by Na of ATP-driven Ca^{2+} pumping reflects a Na-stimulated Ca efflux and militates therefore strongly for the existence of a Na–Ca exchange system at the basolateral membranes. This notion was corroborated by the fact that a Na gradient across the membrane released more calcium from Ca-loaded vesicles than a K gradient if an impermeant anion was used (cyclamate), and that this was not due to different potential gradients across the membrane in the different media. Finally the authors showed that brush border membrane vesicles did not release more Ca into Na- than into K medium. It seems unavoidable, therefore, to assume that both types of Ca transport exist at the basolateral membrane of kidney tubules. Since it is known from micropuncture studies that removal of Na^+ (replaced by Li^+ or choline) or addition of ouabain (in a preparation of hamster kidney in which the blood vessels were perfused and ouabain was present on both sides of the tubular wall) inhibit the transepithelial Ca^{2+} transport *in situ* to a large extent (Ullrich *et al.*, 1976, 1977), one must expect that the bulk of the absorbed Ca^{2+} passes through the Na–Ca exchange pathway and that the ATP-driven pump plays a role only at very low intracellular Ca^{2+} concentration. This looks familiar because the same conclusions were reached by Di Polo and Beaugé (1979) for the concerted action of the two parallel transport systems for Ca^{2+} in nerve (see above).

Miscellaneous Cells

Liver cells

Van Rossum (1970), confirming earlier reports (Dawkins et al., 1959; Judah and Ahmed, 1963; Wallach et al., 1966), showed that by slowing the metabolism in liver slices one obtains cells enriched in calcium. Cells, loaded by cooling to 1°C for 40 min with Ca^{2+}, upon rewarming to 38°C extruded Ca^{2+}, very probably against an electrochemical gradient. This active Ca^{2+} movement was independent of a Na gradient across the membrane and was not abolished by ouabain but reduced by cyanide. The paper also clearly showed that this extrusion mechanism successfully competes with the accumulation of Ca^{2+} by mitochondria unless high phosphate or succinate concentrations unduly stimulate mitochondria. A (Ca + Mg)-ATPase present in purified liver plasma membranes was described by Garnett and Kemp (1975). The activity of the preparation could be further stimulated by Na^+ and K^+.

Ash and Bygrave (1977) showed in a short communication that there is Ca^{2+} transport in the mitochondrial, the plasma membrane and the microsomal fraction of liver cells. These three systems differed in sensitivity to inhibition by ruthenium red. Mitochondria are inhibited by a few tenths of a nanomole per milligram protein, plasma membranes are far from complete inhibition at 20 nmol/mg of protein and microsomes are insensitive up to 30 nmol/mg of protein.

By using La^{3+} during homogenization Van Rossum et al. (1976) were able to prevent Ca^{2+} redistribution during preparation of mitochondria. They showed that when liver slices were loaded by cold storage for 90 min to about three-fold normal total Ca content, mitochondrial Ca rose six-fold. Upon rewarming for 30 min to 38°C, total Ca returned to about twice normal value, whereas mitochondrial Ca concentration returned to 1.6 times the fresh slice value (or 1.3 calculated per kilogram dry material of slice). It looks, therefore, as if mitochondrial uptake of Ca is governed by the total Ca content above a critical total concentration exceeding the physiological value. The physiological value in turn is maintained by extrusion of Ca^{2+} at the plasma membrane (or at membranes of the endoplasmic reticulum which communicates with the extracellular space). This extrusion mechanism requires energy and is not of the Na–Ca exchange type.

Van Rossum et al. (1973) indicated that the Ca^{2+} extrusion mechanism is present in hepatoma cells as well and operates even faster there than in normal liver cells. They also demonstrated in the same study that in liver cells the Ca^{2+} sensitive channel for K described in red cells (see above) exists.

Blood platelets
Human blood platelets, when activated (e.g. by thrombin), show an increase in permeability for Ca^{2+} at the plasma membrane which seems to occur *after* an initial Ca^{2+} release from storage organelles (Massini and Lüscher, 1976). Since this induced leakiness for Ca^{2+} of the plasma membrane seems necessary for the clot retraction, it is highly probable that resting intracellular Ca^{2+} concentration is below that necessary for activation of the contractile protein and that upon activation there is a net inward Ca^{2+} movement. Membrane material from platelets contains a $(Ca + Mg)$-activated ATPase (Käser-Glanzmann *et al.*, 1978). Käser-Glanzmann *et al.* (1979) were able to produce Ca^{2+}-accumulating vesicles from sonicated platelets. The fraction contained both membranes from the dense tubular system and the plasma membrane (28% of a surface label was recovered in the vesicles). The question remains open, therefore, whether the Ca accumulation reflects uptake into internal membrane vesicles or into inside-out plasma membrane vesicles. Käser-Glanzmann *et al.* (1979) showed that the observed Ca^{2+} transport can be stimulated by a cAMP-dependent protein kinase recovered from platelets.

L cells
Lamb and Lindsay (1971) found that in these cultured cells Ca^{2+} efflux shows a pH dependence completely different from that of Ca^{2+} influx, is highly temperature dependent and can be reduced by metabolic inhibitors. They were able to demonstrate clearly that the efflux ceased in ATP depleted cells and could be restored by incorporating ATP (or CTP) into these cells. As they could further rule out a dependence on the Na gradient across the plasma membrane it is quite obvious that L cells have an ATP fuelled, Ca pump and lack the Na–Ca exchange system and are, therefore, very similar to red cells.

Ehrlich ascites tumour cells
Hinnen *et al.* (1979), who observed that addition of glucose reduced Ca^{2+} uptake by these cells, reconstituted plasma membranes into liposomes. This reconstituted system showed an ATP-dependent Ca^{2+} uptake which was not sensitive to removal or addition of

calmodulin. The report eschews specious arguments as to the purity of the plasma membrane preparation used.

Macrophages

Phagocytotic vesicles from macrophages are by definition inside-out vesicles. Lew and Stoessel (1979) studied such vesicles from pulmonary macrophages and found them to take up Ca^{2+}, provided that Mg^{2+} and ATP were present in the medium. In the absence of oxalate the Ca^{2+} accumulated could be released by A 23187. Calmodulin increased the Ca^{2+} uptake 3.5-fold. The K_{Ca} was found to be 0.48 μM. Curiously the authors were unable to demonstrate a corresponding (Ca + Mg)-ATPase activity.

Adipocytes

Pershadsingh et al. (1980a) prepared vesicles from plasma membranes of adipocytes which accumulated Ca^{2+} in an ATP dependent way in the presence of oxalate in the medium. The process was abolished by A 23187. Calmodulin did stimulate the system, particularly when EDTA was used during the vesicle preparation. Its action consisted in shifting the K_{Ca} from above 0.1 μM to below 0.1 μM. In contrast, Ca^{2+} transport by the microsomal fraction of the same cells was not activated by calmodulin. Conversely, the microsomal transport entity was stimulated by 20 mM KCl whereas the plasma membrane system was not.

Pancreas islet cells

Pershadsingh et al. (1980b) reported the presence of a Ca pump in what appears to be a fairly pure plasma membrane preparation from mouse pancreatic islet cells. The material displayed Ca^{2+}-activated ATPase activity which required (at least low) Mg^{2+} concentrations. It showed two different affinities for ATP and had high affinity for Ca^{2+} ($K_{Ca} = 0.1 \mu M$). The same membrane preparation accumulated Ca^{2+} in an ATP-dependent fashion from oxalate containing media. This transport was abolished by A 23187. If the membranes were prepared in the presence of EDTA calmodulin moderately increased the Ca^{2+} uptake.

Summary

In complete cells it is more difficult than in red cells to demonstrate beyond reasonable doubt the presence of an ATP-driven, outwardly directed Ca pump. However, the 13 preceeding examples where, in the reviewer's opinion, evidence is fairly good (although not equally good in all) prove that it does exist. It is even present in cells which certainly are endowed with the Na–Ca exchange mechanism. Its task seems to consist in keeping the intracellular Ca^{2+} concentration low, rather than to deal with heavy loads of Ca^{2+}. It sets the scope for total cellular Ca. Within this scope intracellular Ca-storing organelles can modulate the free ionic Ca^{2+} concentration at short notice according to the needs of the cell or the organism and in response to signals reaching the Ca^{2+} permeability channels of these organelles. By maintaining a very steep Ca^{2+} gradient across the plasma membrane it provides the condition allowing rapid Ca^{2+} entry triggered by external stimuli impinging on the plasma membrane which makes Ca^{2+} itself a signalling agent and justifies its being called a second messenger.

REFERENCES

Andersson, R. G. G. and Norrby, K. (1977). *Virchows Arch. B Cell Path.* **23**, 185-194.

Ansah, T. A. and Katz, S. (1980). *Cell Calcium* **1**, 195-203.

Ash, G. R. and Bygrave, F. L. (1977). *FEBS Lett.* **78**, 166-168.

Au, K. S. (1978a). *Int. J. Biochem.* **9**, 477-480.

Au, K. S. (1978b). *Int. J. Biochem.* **9**, 735-743.

Avioli, L. V. and Birge, S. J. (1978). *In* "Physiology of Membrane Disorders" (Eds E. Andreoli, J. F. Hoffman and D. D. Fanestil) 919-940. Plenum, New York, London.

Avissar, N., de Vries, A., Ben-Shaul, Y. and Cohen, I. (1975). Biochim. Biophys. Acta **375**, 35-43.

Baker, P. F. (1972). *In* "Progress in Biophysics and Molecular Biology" (Eds J. A. V. Butler and D. Noble) Vol. 24, 177-223. Pergamon Press, Oxford.

Baker, P. F. and Glitsch, H. G. (1973). *J. Physiol., Lond.* **233**, 44P-46P.

Baker, P. F. and McNaughton, P. A. (1976a). *J. Physiol., Lond.* **259**, 103-144.

Baker, P. F. and McNaughton, P. A. (1976b). *J. Physiol., Lond.* **260**, 24P-25P.

Baker, P. F. and McNaughton, P. A. (1978). *J. Physiol., Lond.* **276**, 127-150.

Baker, P. F., Blaustein, M. P., Hodgkin, A. L. and Steinhardt, R. A. (1969). *J. Physiol., Lond.* **200**, 431-458.

Balzer, H., Makinose, M. and Hasselbach, W. (1968). *Arch. Pharmakol. exp. Pathol.* **260**, 444-455.

Barrabin, H., Garrahan, P. J. and Rega, A. F. (1980). *Biochim. biophys. Acta* **600**, 796-804.

Bartlett, G. R. (1959). *J. biol. Chem.* **234**, 449-465.

Beaugé, L. A. and Glynn, I. M. (1978). *Nature, Lond.* **272**, 551-552.

Beaugé, L., Di Polo, R., Osses, L., Barnola, F. and Campos, M. (1981). *Biochim. biophys. Acta*, **644**, 147–152.

Beutler, E. (1975). "Red Cell Metabolism", 2nd edn, 149. Grune and Stratton, New York.
Bikle, D. D., Zolock, D. T., Morrissey, R. L. and Herman, R. H. (1978). *J. biol. Chem.* **253**, 484-488.
Blaustein, M. P. (1977). *Biophys. J.* **20**, 79-111.
Blaustein, M. P. and Ector, A. C. H. (1976). *Biochim. biophys. Acta* **419**, 295-308.
Blaustein, M. P. and Oborn, C. J. (1975). *J. Physiol., Lond.* **247**, 657-686.
Blaustein, M. P., Kendrick, N. C., Fried, R. C. and Ratzlaff, R. W. (1977). *In* "Approaches to the Cell Biology of Neurones" (Eds W. M. Cowan and J. A. Ferendelli) 172-194. Soc. Neuroscience Symp., Vol. 2. Society for Neuroscience, Bethesda, Md.
Blaustein, M. P., Ratzlaff, R. W., Kendrick, N. C. and Schweitzer, E. S. (1978). *J. gen. Phsyiol.* **72**, 15-41.
Blum, R. M. and Hoffman, J. F. (1971). *J. memb. Biol.* **6**, 315-328.
Bond, G. H. (1972). *Biochim. biophys. Acta* **288**, 423-433.
Bond, G. H. and Clough, D. L. (1973). *Biochim. biophys. Acta* **323**, 592-599.
Bond, G. H. and Green, J. W. (1971). *Biochim. biophys. Acta* **241**, 393-398.
Bond, G. H. and Hudgins, P. (1978). *Fedn Proc. Fedn Am. Socs exp. Biol.* **37**, 542 Abstr., 313.
Bond, G. H. and Hudgins, P. (1979). *Biochemistry* **18**, 325-331.
Bond, G. H. and Hudgins, P. M. (1980). *Biochim. biophys. Acta* **600**, 781-790.
Bookchin, R. M. and Lew, V. L. (1978). *J. Physiol., Lond.* **284**, 93P.
Bookchin, R. M. and Lew, V. L. (1980). *Nature, Lond.* **284**, 561-563.
Borle, A. B. (1974). *J. memb. Biol.* **16**, 207-220.
Boynton, A. L., Whitfield, J. F. and MacManus, J. P. (1980). *Biochem. biophys. Res. Commun.* **95**, 745-749.
Brinley, F. J., Jr. and Mullins, L. J. (1974). *Ann. N.Y. Acad. Sci.* **242**, 406-432.
Brown, A. M. (1979). *Biochim. biophys. Acta* **554**, 195-203.
Brown, A. M., Ellory, J. C., Young, J. D. and Lew, V. L. (1978). *Biochim. biophys. Acta* **511**, 163-175.
Bucher, U., Coninx, S., Furlan, M., Bürgin, H., Schatzmann, H. J. and Zahler, P. (1978). 17th Congr. Int. Soc. Hematol. Abst., p. 61.
Buckley, J. T. and Hawthorne, (1972). *J. biol. Chem.* **247**, 7218-7223.
Bürgin, H. (1979). *J. vet. Pharmacol. Ther.* **2**, 305-311.
Bürgin, H. and Schatzmann, H. J. (1979). *J. Physiol., Lond.* **287**, 15-32.
Bygrave, F. L. (1978). *Biol. Rev.* **53**, 43-79.
Cantley, L. C., Jr, Josephson, L., Warner, R., Yanagisawa, M., Lechene, C. and Guidotti, G. (1977). *J. biol. Chem.* **252**, 7421-7423.
Cantley, L. C. Jr., Resh, M. D. and Guidotti, G. (1978). *Nature, Lond.* **272**, 552-554.
Carafoli, E. (1974). *Biochem. Soc. Symp.* **39**, 89-109.
Carafoli, E. and Crompton, M. (1978). *In* "Current Topics in Membranes and Transport" (Eds F. Bronner and A. Kleinzeller) Vol. 10, 151-216. Academic Press, New York.
Carafoli, E. and Niggli, V. (1981). *Ann. N.Y. Acad. Sci.* **358**, 159-168.
Caroni, P. and Carafoli, E. (1980). *Nature, Lond.* **283**, 765-767.
Casteels, R. and Raeymaekers, L. (1978). *J. Physiol., Lond.* **285**, 49P.
Casteels, R. and Raeymaekers, L. (1979). *J. Physiol., Lond.* **294**, 51-68.
Casteels, R., Goffin, J., Raeymaekers, L. and Wuytak, F. (1973). *J. Physiol.,*

Lond. **231**, 19P.
Cha, N. Y., Shin, B. C. and Lee, K. S. (1971). *J. gen. Physiol.* **57**, 202-215.
Chambers, E. L. (1980). Abstr. 2nd Intern. Congress Cell Biol. *Eur. J. Cell Biol.* **22**, 476.
Cheung, W. Y. (1980). *Science, N.Y.* **207**, 19-27.
Dean, W. L. and Tanford, Ch. (1978). *Biochemistry* **17**, 1683-1690.
Davis, P. W. and Vincenzi, F. F. (1971). *Life. Sci.* **10**, 401-406.
Dawkins, M. J. R., Judah, J. D. and Rees, K. R. (1959). *J. Pathol. Bacteriol.* **77**, 257-266.
Dedman, J. R., Potter, J. D., Jackson, R. L., Jonson, J. D. and Means, A. R. (1977). *J. biol. Chem.* **252**, 8415-8422.
Di Polo, R. (1978). *Nature, Lond.* **274**, 390-392.
Di Polo, R. and Beaugé, L. (1979). *Nature, Lond.* **278**, 271-273.
Di Polo, R. and Beaugé, L. (1980). *Cell Calcium* **1**, 147-169.
Di Polo, R., Rojas, H. R. and Beaugé, L. (1979). *Nature, Lond.* **281**, 228-229.
Dixon, E. and Winslow, R. M. (1979). *Fedn Proc. Fedn Am. Socs exp. Biol.* **38**, 1127.
Downes, P. and Michell, R. H. (1981). *Nature, Lond.* **290**, 270-271.
Drickamer, L. K. (1975). *J. biol. Chem.* **250**, 1952-1954.
Dunham, E. T. and Glynn, I. M. (1961). *J. Physiol., Lond.* **156**, 274-293.
Eaton, J. W., Skelton, T. D., Swofford, H. S., Koplin, C. and Jacob, H. S. (1973). *Nature, Lond.* **246**, 105-106.
Eaton, J. W., Berger, E., White, J. G. and Jacob, H. S. (1978). *Br. J. Haemat.* **38**, 57-62.
Ellory, J. C. (1977). *In* "Membrane Transport in Red Cells" (Eds J. C. Ellory and V. L. Lew) 363-381. Academic Press, London and New York.
Enyedi, A., Sarkadi, B., Szasz, I., Bot, G. and Gardos, G. (1980a). *Cell Calcium* **1**, 299-310.
Enyedi, A., Szasz, I., Sarkadi, B., Bot, G. and Gardos, G. (1980b). Abstr. Proc. Int. Union Physiol. Sci. **14**, p. 398, No. 1333. Hungarian Physiological Society.
Epel, D. (1980). *Endeavour,* New Series **4**, 26-31.
Farrance, M. L. and Vincenzi, F. F. (1977a). *Biochim. biophys. Acta* **471**, 49-58.
Farrance, M. L. and Vincenzi, F. F. (1977b). *Biochim. biophys. Acta* **471**, 59-66.
Ferreira, H. G. and Lew, V. L. (1975). *J. Physiol., Lond.* **252**, 86P.
Ferreira, H. G. and Lew, V. L. (1976). *Nature, Lond.* **259**, 47-49.
Ferreira, H. G. and Lew, V. L. (1977). *In* "Membrane Transport in Red Cells" (Eds J. C. Ellory and V. L. Lew) 53-91. Academic Press, London and New York.
Flatman, P. and Lew, V. L. (1977). *Nature, Lond.* **267**, 360-362.
Foder, B., Scharff, O. and Tønnesen, P. (1980). *Clin. chim. Acta* **104**, 187-193.
Fortes, P. A., Ellory, J. C. and Lew, V. L. (1973). *Biochim. biophys. Acta* **318**, 262-272.
Gallien-Lartigue, O. (1976). *Cell Tiss. Kinet.* **9**, 533-540.
Garay, R. P. and Garrahan, P. J. (1973). *J. Physiol., Lond.* **231**, 297-325.
Garrahan, P. J. and Rega, A. F. (1978). *Biochim. biophys. Acta* **513**, 59-65.
Garrahan, P. J., Rega, A. F. and Alonso, G. L. (1976). *Biochim. biophys. Acta* **448**, 121-132.
Gardos, G. (1958a). *Acta physiol. Hung.* **15**, 121-125.
Gardos, G. (1958b). *Biochim. biophys. Acta* **30**, 653-654.

Gardos, G., Szasz, I. and Sarkadi, B. (1975). *FEBS Proc.* **35**, 167.

Garnett, H. M. and Kemp, R. B. (1975). *Biochim. biophys. Acta* **382**, 526-533.

Ghijsen, W. E. J. M. and van Os, C. H. (1979). *Nature, Lond.* **279**, 802-803.

Ghijsen, W. E. J. M., De Jong, M. D. and van Os, C. H. (1980). *Biochim. biophys. Acta* **599**, 538-551.

Gietzen, K. and Bader, H. (1980). *IRCS med. Sci.* **8**, 396-397.

Gietzen, K., Mansard, A. and Bader, H. (1980a). *Biochem. biophys. Res. Commun.* **94**, 674-681.

Gietzen, K., Seiler, S., Fleischer, S. and Wolf, H. U. (1980b). *Biochem. J.* **188**, 47-54.

Gietzen, K., Tejčka, M. and Wolf, H. U. (1980c). *Biochem. J.* **189**, 81-88.

Gietzen, K., Wüthrich, W., Mansard, A. and Bader, H. (1981). Abstr. Int. Vinca Symp. Frankfurt, (Eds W. Brade, G. A. Nagel and S. Seeber) 1981, pp. 16–26. S. Karger, Basel.

Glitsch, H. G., Reuter, H. and Scholz, (1970). *J. Physiol., Lond.* **209**, 25-43.

Gmaj, P., Murer, H. and Kinne, R. (1979). *Biochem. J.* **178**, 549-557.

Gopinath, R. M. and Vincenzi, F. F. (1977). *Biochem. biophys. Res. Commun.* **77**, 1203-1209.

Gopinath, R. M. and Vincenzi, F. F. (1979). *Am. J. Hemat.* **7**, 303-312.

Graf, E. and Penniston, J. T. (1981). *J. biol. Chem.* **256**, 1587–1592.

Grand, J. A. and Perry, S. V. (1979). *Biochem. J.* **183**, 285-295.

Greenquist, A. and Shohet, S. B. (1973). *Blood* **42**, 997-1001.

Griffin, H. D. and Hawthorne (1978). *Biochem. J.* **176**, 541-552.

Haaker, H. and Racker, E. (1979). *J. biol. Chem.* **254**, 6598-6602.

Hildmann, B., Schmidt, A. and Murer, H. (1979). *Pflügers Arch. ges. Physiol.* **382**, R 23.

Hinds, T. R., Larsen, F. L. and Vincenzi, F. F. (1978). *Biochem. biophys. Res. Commun.* **81**, 455-461.

Hinds, T. R., Raess, B. U. and Vincenzi, F. F. (1981). *J. membr. Biol.* **58**, 57-65.

Hinnen, R., Miyamoto, H. and Racker, E. (1979). *J. membr. Biol.* **49**, 309-324.

Hoffman, J. F. and Blum, R. M. (1977). *In* "Membrane Toxicity" (Eds M. W. Miller and A. E. Shamoo) 381-405. Plenum, New York.

Holdsworth, E. S. (1965). *Biochem. J.* **96**, 475-481.

Horton, C. R., Cole, W. Q. and Bader, H. (1970). *Biochem. biophys. Res. Commun.* **40**, 505-509.

Jarrett, H. W. and Kyte, J. (1979). *J. biol. Chem.* **254**, 8237-8244.

Jarrett, H. W. and Penniston, J. T. (1977). *Biochem. Biophys. Res. Commun.* **77**, 1210-1216.

Jarrett, H. W. and Penniston, J. T. (1978). *J. biol. Chem.* **253**, 4676-4682.

Jenkins, D. M. G. and Lew, V. L. (1973). *J. Physiol., Lond.* **234**, 41-42P.

Jones, L. R., Besch, H. R. and Watanabe, A. M. (1977). *J. biol. Chem.* **252**, 3315-3323.

Jones, L. R., Besch, H. R., Jr, Fleming, J. W., McConnaughey, M. M. and Watanabe, A. M. (1979). *J. biol. Chem.* **254**, 530-539.

Judah, J. D. and Ahmed, K. (1963). *Biochim. biophys. Acta* **71**, 34-45.

Jundt, H. and Reuter, H. (1977). *J. Physiol., Lond.* **266**, 78P.

Käser-Glanzmann, R., Jakabowa, M., George, J. N. and Lüscher, E. F. (1978). *Biochim. biophys. Acta* **512**, 1-12.

Käser-Glanzmann, R., Gerber, E. and Lüscher, E. F. (1979). *Biochim. biophys. Acta* **558**, 344-347.

Katz, S. and Blostein, R. (1973). *Fedn Proc. Fedn Am. Socs exp. Biol.* **32**, Abstr. 287.
Katz, S. and Blostein, R. (1975). *Biochim. biophys. Acta* **389**, 314-324.
Katz, S. and Remtulla, M. A. (1978). *Biochem. biophys. Res. Commun.* **83**, 1373-1379.
Kawaguchi, T. and Konishi, K. (1980). *Biochim. biophys. Acta* **597**, 577-586.
Kendrick, N. C., Blaustein, M. P., Fried, R. C. and Ratzlaff, R. W. (1977). *Nature, Lond.* **265**, 246-248.
Kinne, R. (1980). Proc. Int. Union. Physiol. Sci., Vol. 14, Abstr. 28 (p. 27).
Kinne-Saffran, E. and Kinne, R. (1974). *J. membr. Biol.* **17**, 263-274.
Kirkpatrick, F. H., Hillman, D. G. and La Celle, P. L. (1975). *Experientia* **31**, 653-654.
Klinger, R., Wetzker, R., Fleischer, I. and Frunder, H. (1980). *Cell Calcium*, **1**, 229-240.
Knauf, P. A., Proverbio, F. and Hoffman, J. F. (1972). Abstr. IV. Int. Biophysics Congress, Moscow, Vol. 3, 104.
Knauf, P. A., Proverbio, F. and Hoffman, J. F. (1974). *J. gen. Physiol.* **63**, 324-336.
Lamb, J. F. and Lindsay (1971). *J. Physiol., Lond.* **218**, 691-708.
Larocca, J. N., Rega, A. F. and Garrahan, P. J. (1981). *Biochim. Biophys. Acta*, **645**, 10–16.
Larsen, F. L. and Vincenzi, F. F. (1979). *Science, N.Y.* **204**, 306-308.
Larsen, F. L., Hinds, T. R. and Vincenzi, F. F. (1978a). *J. membr. Biol.* **41**, 361-376.
Larsen, F. L., Raess, B. U., Hinds, T. R. and Vincenzi, F. F. (1978b). *J. supramol. Struct.* **9**, 269-274.
Lee, K. S. and Shin, B. C. (1969). *J. gen. Physiol.* **54**, 713-728.
Le Peuch, Ch. J., Haiech, J. and Demaille, J. G. (1979). *Biochemistry* **18**, 5150-5157.
Levin, R. M. and Weiss, B. (1978). *Biochim. biophys. Acta* **540**, 197-204.
Levin, R. M. and Weiss, B. (1979). *J. Pharmac. exp. Ther.* **208**, 454-459.
Lew, V. L. and Ferreira, H. G. (1976). *Nature, Lond.* **263**, 336-338.
Lew, V. L. and Ferreira, H. G. (1977), *In* "Membrane Transport in Red Cells" (Eds J. C. Ellory and V. L. Lew) 93-100. Academic Press, London and New York.
Lew, V. L. and Ferreira, H. G. (1978). *In* "Current Topics in Membranes and Transport" (Eds F. Bronner and A. Kleinzeller) Vol. 10, 217-277. Academic Press, New York.
Lew, P. D. and Stoessel, T. P. (1979). *J. biol. Chem.* **255**, 5841-5846.
Lichtner, R. and Wolf, H. U. (1977). *Archs. Pharmac. Suppl.* **297**, R38.
Lichtner, R. and Wolf, H. U. (1980a). *Biochim. biophys. Acta* **598**, 472-485.
Lichtner, R. and Wolf, H. U. (1980b). *Biochim. biophys. Acta* **598**, 486-493.
Loewenstein, W. R. and Rose, B. (1978). *Ann. N.Y. Acad. Sci.* **307**, 285-307.
Luthra, M. G. and Kim, H. D. (1979). *Life Sci.* **24**, 2441-2448.
Luthra, M. G., Hildenbrandt, G. R. and Hanahan, D. J. (1976). *Biochim. biophys. Acta* **419**, 164-179.
Luthra, M. G., Au, K. S. and Hanahan, D. J. (1977). *Biochem. biophys. Res. Commun.* **77**, 678-687.
Luthra, M. G., Watts, R. P., Scherer, K. L. and Kim, H. D. (1981). *Biochim. Biophys. Acta*, in press.

104   H. J. SCHATZMANN

Lynch, Th. J. and Cheung, W. Y. (1979). *Arch. Biochem. Biophys.* **194**, 165-170.

MacIntyre, J. D. and Green, J. W. (1976). *J. gen. Physiol.* **68**, 12a.

MacIntyre, J. D. and Green, J. W. (1978). *Biochim. Biophys. Acta* **510**, 373-377.

Maimon, J. and Pushkin, S. (1978). *J. supramol. Struct.* **9**, 131-141.

Makinose, M. and Boll, W. (1979). *In* "International Bioenergetics Symposium" 89-100. Kobe (Japan) 1978. Academic Press, New York.

Mandersloot, J. G., Roelofsen, B. and De Gier, J. (1978). *Biochim. biophys. Acta* **508**, 478-485.

Massini, P. and Lüscher, E. F. (1976). *Biochim. biophys. Acta* **436**, 652-663.

Mauldin, D. and Roufogalis, B. D. (1980). *Biochem. J.* **187**, 507-513.

Means, A. R. and Dedman, J. R. (1980). *Nature, Lond.* **285**, 73-77.

Melancom, M. J., Jr, and De Luca, H. F. (1970). *Biochemistry* **9**, 1658-1664.

Mircheff, A. K. and Wright, E. M. (1976). *J. membr. Biol.* **28**, 309-333.

Mollman, J. E. and Pleasure, D. E. (1980). *J. biol. Chem.* **255**, 569-574.

Moore, L., Fitzpatrick, D. F., Chen, T. S. and Landon, E. J. (1974). *Biochim. biophys. Acta* **345**, 405-418.

Morcos, N. C. and Drummond, G. I. (1980). *Biochim. biophys. Acta* **598**, 27-39.

Morrissey, R. L., Zolock, D. T., Bikle, D. D., Empson, R. N., Jr, and Bucci, T. J. (1978). *Biochim. biophys. Acta* **538**, 23-33.

Morrissey, R. L., Zolock, D. T., Mellick, P. W. and Bikle, D. D. (1980). *Cell Calcium* **1**, 69-79.

Mualem, S. and Karlish, S. J. D. (1979a). *Nature, Lond.* **277**, 238-240.

Mualem, S. and Karlish, S. J. D. (1979b). *FEBS Lett.* **107**, 209–212.

Mualem, S. and Karlish, S. J. D. (1980). *Biochim. biophys. Acta* **597**, 631-636.

Mualem, S. and Karlish, S. J. D. (1981). *Biochim. biophys. Acta*, **647**, 73–86.

Niggli, V., Penniston, J. T. and Carafoli, E. (1979a). *J. biol. Chem.* **254**, 9955-9958.

Niggli, V., Ronner, P., Carafoli, E. and Penniston, J. T. (1979b). *Archs. Biochem. Biophys.* **198**, 124-130.

Niggli, V., Adunyah, E. S. Penniston, J. T. and Carafoli, E. (1981). *J. biol. Chem.* **256**, 395-401.

Olson, E. J. and Cazort, R. J. (1969). *J. Gen. Physiol.* **53**, 311-322.

Palek, J., Church, A. and Fairbanks, G. (1976). *In* "Membranes and Disease" (Eds L. Bolis, J. F. Hoffman and A. Leaf) 41-59. Raven Press, New York.

Papazian, D., Rahamimoff, H. and Goldin, S. M. (1979). *Proc. natn. Acad. Sci. U.S.A.* **76**, 3708-3712.

Parker, J. C. (1977). *Fedn Proc. Fedn Am. Socs exp. Biol.* **36**, 271.

Parker, J. C. (1978). *J. gen. Physiol.* **71**, 1-17.

Parker, J. C. (1979). *Am. J. Physiol.* **237**, C10-C16.

Parker, J. C. and Welt, L. G. (1972). *Archs intern. Med.* **129**, 320-332.

Parker, J. C., Gitelman, H. J., Glosson, P. S. and Leonard, D. L. (1975). *J. gen. Physiol.* **65**, 84-96.

Parker, J. C., Orringer, E. P. and McManus, Th. J. (1978). *In* "Physiology of Membrane Disorders" (Eds E. Andreoli, J. F. Hoffmann and D. D. Fanestil) 773-799. Plenum Medical, New York and London.

Parkinson, D. K. and Radde, I. C. (1971). *Biochim. biophys. Acta* **242**, 238-246.

Pershadsingh, H. A., Landt, M. and McDonald, J. M. (1980a). *J. biol. Chem.* **255**, 8983-8986.

Pershadsingh, H. A., McDaniel, M. L., Landt, M., Bry, C. G., Lacy, P. E. and McDonald, J. M. (1980b). *Nature, Lond.* **288**, 492-494.
Peterson, S. W., Ronner, P. and Carafoli, E. (1978). *Archs. Biochem. Biophys.* **186**, 202-210.
Pfleger, H. and Wolf, H. U. (1975). *Biochem. J.* **147**, 359-361.
Portzehl, H., Zaoralek, P. and Gaudin, J. (1969). *Biochim. biophys. Acta* **189**, 440-448.
Porzig, H. (1970). *J. membr. Biol.* **2**, 324-340.
Pouchan, M. I., Garrahan, P. J. and Rega, A. F. (1969). *Biochim. biophys. Acta* **173**, 151-154.
Quist, E. E. and Roufogalis, B. D. (1975). *FEBS Lett.* **50**, 135-139.
Quist, E. E. and Roufogalis, B. D. (1977). *J. supramol. Struct.* **6**, 375-381.
Raess, B. U. and Hinds, T. R. (1979). *Pharmacologist* **21**, 277.
Raess, B. U. and Vincenzi, F. F. (1980). *Molec. Pharmacol.* **18**, 253-258.
Rahamimoff, H. and Abramowitz, E. (1978). *FEBS Lett.* **92**, 163-167.
Reeves, J. P. and Sutko, J. L. (1980). *Science, N.Y.* **208**, 1461-1463.
Rega, A. F. and Garrahan, P. J. (1975). *J. membr. Biol.* **22**, 313-327.
Rega, A. F. and Garrahan, P. J. (1978). *Biochim. biophys. Acta* **507**, 182-184.
Rega, A. F. and Garrahan, P. J. (1980). *Biochim. biophys. Acta* **596**, 487-489.
Rega, A. F., Garrahan, P. J. and Wainer, S. R. (1972). *Experientia* **28**, 1158-1159.
Rega, A. F., Richards, D. E. and Garrahan, P. J. (1973). *Biochem. J.* **136**, 185-194.
Rega, A. F., Richards, D. E. and Garrahan, P. J. (1974). *Ann. N.Y. Acad. Sci.* **242**, 317-323.
Reuter, H. and Seitz, N. (1968). *J. Physiol.* **195**, 451-470.
Richards, D. E., Rega, A. F. and Garrahan, P. J. (1977a). *J. membr. Biol.* **35**, 113-124.
Richards, D. E., Vidal, J. C., Garrahan, P. J. and Rega, A. F. (1977b). *J. Membr. Biol.* **35**, 125-136.
Richards, D. E., Rega, A. F. and Garrahan, P. J. (1978). *Biochim. biophys. Acta* **511**, 194-201.
Richardt, H. W., Fuhrmann, G. F. and Knauf, P. A. (1979). *Nature, Lond.* **279**, 248-250.
Rink, T. S., Tsien, R. Y. and Warner, A. E. (1980). *Nature, Lond.* **283**, 658-660.
Riordan, J. R. and Passow, H. (1973). *In* "Comparative Physiology." (Eds L. Bolis, K. Schmidt-Nielsen and S. H. P. Maddrell) 543-581. North-Holland, Amsterdam.
Robinson, J. D. (1976). *Archs Biochem. Biophys.* **176**, 366-374.
Robinson, J. D. (1978). *FEBS Lett.* **87**, 261-264.
Roelofsen, B. (1977). 11th FEBS Meeting, Copenhagen. Vol. 45, Symp. A4, Ed P. Nicholls, pp. 183-190.
Roelofsen, B. and Schatzmann, H. J. (1977). *Biochim. biophys. Acta* **464**, 17-36.
Roelofsen, B. and van Deenen, L. L. M. (1973). *Eur. J. Biochim.* **40**, 245-257.
Romero, P. J. (1981) *In* "Advances in Physiological Sciences" (Eds S. R. Hollan, G. Gardos and B. Sarkadi) Vol. 6, 189-194. Pergamon Press, Oxford.
Romero, P. J. and Whittam, R. (1971). *J. Physiol., Lond.* **214**, 481-507.
Ronner, P., Gazzotti, P. and Carafoli, E. (1977). *Archs Biochem. Biophys.* **179**, 578-583.
Rorive, G. and Kleinzeller, A. (1972). *Biochim. biophys. Acta* **274**, 226-239.

Rosenthal, A. S., Kregenow, F. J. M. and Moses, H. L. (1970). *Biochim. Biophys. Acta* 196, 254-262.
Rossi, J. P. F. C., Garrahan, P. J. and Rega, A. F. (1978). *J. membr. Biol.* 44, 37-46.
Rossi, J. P. F. C., Rega, A. F. and Garrahan, P. J. (1980). Abstr. IX. Reunion Ann. Soc. Arg. Biophys., Bermejo Mendoza. In press.
Roufogalis, B. D. (1979). *Can. J. Physiol. Pharmac.* 57, 1331-1349.
Sarkadi, B. (1980). *Biochim. biophys. Acta,* 604, 159-190.
Sarkadi, B. and Tosteson, D. C. I. (1979). *In* "Transport Across Single Biological Membranes" (Eds G. Giebisch, D. C. Tosteson and H. H. Ussing) 117-160. Springer-Verlag, Berlin.
Sarkadi, B., Szasz, I. and Gardos, G. (1976). *J. membr. Biol.* 26, 357-370.
Sarkadi, B., Szasz, I. Gerloczy, A. and Gardos, G. (1977). *Biochim. biophys. Acta* 464, 93-107.
Sarkadi, B., MacIntyre, J. D. and Gardos, G. (1978). *FEBS Lett.* 89, 78-82.
Sarkadi, B., Schubert, A. and Gardos, G. (1979a). *Experientia* 35, 1045-1047.
Sarkadi, B., Szasz, I. and Gardos, G. (1979b). XI Int. Congress Biochem. Abstr. 116.
Sarkadi, B., Szasz, I. and Gardos, G. (1979c). *J. membr. Biol.* 46, 183-184.
Sarkadi, B., Enyedi, A. and Gardos, G. (1980a). *Cell Calcium* 1, 287-297.
Sarkadi, B., Szasz, I. and Gardos, G. (1980b). *Biochim. biophys. Acta* 598, 326-338.
Schanne, F. A. X., Kane, A. B., Young, E. E. and Farber, J. L. (1979). *Science, N.Y.* 206, 700-702.
Scharff, O. (1972). *Scand. J. clin. Lab. Invest.* 30, 313-320.
Scharff, O. (1976). *Biochim. biophys. Acta* 443, 206-218.
Scharff, O. (1978). *Biochim. biophys. Acta* 512, 309-317.
Scharff, O. (1979). *Analyt. chim. Acta* 109, 291-305.
Scharff, O. (1980). *In* "Membrane Transport in Erythrocytes" (Eds U. V. Larsen, H. H. Ussing and J. O. Wieth) 236-254. Alfred Benzon Symp. No. 14. Munskgaard, Copenhagen.
Scharff, O. and Foder, B. (1975). *Scand. J. clin. Lab. Invest.* 35, 583-589.
Scharff, O. and Foder, B. (1978). *Biochim. biophys. Acta* 509, 67-77.
Schatzmann, H. J. (1966). *Experientia* 22, 364-368.
Schatzmann, H. J. (1969). *In* "A Symposium on Calcium and Cellular Function" (Ed. A. W. Cuthbert) 85-95. Biological Council Symposia on Drug Action. MacMillan, London.
Schatzmann, H. J. (1970). *Experientia* 26, 687.
Schatzmann, H. J. (1973). *J. Physiol.* 235, 551-569.
Schatzmann, H. J. (1974). *Nature, Lond.* 248, 58-60.
Schatzmann, H. J. (1975). *In* "Current Topics in Membranes and Transport" (Eds F. Bronner and A. Kleinzeller) Vol. 6, 126-168. Academic Press, New York.
Schatzmann, H. J. (1977). *J. membr. Biol.* 35, 149-158.
Schatzmann, H. J. and Bürgin, H. (1978). *Ann. N.Y. Acad. Sci..* 307, 125-147.
Schatzmann, H. J. and Roelofsen, B. (1977). *In* "Biochemistry of Membrane Transport", (Eds G. Semenza and E. Carafoli) 389-400. FEBS Symposium No. 42. Springer-Verlag, Berlin.
Schatzmann, H. J. and Rossi, G. L. (1971). *Biochim. biophys. Acta* 241, 379-392.

Schatzmann, H. J. and Scheidegger, H. R. (1975). *Experientia* **31**, 1260-1261.
Schatzmann, H. J. and Tschabold, M. (1971). *Experientia* **27**, 59-61.
Schatzmann, H. J. and Vincenzi, F. F. (1969). *J. Physiol., Lond.* **201**, 369-395.
Seamon, K. B. (1980). *Biochemistry* **19**, 207-215.
Shikegawa, M. and Pearl, L. J. (1976). *J. biol. Chem.* **251**, 6947-6952.
Simons, T. J. B. (1976a). *J. Physiol., Lond.* **256**, 209-225.
Simons, T. J. B. (1976b). *J. Physiol., Lond.* **256**, 227-244.
Spencer, R., Charman, M., Wilson, P. and Lawson, E. (1976). *Nature, Lond.* **263**, 161-163.
Steck, R. L., Weinstein, R. S., Straus, J. H. and Wallach, D. F. H. (1970). *Science, N.Y.* **168**, 255.
Stieger, J. and Luterbacher, S. (1981). *Biochim. biophys. Acta* **641**, 270-275.
Suki, W. N. (1979). *Am. J. Physiol.* **237**, F1-F3.
Suki, W. N. and Rouse, D. (1980). *In* "Advances in Physiological Sciences" (Ed. L. Takacs) Vol. 11, 457-459. Pergamon Press, Oxford.
Szasz, I., Hasitz, M., Sarkadi, B. and Gardos, G. (1978a). *Molec. Cell Biochem.* **22**, 147-152.
Szasz, I., Sarkadi, B., Schubert, A. and Gardos, G. (1978b). *Biochim. biophys. Acta* **512**, 331-340.
Taverna, R. D. and Hanahan, D. J. (1980). *Biochem. biophys. Res. Commun.* **94**, 652-659.
Taylor, A. and Windhager, E. E. (1979). *Am. J. Physiol.* **236**, F505-F512.
Ting, A., Lee, J. W. and Vidaver, G. A. (1979). *Biochim. biophys. Acta* **555**, 239-248.
Trumble, W. R., Reeves, J. P. and Sutko, J. L. (1979). *Physiologist* **22**, 125.
Trumble, W. R., Sutko, J. L. and Reeves, J. P. (1980). *Life Sci.* **27**, 207-214.
Ullrich, K. J., Rumrich, G. and Klöss, S. (1976). *Pflügers Arch. ges. Physiol.* **364**, 223-228.
Ullrich, K. J., Capasso, G., Rumrich, G., Papavassiliou, F. and Klöss, S. (1977). *Pflügers Arch. ges. Physiol.* **368**, 245-252.
Van Rossum, G. D. V. (1970). *J. gen. Physiol.* **55**, 18-32.
Van Rossum, G. D. V., Smith, K. P. and Morris, H. P. (1973). *Cancer Res.* **33**, 1086-1091.
Van Rossum, G. D. V., Smith, K. P. and Beeton, P. H. (1976). *Nature, Lond.* **260**, 335-337.
Varghese, S. and Brown-Cunningham, E. (1980). *Biochim. biophys. Acta* **596**, 468-471.
Vincenzi, F. F. (1968). *Proc. west. Pharmacol. Soc.* **11**, 58-60.
Vincenzi, F. F. and Hinds, T. R. (1980). *In* "Calcium and Cell Function" (Ed. W. Y. Cheung). Academic Press, New York. **1**, 127–165.
Vincenzi, F. F. and Hinds, T. R. (1980). *In* "Calcium Binding Proteins as Cellular Regulators" (Ed. W. Y. Cheung). Academic Press, New York. In press.
Vincenzi, F. F. and Larsen, F. L. (1980). *Fedn Proc. Fedn Am. Socs exp. Biol.* **39**, 2427-2431.
Vincenzi, F. F., Hinds, T. R. and Raess, B. U. (1980). *Ann. N.Y. Acad. Sci.* **356**, 232–244.
Wallach, S., Reizenstein, D. L. and Bellavia, J. V. (1966). *J. gen. Physiol.* **49**, 743-751.
Wang, J. H. and Desai, R. (1977). *J. biol. Chem.* **252**, 4175-4184.
Wassermann, R. H., Corradino, R. A. and Taylor, A. N. (1968). *J. biol. Chem.*

243, 3978-3986.

Wassermann, R. H., Corradino, R. A., Taylor, A. N. and Morrissey, R. L. (1971). *In* "Cellular Mechanisms for Calcium Transfer and Homeostasis" (Eds. G. Nichols Jr and R. H. Wassermann) 293-311. Academic Press, New York.

Watson, E. L., Vincenzi, F. F. and Davis, P. W. (1971). *Life Sci.* 10, 1399-1404.

Weidekamm, E. and Brdiczka, D. (1975). *Biochim. biophys. Acta* 401, 51-58.

Weiner, M. L. and Lee, K. S. (1972). *J. gen. Physiol.* 59, 462-475.

White, M. D. and Ralston, G. B. (1980). *Biochim. biophys. Acta* 596, 472-475.

Whitfield, J. F., Boynton, A. L., MacManus, J. P., Sikorska, M. and Tsang, B. K. (1979). *Molec. Cell. Biochem.* 27, 155-179.

Whitfield, J. F., Boyton, A. L., MacManus, J. P., Rixon, R. H., Sikorska, M., Tsang, B. and Walker, P. R. (1980). *Ann. N. Y. Acad. Sci.* 339, 216-262.

Wierichs, R. and Bader, H. (1980). *Biochim. biophys. Acta* 596, 325-328.

Wiley, J. S. (1976). *In* "Membranes and Disease" (Eds L. Bolis, J. F. Hoffman and A. Leaf) 89-94. Raven Press, New York.

Wiley, J. S. and Gill, F. M. (1976). *Blood* 47, 197-210.

Wiley, J. S. and Shaller, C. C. (1977). *J. clin. Invest.* 59, 1113-1119.

Wins, T. and Schoffeniels, E. (1966). *Biochim. biophys. Acta* 120, 341-350.

Wolf, H. U. (1973a). *Experientia* 29, 241.

Wolf, H. U. (1973b). Habilitationsschrift, Universität Mainz.

Wolf, H. U. and Dieckvoss, G. (1976). *Arch. Pharmac.* Suppl. 293, R43.

Wolf, H. U. and Knipser, W. (1975a). *Experientia* 31, 726-729.

Wolf, H. U. and Knipser, W. (1975b). *Hoppe-Seyler's Z. physiol. Chem.* 356, 290-295.

Wolf, H. U., Dieckvoss, G. and Lichtner, R. (1977). *Acta biol. med. germ.* 36, 847-858.

Wolff, D. J., Poirier, P. G., Brostrom, C. O. and Brostrom, M. A. (1977). *J. biol. Chem.* 252, 4108-4117.

Wüthrich, A. and Schatzmann, H. J. (1980). *Cell Calcium* 1, 21-35.

Wüthrich, A., Schatzmann, H. J. and Romero, P. (1979). *Experientia* 35, 1589-1590.

Wuytack, F. and Casteels, R. (1980). *Biochim. biophys. Acta* 595, 257-263.

Wuytack, F., Landon, E., Fleischer, S. and Hardman, J. G. (1978). *Biochim. biophys. Acta* 540, 253-269.

Wuytack, F., De Schutter, G. and Casteels, R. (1980). *Biochem. J.* 190, 827-831.

Wuytack, F., De Schutter, G. and Casteels, R. (1981). *Biochem. J.* in press.

3

The Transport of Calcium Across the Inner Membrane of Mitochondria

ERNESTO CARAFOLI

Laboratory of Biochemistry, Swiss Federal Institute of Technology (ETH), Zurich, Switzerland

HISTORICAL BACKGROUND

A very remarkable observation made by Chance in 1955 can be considered as the first indication that mitochondria can actively take up Ca^{2+}. He observed that Ca^{2+}, unique among uncouplers, did not stimulate respiration permanently, but temporarily. He also observed that the duration of the stimulation was directly proportional to the amount of Ca^{2+} added, and suggested that Ca^{2+} was somehow "consumed" during the uncoupling phase. It was a remarkable suggestion, and even if it stopped short of concluding that Ca^{2+} was taken up by mitochondria in an energy-linked process, it seems clear in retrospect that Chance had understood the essence of what was happening in his experiment. That mitochondria could absorb large amounts of Ca^{2+} from the medium had already been observed by Slater and Cleland two years before (Slater and Cleland, 1953). However, the observation that the absorption took place at $0°$ C had led them to conclude that the process was passive, and not dependent on respiration. Looking back to those early days in the light of what is known today, it is clear that the observation and the suggestion of Chance were probably ahead of the times. Indeed, some years were to elapse before the phenomenon of energy-linked mitochondrial Ca^{2+} uptake could be firmly established experimentally, and reasonably well interpreted conceptually.

THE DISCOVERY OF THE PROCESS AND THE EARLY
OBSERVATIONS ON THE "MASSIVE" UPTAKE OF CALCIUM
BY MITOCHONDRIA

A series of fundamental observations were made in the period 1959–1962. Saris (1959) studied the pH decrease induced by the addition of Ca^{2+} to mitochondria, and concluded that the phenomenon was due to the formation of H^+ and Ca^{2+} gradients produced by the uptake of Ca^{2+} by mitochondria. Vasington and Murphy (1961) observed uptake of Ca^{2+} coupled to oxidative phosphorylation in isolated mitochondria, and De Luca and Engstrom (1961) demonstrated that coupled respiration was not an absolute requirement for Ca^{2+} uptake, provided that external ATP was present. The concept that Ca^{2+} uptake could be driven alternatively by respiration or ATP hydrolysis was later supported by evidence provided by Brierley *et al.* (1964) and Lehninger *et al.* (1963). The most comprehensive of the early studies of the process, however, is that published by Vasington and Murphy in 1962. They established that the uptake of Ca^{2+} required respiration, ATP (or ADP), Mg^{2+}, and phosphate, was inhibited by uncoupling agents, but *not* by the phosphorylation inhibitor oligomycin. During the uptake of Ca^{2+}, no phosphorylation of ADP took place. This last observation clearly established the process of Ca^{2+} uptake as an alternative to ADP phosphorylation in the utilization of respiratory energy, whereas the insensitivity to oligomycin indicated that the energy used for the uptake reaction was harvested at a "point" prior to the synthesis of ATP. Both observations were of fundamental importance, and had to be incorporated in all future formulations of the mechanism of the reaction as essential ingredients. The amounts of Ca^{2+} accumulated in the experiments of Vasington and Murphy were very large (up to $2.6\,\mu mol/mg$ of protein), implying that most of the Ca^{2+} was removed from solution in the very limited intramitochondrial aqueous space. Indeed, Lehninger *et al.* (1963) soon found that inorganic phosphate was accumulated by mitochondria together with Ca^{2+}, an observation that rationalized the requirement for phosphate of the Ca^{2+} uptake process, and offered a plausible mechanism for the storage of the "massive" amounts of Ca^{2+} inside mitochondria. Ca^{2+} and phosphate precipitate inside mitochondria, and are easily visualized in the electron microscope in the form of electron-opaque masses that may fill up most of the matrix space (Greenawalt *et al.*, 1964). The requirement for ATP (ADP), and the fact that ATP hydrolysis was apparently not involved (oligomycin did not eliminate the effect of ATP), was rationalized by the finding, made by Carafoli *et al.* in 1965, that ATP (or ADP) was taken up by mitochondria together with Ca^{2+} and phosphate. The conclusion was reached that adenine nucleotides, which are known to be absorbed by growing hydroxylapatite crystals, precipitate inside mitochondria together with Ca^{2+} and phosphate, somewhat stabilizing the dense granule, or even "priming" its deposition. It is of great interest that the granules precipitated inside mitochondria do not show the typical X-ray diffraction pattern of crystalline hydroxylapatite, but remain amorphous for indefinite periods of time. Evidently, Ca^{2+}-loaded mitochondria contain substances that prevent the crystalline transformation that would otherwise normally occur in the intramitochondrial environment. Very recent work by Lehninger and his associates has tentatively identified this substance as phosphocitrate (Lehninger, 1981).

Rossi and Lehninger (1963) measured the Ca^{2+}/phosphate accumulation ratios, and found them to average 1.67:1. Since this ratio is precisely that of hydroxylapatite, it was concluded that the precipitate inside mitochondria consisted of (amorphous) hydroxylapatite. Rossi and Lehninger (1963) also measured the ratio between the "extra" oxygen consumption induced by the addition of Ca^{2+}, and the amount of the latter (and of phosphate) accumulated. The finding was that 1.67 Ca^{2+} and 1.0 phosphate were transported into mitochondria as a pair of electrons travelled across each one of the three coupling regions of the respiratory chain. The fact that the $P_i : 2e^-$ ratio was an integral number, whereas the $Ca^{2+} : 2e^-$ ratio was not, led Rossi and Lehninger to suggest that the uptake of phosphate was the primary event, the uptake of Ca^{2+} following for reasons of charge compensation. The reasoning was naturally influenced by the view, prevailing at the time, of a strict chemical coupling between electron transport and phosphorylation. As will be discussed later, with the general acceptance of the principles of the chemiosmotic theory it has become clear that the primary event in the process is the uptake of Ca^{2+}.

THE "LIMITED LOADING" OF MITOCHONDRIA WITH CALCIUM

One negative aspect of the experiments in which mitochondria are made to accumulate massive amounts of Ca^{2+} is the irreversible functional and structural damage to the organelles. The details of the accumulation process and its mechanism are thus better studied under conditions of "limited loading", in which mitochondria are exposed to much lower, and presumably non-injuring, Ca^{2+} concentrations. Interestingly, the study by Chance (1955) quoted at the outset of this chapter employed "limited loading" conditions, and it indeed led to suggestions that were remarkably accurate, as later work has shown. It can be mentioned here that two atoms of Ca^{2+} were required in the experiment of Chance to produce the same amount of "extra" oxygen consumption as one molecule of ADP, an interesting observation in view of the latter conclusion by Rossi and Lehninger (1963, see above) that 1.67 atoms of Ca^{2+} are accumulated (or 1 ADP phosphorylated) at each one of the coupling sites.

The $Ca^{2+}:O$ ratios, and the effects of the addition of limited amounts of Ca^{2+}, were investigated in detail by Rossi and Lehninger (1964). They observed that the mitochondrial respiration remained in the activated state until only about 1–2 μM Ca^{2+} remained in the extramitochondrial medium. This finding, in fact, indicated that below 1–2 μM external Ca^{2+} the system for the uptake of Ca^{2+} ceased to be activated, i.e. it showed that the apparent K_m of the uptake system was in this concentration region. Rossi and Lehninger (1964) found that phosphate was not necessary for the Ca^{2+}-induced activation of respiration. However, they found that in the absence of phosphate the capacity of mitochondria to accumulate Ca^{2+} was rather limited (about 100 nmol/mg of protein), a finding that confirmed the "Ca^{2+} trapping" role of phosphate in the "massive loading" experiments. In the presence of phosphate, at least when its concentration was 2.0 mM or higher, Ca^{2+} was apparently not retained by mitrochondria, but lost continuously to the medium, where it produced an indefinite stimulation of the respiration. The interpretation offered by Rossi and Lehninger (1964) was that the loss of Ca^{2+} was due to a deleterious effect of phosphate on mitochondria. Indeed, the indefinite stimulation of respiration induced by phosphate was prevented by agents like ADP and Mg^{2+}, which are known to protect mitochondria against structural damage. That the combined accumulation of Ca^{2+} and phosphate may damage mitochondria is an accepted concept that can, for instance, be verified by electron microscopy. It is important to realize, however, that the indefinite stimulation of respiration implies that the energy-linked re-accumulation of the lost Ca^{2+} proceeds continuously. That is, mitochondria are still capable of a certain degree of energy coupling under conditions where phosphate induces release of the accumulated Ca^{2+}.

Another important finding which was made using conditions of "limited loading" was that during the accumulation of Ca^{2+}, H^+ are ejected to the medium. It will be recalled that the acidification of the outside medium upon addition of Ca^{2+} to mitochondria had been observed already in the early work by Saris (1959, 1963) and found to correspond to a H^+/Ca^{2+} ratio of about 1 in the absence of phosphate. It was soon found by other investigators (Chappell et al., 1963; Engstrom and De Luca, 1963; Rasmussen et al., 1965; Chance, 1965; Rossi et al., 1967) that the H^+/Ca^{2+} ratio, as an expression of the electrochemical balance during Ca^{2+} uptake, may vary considerably depending on the experimental conditions. In particularly, it may be depressed to values lower than 0.3 in the presence of permeant anions like acetate. The counterpart to the H^+ ejection was an increase in the titratable alkalinity of mitochondria ($OH^-:Ca^{2+}$ about 1, Chance and Mela, 1966a b; Gear et al., 1967; Addanki et al., 1968), which is reflected into the fact that the energy-linked uptake of Ca^{2+} may be accompanied by the uptake of those anions which are able to penetrate electroneutrally and to donate H^+ to the matrix, e.g. phosphate, acetate, bicarbonate (Chappell et al., 1963; Lehninger et al., 1974).

In concluding this presentation of the early history of the Ca^{2+} uptake process, one more comment is in order on a paper published by Drahota et al. in 1965. The emphasis of the early research had been placed exclusively on the Ca^{2+} uptake reaction, almost as if the opposite process, i.e. the release of the accumulated Ca^{2+}, never occurred. However, release of Ca^{2+} in vivo is obviously a necessity, since mitochondria do not calcify in situ, at least in soft tissues. As will be discussed in later sections, the process of Ca^{2+} release has now become a central issue in the field. What was unusual in the paper by Drahota et al. (1965) was that they concluded that the accumulated Ca^{2+} was maintained in mitochondria in a dynamic steady state, with continuous efflux counterbalanced by continuous energy-linked uptake. They showed that the efflux of Ca^{2+} is not simply a passive diffusion, but has the characteristics of an enzyme-controlled process.

INHIBITION OF THE CALCIUM UPTAKE PROCESS

As expected of an energy-dependent process, the uptake of Ca^{2+} by mitochondria is inhibited by agents that interfere with the energy flow, like the uncouplers, or, depending on whether the energy chain inhibitors or by oligomycin. Clearly, in these cases the inhibition is not due to specific interactions with the transport system. Specific inhibitors, however, have also been described. La^{3+} and other lanthanides (Mela, 1968; Lehninger and Carafoli, 1971) are active at very low concentrations, in the micromolar range. They apparently act by interfering with (a) specific component(s) of the uptake system (Niggli *et al.*, 1978). The inhibition is ionic radius related (Tew, 1977) and is maximal with Tm^{3+} (Crompton *et al.*, 1978b). Another very useful inhibitor is ruthenium red (Moore, 1971; Vasington *et al.*, 1972), a hexavalent cation which also interferes specifically with (a) component(s) of the Ca^{2+} uptake system (Niggli *et al.*, 1977). As will be discussed in later sections, ruthenium red has now become a fundamental tool for separating functionally the pathway for Ca^{2+} uptake from that for Ca^{2+} release.

A specific inhibition of the Ca^{2+} uptake reaction is produced by Mg^{2+}, and the case here is more complex. The inhibition is much more evident in heart (Jacobus *et al.*, 1975; Sordahl, 1974) than in liver mitochondria (Jacobus *et al.*, 1975; Åkerman *et al.*, 1977). In the presence of Mg^{2+}, the kinetics of the uptake reaction, which is normally hyperbolic, becomes markedly sigmoidal (Crompton *et al.*, 1976a; Bragadin *et al.*, 1979; Affolter and Carafoli, 1981). An important aspect of the Mg^{2+} inhibition is that it is seen at concentrations which are normally present in most cytosols. This indicates a physiological role for Mg^{2+}, possibly as a mechanism for controlling the *in vivo* balance for Ca^{2+} between mitochondria and the cytosol (see below).

THE MECHANISM OF THE CALCIUM UPTAKE REACTION

The early attempts to explain the mechanism of the Ca^{2+} uptake process were based on the postulates of the so-called chemical coupling hypothesis (Rasmussen *et al.*, 1965). After the introduction and the general acceptance of the chemiosmotic theory (Mitchell, 1966), it became clear that the interpretation of the Ca^{2+} uptake process had also to be sought along the guidelines of the chemiosmotic principles. In 1970, two groups found that Ca^{2+} uptake could be driven, in the absence of respiration or added ATP, by diffusion potentials created by H^+ permeation in the presence of protonophoric uncouplers, by the permeant anion thiocyanate or by the exit of K^+ in the presence of the specific ionophore valinomycin (Selwyn *et al.*, 1970; Scarpa and Azzone, 1970). These important observations indicated very clearly that the uptake of Ca^{2+} is an electrophoretic process. Confirmation and extension of this concept came from the work of Rottenberg and Scarpa (1974), who compared the distribution of Ca^{2+}

between energized mitochondria and medium with that of Rb^+, which is known to distribute, in the presence of valinomycin, according to the membrane potential, and to attain electrochemical equilibrium according to the Nernst equation (1). Rottenberg and Scarpa (1974) demonstrated that the distribution of Ca^{2+} (in the presence of the permeant anion acetate to minimize Ca^{2+} binding to anionic sites on the matrix face of the inner membrane) was also determined by the membrane potential, and occurred with a net charge transfer of 2.

$$E = \frac{RT}{nF} \cdot \ln \frac{[Rb^+]_i}{[Rb^+]_o} \qquad (1)$$

$$\frac{[^{86}Rb^+]_i}{[^{86}Rb^+]_o} = 13.3 \qquad \text{log of the ratio} = 1.24$$

$$\frac{[^{45}Ca^{2+}]_i}{[^{45}Ca^{2+}]_o} = 230 \qquad \text{log of the ratio} = 2.36$$

$$\log \frac{[Ca^{2+}]_i}{[Ca^{2+}]_o} = 2 \log \frac{[Rb^+]_i}{[Rb^+]_o}$$

The Rb^+ distribution technique was also used by Heaton and Nicholls (1976), who supported the conclusions of Rottenberg and Scarpa. One problem, in experiments of this type, is that the actual activity of Ca^{2+} in the matrix is not known with precision. Permeable anions that donate H^+ to the matrix certainly limit the binding of Ca^{2+} inside mitochondria, and they have been used by both Rottenberg and Scarpa (1974) and Heaton and Nicholls (1976). Some Ca^{2+}, however, is invariably bound, and its extent depends on the experimental conditions used (Gunter and Puskin, 1976). For this reason Ca^{2+} distribution data, even in the presence of acetate, are never absolutely quantitative.

Results suggesting the alternative possibility that the uptake route functions with a net charge transfer of 1, i.e. with partial compensation of the Ca^{2+} charges, have been presented by Åkerman (1978), by Reed and Bygrave (1975) and by Moyle and Mitchell (1977a, b). The first two authors have suggested a Ca^{2+}/H^+ antiporter, but the data on which the proposal was based are open to a number of other interpretations. Moyle and Mitchell (1977a, b) have proposed a Ca-phosphate symport, or a Ca-β-hydroxybutyrate (when the latter is the supporting substrate) symport. Recent evidence (Reynafarje and Lehninger, 1977; Crompton et al., 1978c), however, has excluded the direct participation of inorganic phosphate in the uptake of Ca^{2+}.

The monocarboxylate (β-hydroxybutyrate) possibility has not yet been fully explored.

The fluctuations of the transmembrane potential during the influx of Ca^{2+} have been recently measured directly by Åkerman (1978) using the saffranine method, and by Lötscher *et al.* (1980a) using an electrode sensitive to the lipophylic cation tetraphenylphosphonium.

The problem of the charge transfer during Ca^{2+} uptake has recently been re-investigated by Crompton and Heid (1978). They have chosen a method in which no assumptions were made on the actual amount of ionized matrix Ca^{2+}. They determined the extramitochondrial free Ca^{2+} required to maintain a constant intramitochondrial free Ca^{2+} as the transmembrane potential was varied with uncouplers. Their values ($n = 1.8-2.1$) are of particular interest, since they are obtained under experimental conditions that eliminate all problems arising from the possible, and not quantifiable, binding (or precipitation) of Ca^{2+} in the intramitochondrial space. The older data by Rottenberg and Scarpa (1974) and Heaton and Nicholls (1976) are thus fully legitimized. It can thus be stated that at the present state of knowledge the bulk of the available evidence strongly favours the view that the uptake of Ca^{2+} is a purely electrophoretic process. As will be discussed in detail later, this conclusion has been instrumental in the development of the concept of separate Ca^{2+} influx and efflux routes in mitochondria.

The influx of Ca^{2+} into mitochondria exhibits saturation kinetics, and can be competitively inhibited by Sr^{2+} (Carafoli, 1965), which can be also actively accumulated. Half maximal velocity of the uptake reaction is obtained at about 13 μM Ca^{2+} (Crompton *et al.*, 1976a) in heart mitochondria, but possibly at lower values in other mitochondrial types (Reed and Bygrave, 1975; Hutson *et al.*, 1976; Heaton and Nicholls, 1976; see Carafoli and Crompton, 1978, for a comprehensive discussion of the matter of the affinity of mitochondria for Ca^{2+}). The K_m of the influx system for Ca^{2+} has been determined using several methods, in several mitochondrial types, and under different experimental conditions, and has sometimes yielded figures well in excess of the 5–13 μM mentioned above (Scarpa and Graziotti, 1973). This appears to be due to the inclusion of Mg^{2+} in the reaction medium to minimize nonspecific Ca^{2+} binding to the mitochondrial surface. As mentioned above, Mg^{2+} competes with Ca^{2+} for the transport system, especially at the very low Ca^{2+}/Mg^{2+} ratios normally employed in experiments of this type. This is the reason for the finding made by several authors that the kinetics of the Ca^{2+} influx reaction has a marked sigmoidal character: evidently, measurements in the presence of Mg^{2+} do not reflect the true affinity of the influx system for Ca^{2+}. It is now clear (Crompton et al., 1976a) that the kinetics of the uptake system, at least in heart mitochondria, is hyperbolic in the absence of Mg^{2+}, and reaches half maximal rate at about 10 μM Ca^{2+}. Other authors, however, have observed sigmoidal kinetics of the uptake reaction in liver mitochondria also in the absence of Mg^{2+} (Reed and Bygrave, 1975), and have interpreted the data as meaning that the hypothetical Ca^{2+} carrier has two binding sites for Ca^{2+} (Spencer and Bygrave, 1973; Reed and Bygrave, 1975). Recent data (see above, Bragadin et al., 1979; Affolter and Carafoli, 1981), however, have shown that also the kinetics of the influx reaction in liver mitochondria is hyperbolic, provided that the measurement is carried out under appropriate nonlimiting conditions.

   The maximal velocity of the influx system is about 0.1 μmol Ca^{2+} /mg of protein/min at 25° C, and twice this value at the physiological temperature of 38° C (heart mitochondria, Crompton et al., 1976a). One interesting finding (Crompton et al., 1976a) is that phosphate stimulates the maximal rate of the influx reaction, to values that can be as high as 0.6 μmol/mg of protein/min at 25° C (Vercesi et al., 1978). As mentioned before, phosphate increases very substantially the total amount of Ca^{2+} that can be accumulated. The effect of phosphate on the rate of Ca^{2+} entry is evidently due to a different phenomenon and is not, at the moment, completely understood. The entry of phosphate restores the Δ pH, which decreases as the result of the penetration of the positively charged Ca^{2+}, and of the net H^+ extrusion by the respiratory chain (Lötscher et al., 1980a). This may remove a limiting factor in the Ca^{2+} influx reaction. The precipitation of hydroxylapatite in the intramitochondrial space, however, may add a complicating factor to the picture, since it involves net acidification

$$3Ca^{2+} + 2HPO_3^{2-} \rightarrow Ca_3(PO_4)_2 + 2H^+$$

THE PROBLEM OF THE MITOCHONDRIAL
CALCIUM CARRIER

The kinetic parameters described above have in general been taken to indicate that the process of Ca^{2+} uptake is mediated by a specific carrier (Lehninger and Carafoli, 1969). Efforts aimed at the isolation of such a carrier have produced a variety of fractions which are able to bind Ca^{2+} with high affinity. Two of them have been characterized to a very extensive degree, and have permitted experiments that indicate rather compellingly their involvement in the Ca^{2+} uptake process. They will be described in some detail, but it must be emphasized here that the kinetic parameters of the uptake process mentioned in the preceeding sections (saturation kinetics, competitive inhibition by other divalent cations like Sr^{2+}, specific inhibition by La^{3+} and by ruthenium red) could also be explained by a superficial Ca^{2+} receptor, not functioning as a transmembrane Ca^{2+} carrier. The existence of such a carrier, therefore, remains for the moment only a postulation.

One of the two Ca^{2+} binding fractions that have been extensively purified is a glycoprotein of about 30 000 mol. wt (Sottocasa *et al.*, 1972; Carafoli and Sottocasa, 1974), which is contained in the inner and outer membrane, in the intermembrane space, but not in the matrix. A fraction of it is apparently bound only weakly to the membrane domain, and it can be liberated by osmotic shocks. However, a more drastic treatment is necessary to extract from the membrane the remainder of it. The glycoprotein contains sialic acid, hexosamines, neutral sugars, and a considerable amount of phospholipids. It binds Ca^{2+} at two classes of sites, differing widely in their affinity for it. The high-affinity class binds 2–3 moles of Ca^{2+} per mole of protein, with a K_d of the order of 1 μM, and is inhibited by La^{3+} and by ruthenium red.

The involvement of the glycoprotein in the process of Ca^{2+} uptake is indicated by several lines of evidence. Mitochondria that are unable to interact with Ca^{2+} with high affinity and to transport it (blowfly flight muscle, Carafoli *et al.*, 1971; yeast, Carafoli *et al.*, 1970) contain a glycoprotein that has apparently lost the high affinity Ca^{2+} binding sites (Carafoli *et al.*, 1973a). A specific antibody prepared by Sottocasa and his coworkers inhibits the energy-linked transport of Ca^{2+} without affecting other mitochondrial functions (Panfili *et al.*, 1976) (rather interestingly, the antiglycoprotein antibody inhibits also the Ca^{2+} efflux induced by the oxidation of intramitochondrial pyridine nucleotides, Panfili *et al.*, 1980, see below). Experiments with glycoprotein-depleted mitochondria and pure glycoprotein (Sandri *et al.*, 1979) have successfully reconstituted Ca^{2+} uptake. It thus appears very probable that the glycoprotein is a component of the electrophoretic Ca^{2+} import system. What role the protein plays in the process, however, is an open question. Reconstruction experiments by Prestipino *et al.* (1974) using planar lipid bilayers, as well as experiments on Ca^{2+} extraction into apolar bulk phases by the glycoprotein (Carafoli, 1975a) have led to the suggestion (Carafoli, 1975a) that the glycoprotein may function as a superficial Ca^{2+} receptor, which becomes associated with the membrane in the presence of Ca^{2+}, and operates in series with other transmembrane components responsible for its final translocation.

The other fraction that has been characterized in detail is a protein of 3000 mol. wt which has been named "calciphorin" (Jeng and Shamoo, 1980a, b). It has been purified from calf heart mitochondria with the help of Mn^{2+}, a paramagnetic analogue of Ca^{2+} that is also bound by mitochondria with high affinity, and transported into the matrix. The binding of Mn^{2+} has been used as a "marker" to follow the protein throughout the purification scheme. The purified protein binds Ca^{2+} with high affinity and high specificity, and can extract it into organic phases. Since the bulk-phase transfer of Ca^{2+} is greatly stimulated by lipophilic anions, it has been concluded that calciphorin mediates an electrophoretic process, probably corresponding to the electrophoretic mitochondrial uptake import. In agreement with this suggestion, low concentrations of ruthenium red and La^{2+} inhibit the calciphorin-mediated extraction of Ca^{2+} into organic phases.

At the present moment, no conclusions can be drawn on the (respective) roles of the two purified Ca^{2+} binding proteins, nor on their possible interplay. Calciphorin is very hydrophobic, and it is therefore a more likely candidate for a transmembrane role than the glycoprotein, which has a pronounced hydrophobic character. Possibly, the two proteins operate in series, the glycoprotein as a superficial receptor as suggested by Carafoli (1975a), and calciphorin as a transmembrane Ca^{2+} channel.

THE PROBLEM OF THE REVERSIBILITY OF THE CALCIUM UPTAKE PROCESS

The electrical potential across the inner membrane of mitochondria *in vivo* is probably comparable to that normally measured in respiring isolated mitochondria (see Scott and Nicholls, 1980, for a recent discussion of the matter). If the electrophoretic Ca^{2+} uptake uniporter, which catalyses the transfer of two positive charges, would reach equilibrium *in vivo* against a potential of 150–180 mV, a transmembrane gradient of ionized Ca^{2+} of 10^5-10^6 would develop. Precise measurements of the ionized Ca^{2+} concentrations are difficult, and available only for some cytosol types (see Carafoli and

Crompton , 1978, for a review), but it is generally agreed that they oscillate in the $0.1-1.0 \mu M$ range. The concentration of ionized Ca^{2+} in the mitochondrial matrix has never been measured directly, but can be estimated indirectly from the level of activity of some Ca^{2+} requiring matrix enzymes (Carafoli et al., 1977). It probably approaches $1 \mu M$. It is thus evident that the gradient of ionized Ca^{2+} across the inner membrane is less than 10^6, and in all likelihood even less than 10^3. This is confirmed by data of Puskin et al. (1976), who have used Mn^{2+} as a paramagnetic Ca^{2+} analogue. The use of Mn^{2+} permits free and bound intramitochondrial Ca^{2+} to be estimated by EPR. At low values of transmembrane potential, and high concentrations of extramitochondrial Mn^{2+}, the gradients observed were in agreement with those predicted by the Nernstian equilibrium. As the transmembrane potential increased, Puskin et al. (1976) found considerable deviations from the Nernstian equilibrium, and at the physiological transmembrane potential of 160 mV they measured gradients which were only a small fraction of the predicted 10^6.

The finding that the gradient of ionized Ca^{2+} is far lower than 10^6 is a compelling argument for the conclusion that the electrophoretic uptake uniporter does not reach equilibrium. This has led to the logical suggestion that in energized mitochondria which take up Ca^{2+}, efflux of the latter is continuously catalysed by systems independent of the uptake uniporter (Carafoli and Crompton, 1976; Puskin et al., 1976). The uniporter would thus operate essentially as an inwardly directed one-way pump driven by the membrane potential; the release system would return Ca^{2+} to the extramitochondrial space in the presence of a fully maintained transmembrane potential.

Research on the separate way(s) that mediate Ca^{2+} efflux has been very active during the last few years, and detailed accounts of it will be given in the following sections. At this point, however, it is appropriate to mention some older observations which were very suggestive of the existence of separate uptake and release pathways, and which have provided essential tools for the verification of the concept. Among them, one can mention again the previously quoted observation by Drahota et al. (1965) that the accumulated Ca^{2+} is not sequestered irreversibly in mitochondria, but is maintained in a dynamic steady state, in which energy-linked uptake is balanced by what the authors interpreted as a continuous efflux. Stücki and Ineichen (1974) demonstrated that mitochondria continuously lost Ca^{2+}, albeit at a slow rate, (about 0.05 mol/mg of protein/s) during the energized state when the membrane potential, negative inside, would naturally prevent any leaks on the reversed uptake uniporter. Then in 1973 Rossi *et al.* demonstrated that Ca^{2+} release from mitochondria could be induced in the presence of ruthenium red. Since under these conditions the uptake uniporter is blocked by the inhibitor, this observation provided the first *direct* demonstration of the existence of an independent Ca^{2+} release pathway. The insensitivity of Ca^{2+} efflux to ruthenium red has now become the principal parameter in establishing the existence of independent release pathways.

The general concept that has emerged during the last two or three years is that the notion of mitochondrial energy-linked Ca^{2+} uptake is more conveniently substituted by the comprehensive idea of an energy dissipating "mitochondrial Ca^{2+} cycle" (Carafoli, 1979), in which the energy-driven penetration of Ca^{2+} into the organelle is counterbalanced by a continuous efflux, independent of the transmembrane potential.

THE RELEASE OF CALCIUM FROM MITOCHONDRIA

Several Ca^{2+}-releasing routes, which do not mediate uptake of the cation, have been described. Whether they all have physiological significance, and whether more than one coexists in the same mitochondrion, is an open question, but the concept is now emerging that they may be differently represented in different tissues. At the moment, the most thoroughly characterized releasing system, and the one for which a physiological role is very probable, is a specific $Ca^{2+}–Na^+$ exchange.

At the end of the description of the various independent release pathways, the possibility of Ca^{2+} release through a reversal of the uptake uniporter will be discussed. This is necessary, because many of the conditions that have been proposed as Ca^{2+} releasers have frequently been accused of irreversibly damaging mitochondria, i.e. of collapsing the membrane potential.

THE SODIUM-PROMOTED CALCIUM RELEASE PATHWAY

In 1974, Carafoli and his coworkers showed that the addition of Na^+ to energized heart mitochondria after small amounts of Ca^{2+} had been accumulated induced a rapid efflux of Ca^{2+}. With the partial exception of Li^+, all other monovalent cations had no effect. Two aspects of the experiment by Carafoli *et al.* (1974b) are worth emphasizing here because they indicate very clearly that a separate Ca^{2+} release pathway was in operation: (a) mitochondria were continuously energized throughout the experiment (i.e. the transmembrane potential was retained) and (b) ruthenium red did not prevent the release of Ca^{2+}, it actually made it considerably faster. Also of interest was the fact that the inducer of Ca^{2+} release was Na^+, a substance normally present in the cytosol, where its concentration could conceivably be modulated. It thus appeared likely from the outset that the phenomenon might have been physiologically meaningful.

In 1976 Crompton *et al.* (1976b) established that Ca^{2+} release could be induced, under appropriate conditions, by concentrations of Na^+ that were lower than those employed in the original experiment of Carafoli *et al.* (1974b). They could also demonstrate that the rate of Ca^{2+} release was linked to the concentration of added Na^+ by a marked sigmoidal dependance, which gave Hill coefficients of 2–3, and thus suggested the involvement of two or more Na^+ in the process. Half-maximal release rate was obtained with 6–8 mM Na^+, corresponding to about 0.2–0.3 nmol Ca^{2+} released/mg of protein/s. Crompton *et al.* (1978b), Nicholls (1978a) and Al Shaikhaly *et al.* (1979) showed that the Na^+-dependent efflux was present in a large series of tissues, being maximally active in heart and adrenal cortex mitochondria. Crompton *et al.* (1978b), however, found it to be practically inactive in other mitochondria, including those from liver and kidney, (recent work in which kidney cortex and medulla were separated has shown that the Na^+ pathway is relatively active in the latter tissue (Haworth *et al.*, 1980)). Crompton *et al.* (1976b) suggested that the release reaction was mediated by a Na^+–Ca^{2+} exchange carrier which operated in parallel with a Na^+/H^+ antiporter (Mitchell and Moyle, 1967). The Na^+ that had entered mitochondria in exchange for Ca^{2+} would return to the extramitochondrial space in exchange for H^+, producing in the end a Ca^{2+}/H^+ exchange, as normally observed in experiments of Ca^{2+} release from mitochondria (Fig. 1).

Nicholls (1978b) ruled out the alternative possibility of a primary Ca^{2+}/H^+ exchange, coupled to Na^+ influx to eliminate the pH gradient created by the exit of Ca^{2+}: dissipation of the pH gradient by K^+ influx in the presence of nigericin did not induce Ca^{2+} release. As mentioned, the Na^+/Ca^{2+} carrier is insensitive to ruthenium red, but is still sensitive to lanthanides, albeit at higher concentrations than the uptake uniporter. Crompton *et al.* (1978a) have found that the maximally effective lanthanide is La^{3+}, and it will be recalled

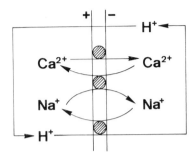

Fig. 1. *The Na^+/Ca^{2+} cycle of heart mitochondria.*

that in the case of the uniporter maximal sensitivity was obtained, instead, with Tm^{3+}. The lanthanide series thus provides a useful pool in distinguishing between the uptake and the Na^+-promoted routes. Very recently, it has been found that the Na^+/Ca^{2+} antiporter is stimulated by K^+ (Crompton et al., 1980), and inhibited substantially by physiological concentrations of Mg^{2+} (Clark and Roman, 1980).

In agreement with the expected properties of a Ca^{2+}/Na^+ carrier, Ca^{2+}/Ca^{2+} exchanges could also be demonstrated (Crompton et al., 1977) in heart mitochondria, with the same La^{3+} sensitivity as the Na^+/Ca^{2+} exchange. In the case of the Ca^{2+}/Ca^{2+} exchange, half maximal rate of the carrier was achieved with 13 μM (external) Ca^{2+}. It was also established that the stoichiometry of the Ca^{2+}/Ca^{2+} exchange was 1. The Na^+/Ca^{2+} ratio on the other hand could not be determined directly but has recently been inferred to be $2:1$, since the exchange appears to proceed electroneutrally (Affolter and Carafoli, 1980).

THE RELEASE OF CALCIUM FROM SODIUM-INSENSITIVE MITOCHONDRIA

The release of Ca^{2+} from all mitochondria, whether Na^+ sensitive or not, leads eventually to the re-uptake of H^+. This is required to permit the re-uptake of the lost Ca^{2+}, and thus the continuous "cycling" of Ca^{2+} across the inner membrane. The exchange of Ca^{2+} for H^+ is not *direct* in Na^+ sensitive mitochondria (see above), and the question now arises of whether such a direct exchange exists in other mitochondria, e.g. of liver. This has been proposed by Fiskum and Lehninger (1979) on the basis of experiments in which release of Ca^{2+}

was induced by acetoacetate in the presence of ruthenium red, and was associated with the re-uptake of $2H^+/Ca^{2+}$. An important problem, however, is that of the relationship of this hypothetical Ca^{2+}/H^+ antiporter to the various mechanisms that have been described for releasing Ca^{2+} from liver and other Na^+ insensitive mitochondria. Of these mechanisms, some have been characterized in sufficient detail to warrant their discussion here, but it is well to remember that for none the level of understanding has reached that of the Na^+/Ca^{2+} antiporter. A related problem is that of the presence of any of the alternative Na^+-independent mechanisms for Ca^{2+} release also in mitochondria that are sensitive to Na^+.

FATTY ACIDS AS CALCIUM-RELEASING AGENTS

Fatty acids have long been known as uncouplers of oxidative phosphorylation. Due to their potential-collapsing effect, they are thus expected to induce release of Ca^{2+} on the reversed uptake uniporter, as any other uncoupler would do. Some recent studies, however, have indicated that fatty acids may induce a massive Ca^{2+} release even when added at sub-uncoupling concentrations. More importantly, the release of Ca^{2+} observed under these conditions is insensitive to ruthenium red, a clear indication for the operation of a distinct release pathway. Carafoli et al. (1973b) and Malmström and Carafoli (1975a) have discovered that prostaglandins induce a ruthenium red-dependent release of Ca^{2+}, at concentrations that do not uncouple oxidative phosphorylation, but which are far in excess of those that would indicate a hormonal effect. The suggestion was made of an ionophoric effect of prostaglandins, who would penetrate across the inner membrane into the matrix in the protonated form, but would *only* cross it back as Ca^{2+} complexes.

A comprehensive study of the Ca^{2+} releasing effect of fatty acids in kidney and liver mitochondria has recently been carried out by Roman et al. (1979). It was found that both unsaturated and saturated fatty acids of medium chain length (C_{12} and C_{14}) promoted a rapid and extensive release of Ca^{2+} which was insensitive to ruthenium red, but inhibited by Na^+ (and Li^+) ions. The effect of fatty acids was independent of their uncoupling activity: no release was seen in the presence of Na^+, which is known to have no effects on the uncoupling ability of fatty acids. In parallel experiments, extensive mitochondrial swelling was observed in the presence of fatty acids and Ca^{2+}, but only in media containing Rb^+ (or Cs^+). On the basis of their observations, Roman *et al.* (1979) proposed a mechanism which is essentially an electroneutral Ca^{2+}/H^+ exchange (see above) in which fatty acids act as mobile Ca^{2+} and H^+ carriers. They interpreted the inhibition by Na^+ with a Na^+/Ca^{2+} competition for the binding fatty acids at the inner side of the inner membrane, and suggested that the accumulation of H^+ in the matrix induces a K^+/H^+ exchange which is eventually responsible for the observed swelling in K^+-containing media.

Whether the fatty acid-induced Ca^{2+} release has physiological significance is impossible to say at the moment. One problem here is that of reconciling the necessity of Ca^{2+} specificity in the releasing effect with the apparent unspecificity of fatty acids in complexing cations and in moving them into apolar environments. Nevertheless, it is interesting that Na^+, albeit at high concentrations, inhibits the release of Ca^{2+}, at variance with its well documented stimulating effect in heart mitochondria. In kidney cells, were the extrusion of Ca^{2+} proceeds in large part by a Na^+/Ca^{2+} exchange mechanism, the inhibition by Na^+ of mitochondrial Ca^{2+} efflux could provide a negative feedback loop between the transport of Ca^{2+} in mitochondria and that across the plasma membrane.

CALCIUM RELEASE BY ALTERATIONS OF THE REDOX RATIO OF INTRAMITOCHONDRIAL PYRIDINE NUCLEOTIDES

In 1978, Lehninger *et al.* observed that liver mitochondria released the accumulated Ca^{2+} when exposed to oxaloacetate, and re-accumulated it when β-hydroxybutyrate was added. These two substrates are known to have opposing effects on the state of oxidation of intramitochondrial pyridine nucleotides. Indeed, Lehninger *et al.* (1978) observed that the Ca^{2+} release induced by oxaloacetate was accompanied by extensive oxidation of NADH, whereas the re-uptake induced by β-hydroxybutyrate was accompanied by the re-reduction of NAD^+. Various substrates could be used to induce oxidation, or re-reduction of the pyridine nucleotides, with essentially the same effects on the movements of Ca^{2+}. It is important to note that mitochondria remained energized throughout the entire Ca^{2+} uptake/release cycle under the conditions employed by the authors (succinate, or other pyridine nucleotide-independent substrates were present). The release could thus not be attributed to a reversal of the electrophoretic uniporter: indeed, it was found to be insensitive to ruthenium red.

The phenomenon described by Lehninger *et al.* (1978) has recently been studied in a number of other Laboratories. Lötscher *et al.* (1979b) have used hydroperoxides to induce oxidation of intramitochondrial NAD(P)H, and have found (Lötscher *et al.* 1980a) that in the presence of Ca^{2+} pyridine nucleotides are hydrolysed inside mitochondria, and nicotinamide is released to the external medium. In the presence of added nicotinamide, however, the hydrolysis of intramitochondrial pyridine nucleotides is reversible. Lötscher *et al.* (1980a) have also found that the release process is electroneutral, and have been able to establish, by measuring directly the transmembrane potential, that it does not damage mitochondria irreversibly. This is at variance with the conclusion by Nicholls and Brand (1980) that the oxidation of pyridine nucleotides sensitizes mitochondria to damage resulting from the accumulation of Ca^{2+} and phosphate. The basis for the conclusion of Nicholls and Brand was the finding that the Ca^{2+} efflux promoted by acetoacetate, which induces oxidation of pyridine nucleotides, was inhibited by ATP plus oligomycin. ATP plus oligomycin, however, which are commonly assumed to protect mitochondria against structural damage, did not prevent the acetoacetate-induced oxidation of pyridine nucleotides. According to Nicholls and Brand (1980) acetoacetate addition results in Ca^{2+} release when the amount of accumulated Ca^{2+} exceeds the level of about 130 nmol/mg of protein, and is accompanied by the complete collapse of the transmembrane potential. As a result, Ca^{2+} flows out on the reversed uptake uniporter. Dissociation between oxidation of pyridine nucleotides and release of Ca^{2+} has been recently reported also by Wolkowicz and McMillin-Wood (1980), who have been able to prevent the latter, but not the former, by the addition of lactate.

It is clear, in summary, that important questions still remain open on the matter of the mechanism and of the physiological significance of the release of Ca^{2+} induced by the oxidation of intramitochondrial pyridine nucleotides. One additional factor of complication is added by the above mentioned recent finding of Panfili *et al.* (1980) that the antibody against the mitochondrial Ca^{2+}-binding glycoprotein (see above) inhibits this type of release.

THE RELEASE OF CALCIUM INDUCED BY INORGANIC PHOSPHATE

When accumulated by mitochondria together with massive amounts of Ca^{2+}, phosphate may induce irreversible structural damage and lead to the release of Ca^{2+}. This has been recognized very early by workers in the field (see above) and is generally accepted. Very recently, however, a series of studies in which very limited amounts of Ca^{2+} have been employed have indicated that phosphate may also release Ca^{2+} through a mechanism that is not necessarily associated with mitochondrial damage. The first indications for such a possibility came from the finding that *inverted* liver submitochondrial vesicles could only accumulate Ca^{2+} when phosphate was present and also accumulated (Wehrle and Pedersen, 1979; Lötscher *et al.*, 1979a). Direct tests of these indications have now been provided by two studies on intact liver mitochondria. In one of them (Siliprandi *et al.*, 1979) it was found that 2 mM phosphate accelerated the release of accumulated Ca^{2+}, which was accompanied by the release of mito- chondrial phosphate. The finding led the authors to suggest a $Ca^{2+}-P_i$ cotransport functioning in the direction of Ca^{2+} release. The critical factor in these experiments was apparently the loss of Mg^{2+} from mitochondria which led to a decrease of the transmembrane potential: prevention of Mg^{2+} loss by the local anestetic tetracaine also pre- vented the release of Ca^{2+}. One unexpected feature of the study was the finding that the release of Ca^{2+}, despite the fact that it occurred simultaneously with the collapse of the transmembrane potential, was increased greatly by ruthenium red. This indicates that a separate release pathway was in operation, and that the collapse of the potential may have been secondary to the re-uptake of the lost Ca^{2+} (the "Ca^{2+} cycle").

In the second study (Roos *et al.*, 1980) liver mitochondria depleted of inorganic phosphate have been used. It was found that phosphate accelerated, as expected, the uptake of Ca^{2+} and was taken up with it, but then spontaneously induced its release through a process that was insensitive to ruthenium red. During the release, phosphate was also lost from mitochondria, but it could be directly demonstrated that the effluxes of the two species did not coincide in time. Phos- phate was released first, thus ruling out a common Ca^{2+}–phosphate symporter. Roos *et al.* (1980) could also establish through direct measurements that during the entire period of Ca^{2+} release the transmembrane potential did not vary significantly, which was a clear indication that the release phenomenon was not linked to gross mitochondrial damage. Direct measurements of reduced and oxidized intramitochondrial pyridine nucleotides have shown that the phosphate-promoted release of Ca^{2+} was not linked to modifications in the oxidation state of pyridine nucleotides.

It is difficult to evaluate at the present state of knowledge what the physiological significance of the phosphate-promoted release might be. Phosphate is normally present in the cytosol in sufficient amounts to activate the release pathway. One critical problem, however, is that of its regulation to insure a modulated Ca^{2+}-releasing effect. As for the mechanism of the phenomenon, a Ca^{2+}–phosphate symporter, which would have been a convenient explanation, is unfortunately ruled out by the experiment of Roos et al. (1980). The possibility of a phosphate-activated Ca^{2+}/H^+ exchange process (see above) seems a likely possibility.

THE PROBLEM OF THE RELEASE OF CALCIUM THROUGH THE REVERSED UPTAKE UNIPORTER

The discussion on the separate pathways for Ca^{2+} release has proceeded from the concept that the uptake pathway, being driven by an essentially constant potential of 160–180 mV, operates as a one-way pump. A prerequisite to permit Ca^{2+} efflux through it is then the substantial lowering of the transmembrane potential. Several conditions have been described in the literature in which Ca^{2+} release has been promoted by its collapse, and it follows from what has been discussed above that a good diagnostic tool would be ruthenium red. Its effect has been tested only in a minority of cases, but one would expect that it blocks the release consequent upon a lowering of the transmembrane potential.

The transmembrane potential can be lowered by protonophoric uncouplers, both natural and artificial. Fatty acids, when added in large amounts, are examples of natural compounds that fall in this category. As pointed out by Nicholls and Crompton (1980), agents that increase the ΔH^+ across the membrane without lowering the total protonmotive forces also decrease the transmembrane potential. This is the case for N-ethylmaleimide, which blocks phosphate influx and the consequent dissipation of the ΔH^+.

A third class of substances that induce Ca^{2+} efflux via reversal of the uptake uniporter is represented by those that induce structural damage to mitochondria. It is not always easy to evaluate when membrane damage has actually occurred, and sometimes the same substance may induce damage-related Ca^{2+} release, or activate a specific release pathway, depending on its concentration. As discussed above, this is the case for fatty acids and inorganic phosphate. Useful parameters for assessing structural damage are the decreased light scattering of mitochondria, the loss of normally excluded matrix components like adenine nucleotides and the increased permeability to substances that are normally impermeant, like NADH (see Nicholls and Brand, 1980, for a discussion). The compounds that induce this type of release are obviously many, but some special cases can perhaps be mentioned here. One is phosphoenylpyruvate (Chudapongse and Haugaard, 1973; Peng et al., 1974; Sul et al., 1976; Roos et al., 1978), which induces a ruthenium red-sensitive Ca^{2+} release in liver mitochondria secondary to a lowering of the membrane potential. The mechanism by which this lowering occurs, however, is obscure.

The other case to be mention is Ca^{2+} itself, which has been shown to induce a conformational transition of beef heart mitochondria when added at low concentrations (100 nmol/mg of protein) (Hunter et al., 1976; Hunter and Haworth, 1979). The phenomenon is accompanied by a nonspecific increase in the permeability of the inner membrane, which leads to the release of accumulated (and endogenous) Ca^{2+}, and is potentiated by substances like arsenate, phosphate and fatty acids. Unexpectedely, however, this release of Ca^{2+} is insensitive to ruthenium red. As expected for a phenomenon linked to membrane alterations, agents like Mg^{2+}, bovine serum albumin and ADP, have pronounced inhibitory effects. Interestingly, the release is also inhibited by keeping the intramitochondrial NAD^+ reduced, which raises questions on its relationship to the release induced by the oxidation of mitochondrial pyridine nucleotides.

MITOCHONDRIA AS INTRACELLULAR CALCIUM REGULATORS

The problem of the role of mitochondria in the regulation of cytosolic Ca^{2+} has received considerable attention since the early days of the field, (for reviews see Lehninger, 1970; Carafoli, 1974; Bygrave, 1977). Its importance has grown in parallel with the growth in importance of Ca^{2+} as an intracellular metabolic regulator, but for a considerable length of time, much research and discussion has been devoted to a definition of the role of mitochondria in the regulation of Ca^{2+} in heart (Carafoli, 1975b). This has evidently been due to the existence in heart cells of a well-known Ca^{2+} regulated process (the contraction and relaxation of myofibrils) with Ca^{2+} requirements that are clearly defined kinetically. Accessory factors have undoubtedly been the lack of a physiologically feasible model for the release of Ca^{2+} from sarcoplasmic reticulum, the organelle universally regarded as the chief regulator of contraction related Ca^{2+}

in heart, and the great abundance of mitochondria in heart cells. In retrospect, it appears now clear that the choice of heart contraction and relaxation as a possible target for mitochondrial Ca^{2+} regulation has been less than ideal, although it has undoubtedly helped defining the characteristics, kinetic and otherwise, of both the uptake and the release of Ca^{2+} by heart mitochondria very precisely. The information gathered has made evident that the affinity of mitochondria for Ca^{2+} and their maximal velocity of uptake and (Na^+-induced) release are not adequate for the beat-to-beat control of Ca^{2+} movements. The Na^+-induced release, however, may well mediate the positive inotropic response of cardiac glycosides, which are known to increase the intracellular Na^+ concentration (Crompton et al., 1976b).

In addition to myofibrils contraction and relaxation, a vast number of Ca^{2+}-dependent activities exist in heart and cells of all types, and they may have Ca^{2+} requirements that are less unfavourable to mitochondria, in terms of kinetics, than the contraction and relaxation of myofibrils. It must be emphasized here that mitochondria, in terms of total capacity, represent the largest Ca^{2+} reservoir in cells: as an example, the case of heart may be offered, where mitochondria in situ probably store more than 1 μmol of Ca^{2+}/g of tissue (Scarpa and Graziotti, 1973; Carafoli et al., 1977). It must also be stressed that the Na^+-induced Ca^{2+} release, which has a very high degree of physiological probability, is characterized by a very marked sigmoidal dependence of the rate of Ca^{2+} efflux on the concentration of added Na^+. Since this dependence occurs in the physiological Na^+ range, it is easy to see how even a comparatively minor fluctuation of Na^+ in the ambient surrounding mitochondria may result in the release of considerable amounts of Ca^{2+}.

The discussion above privileges essentially the release route as the mechanism for the regulation of the Ca^{2+} balance between mitochondria and the cytosol. This is a logical prediction in heart and the other tissues where the Na^+ route offers the possibility of such a regulation. In other tissues, however, the steady state balance between mitochondria and the cytosol may be influenced by controlling the activity of the influx uniporter, rather than that of the (unidentified) release pathway(s). Recent experiments by Nicholls (1978a) have shown that respiring liver mitochondria maintain a steady-state external free Ca^{2+} concentration of about 1 μM, independently of the Ca^{2+} accumulated in the matrix until about 60 nmol/mg of protein, and independently of the membrane potential, if the latter does not fall below a critical value. (If this happens, the

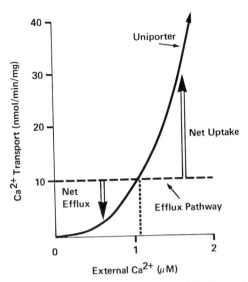

Fig. 2. *A schematic representation of the kinetics of the Ca²⁺ influx and efflux pathways in liver mitochondria (courtesy of Dr D. Nicholls).*

distribution of Ca^{2+} appears to be determined by the thermodynamic equilibration of Ca^{2+} through the uptake uniporter.) Nicholls (1978a) has proposed a kinetic regulation of the external Ca^{2+} by liver mitochondria based on the superimposition of an essentially constant efflux rate over a rate of the uptake uniporter which will be determined by the free Ca^{2+} concentration in the cytosol (Fig. 2). As the external free Ca^{2+} concentration approaches the "set-point" of about 1 μM, the rates of influx and efflux become closer, until at steady state they correspond, in Nicholls' calculations, to the energy-dissipating cycling of 5 nmol Ca^{2+}/mg of protein/min. Decreasing of the external free Ca^{2+} below the "set-point" results in the temporary predominance of the efflux pathway, until the "set-point" is attained again. The opposite would happen if the external free Ca^{2+} increases above the "set-point": the uniporter would become activated, returning the situation to normal (Fig. 3).

The concepts discussed above offer reasonable models for the regulation of the Ca^{2+} balance between mitochondria and medium in both Na^+-sensitive and insensitive mitochondria. Additional possibilities are offered by the activity of certain hormones, which have been shown, during the last few years, to influence the Ca^{2+} transporting ability of mitochondria. Among the hormones that have been shown to influence the Ca^{2+} transporting system of

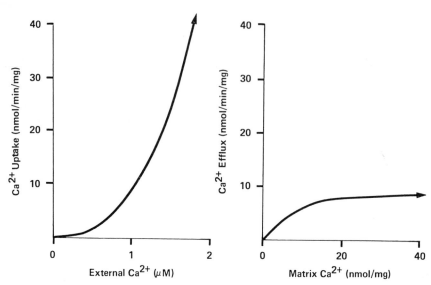

Fig. 3. *Steady-state Ca²⁺ regulation by the concerted operation of the infux and efflux pathways in liver mitochondria (courtesy of Dr D. Nicholls).*

mitochondria are steroids (Kimberg and Goldstein, 1966, 1967; Kimura and Rasmussen, 1977), insulin and glucagon (Dorman *et al.*, 1975; Yamazaki, 1975; Hughes and Barritt, 1978). Earlier reports of a direct stimulatory effect of cAMP on the release of Ca²⁺ from various mitochondrial types (Borle, 1974; Matlib and O'Brien, 1974) have not been confirmed in other Laboratories (Scarpa *et al.*, 1976). Very recent work, however (Arshad and Holdsworth, 1980) has shown that cAMP may yet have, under particular experimental conditions, some effect.

The problem, however, is that of transferring the information discussed in the preceeding sections to the *in vivo* situation. This is difficult, given the impossibility of reproducing completely the composition, ionic and otherwise, of the cytosol in *in vitro* experiments, and also considering that the extraction of mitochondria from their native cytosolic environment may alter their transport abilities. Recently, however, studies from the laboratories of Nicholls (Scott and Nicholls, 1980; Nicholls, 1981) and of Lehninger (Becker *et al.*, 1980) have employed conditions, and obtained results, that can be related almost directly to the *in vivo* situations. Nicholls has devised methods that have enabled him to estimate simultaneously the membrane potentials across the membrane of isolated nerve endings and across the inner membrane of mitochondria contained in the nerve ending cytoplasm. He has then been able to establish that the level of free Ca^{2+} within the nerve ending cytoplasm correlates directly with the mitochondrial membrane potential, and is normally maintained at a steady-state level below 1 μM. Becker *et al.* (1980) have tried to simulate *in vitro* the conditions of liver cytoplasm, including the addition of isolated liver endoplasmic reticulum vesicles. They have also used liver cells treated with digitonin to abolish the impermeability barriers of the plasma membrane, and have found that respiring mitochondria maintain the extramitochondrial free Ca^{2+} concentration at a "set-point" of about 0.5 μM. The addition of liver microsomes lowers the cytosolic steady state level to about 0.2 μM, indicating clearly that mitochondria and endoplasmic reticulum cooperate in the control of the intracellular Ca^{2+} homoeostasis, the latter organelle being evidently able to exert its effect at lower cytosolic concentrations of free Ca^{2+} than mitochondria.

An integrated picture of the overall regulation of intracellular Ca^{2+}, as performed by the different cellular membranes, has recently been attempted for the case of heart by Carafoli *et al.* (1982). In heart, the kinetic parameters of Ca^{2+} transport (uptake) are known for the four systems that contribute to the cellular homoeostasis of Ca^{2+}: the Ca^{2+} ATPase of sarcoplasmic reticulum, the electrophoretic uptake system in mitochondria, the Ca^{2+} ATPase, and the Na^+/Ca^{2+} exchange, in sarcolemma. Table I shows that, under conditions approaching the physiological ones, only minor fractions of the intracellular Ca^{2+} are handled by the plasma membrane and the mitochondria, the most active organelle being the sarcoplasmic reticulum. Under conditions of relative intracellular Ca^{2+} overload, however, the role of mitochondria becomes proportionally more important, and may equal or exceed that of sarcoplasmic reticulum. The picture that emerges is that of mitochondria as long-term Ca^{2+} buffers, which can handle very large amounts of Ca^{2+} whenever the need arises. In so doing, evidently, they serve an essential function, since they are responsible for maintaining the Ca^{2+} levels in the cell within limits compatible with the demands of the regulatory function of Ca^{2+}.

One last aspect of the process of mitochondrial Ca^{2+} transport must be mentioned in closing this chapter. The mitochondrial matrix contains some important reactions that are precisely regulated by Ca^{2+} in the micromolar activity level (Denton *et al.*, 1972, 1978; Cooper *et al.*, 1974; McCormack and Denton, 1979). The maintenance of the matrix Ca^{2+} activity of about this level, and of regulating it with precision, is thus of the greatest importance to the

Table I. An integrated view of the transport of Ca^{2+} in heart cells.

Membrane system	Ca^{2+} transporting area of heart cells (m^2/g)	(% of total area, a)	Ca^{2+} uptake					
			At 10 μM Ca^{2+}			At 1 μM Ca^{2+} + 1–3 mM Mg^{2+}		
			Rate (b) (nmol/mg of protein/s)	"Total" uptake (a × b) (nmol/mg of proteins)	% of total uptake	Rate (b) (nmol/mg of protein/s)	"Total" uptake (a × b) (nmol/mg of proteins)	% of total uptake
Sarcolemma	0.10	0.8						
Ca^{2+}–ATPase			1.5	1.2	0.4	1.5	1.2	0.7
Na^+–Ca^{2+} exchange			20	16	3.1	10	8	4.9
Sarcoplasmic reticulum	1.5	12.1	20	242	46.5	12	145	88
Mitochondria	10.6	87	3	261	50	0.12	10.4	6.3

The data on the relative distribution of the Ca^{2+}-transporting membrane area are taken from Carafoli and Crompton (1978). For mitochondria, only the inner membrane has been considered. Uptakes are given at 38° C. In the case of mitochondria, the rates given by Carafoli and Crompton (1978) have been doubled, on the assumption that the inner membrane represents 50% of the total protein of heart mitochondrial preparations. For sarcoplasmic reticulum, the rates have been calculated from the data given by Will et al. (1976). The following additional assumptions have been made: (a), that the degree of contamination of all preparations with extraneous organelles is the same; (b), that the protein content for the unit of membrane area is the same in all organelles (From Carafoli et al., 1982).

functioning of the citric acid cycle, and of other reactions that deliver reducing equivalents to the respiratory chain (Malmström and Carafoli, 1976; Otto and Ontko, 1978). Whereas other organelles contribute to the regulation of cytosolic Ca^{2+}, it is evident that only the mitochondrial Ca^{2+} transport activity is responsible for the regulation of intramitochondrial Ca^{2+}. As suggested by Denton and McCormack (1980, see also Carafoli, 1980) the most important function of the mitochondrial Ca^{2+} transport system could then well be that of regulating Ca^{2+} inside the mitochondrion itself.

REFERENCES

Addanki, S., Cahill, F. D. and Sotos, J. F. (1968). *J. biol. Chem.* **243**, 2337–2348.

Affolter, H. and Carafoli, E. (1980). *Biochim. biophys. Res. Commun.* **95**, 193–196.

Affolter, H. and Carafoli, E. (1981). *Eur. J. Biochem.* **119**, 199–201.

Åkerman, K. E. O. (1978). *Biochim. biophys. Acta* **502**, 359–366.

Åkerman, K. E. O., Wikstrom, M. K. F. and Saris, N. E. (1977). *Biochim. biophys. Acta* **464**, 287–294.

Al Shaikhaly, M. H. M., Nedergaard, J. and Cannon, B. (1979). *Proc. natn. Acad. Sci. U.S.A.* **76**, 2350–2353.

Arshad, J. H. and Holdsworth, E. S. (1980). *J. Membr. Biol.* **57**, 207–212.

Becker, G. L., Fiskum, G. and Lehninger, A. C. (1980). *J. biol. Chem.* **255**, 9009–9012.

Borle, A. (1974). *J. Membr. Biol.* **16**, 221–236.

Bragadin, M., Pozzan, T and Azzone, G. T. (1979). *Biochemistry* **18**, 5972–5978.

Brierley, G. P., Murer, E. and Bachmann, E. (1964). *Archs. Biochem. Biophys.* **105**, 89–102.

Bygrave, F. L. (1977). *Curr. Top. Bioenerg.* **6**, 259–318.

Carafoli, E. (1965). *Biochim. biophys. Acta* **97**, 99–106.

Carafoli, E. (1974). *Biochem. Soc. Symp.* **39**, 89–109.

Carafoli, E. (1975a). *Molec. cell. Biochem.* **8**, 133–140.

Carafoli, E. (1975b). *J. molec. cell. Cardiol.* **7**, 83–89.

Carafoli, E. (1979). *FEBS Lett.* **104**, 1–5.

Carafoli, E. (1980). *In* "Exercise Bioenergetics and Gas Exchange" (Eds. P. Cerretelli and B. J. Whipp). 3–12. Elsevier/North-Holland, Amsterdam.

Carafoli, E. and Crompton, M. (1976). *In* "Calcium in Biological Systems" (Ed. C. J. Duncan) 89–115. Symp. No. 30 Soc. Exptl Biol. Cambridge University Press.

Carafoli, E. and Crompton, M. (1978). *Curr. Top. Membr. Transp.* **10**, 152–216.

Carafoli, E. and Sottocasa, G. L. (1974). *In* "Dynamics of Energy-Transducing Membranes" (Eds L. Ernster, R. W. Eastabrook and E. C. Slater) 455–469. Elseveir, Amsterdam.

Carafoli, E., Rossi, C. S. and Lehninger, A. L. (1965). *J. biol. Chem.* **240**, 2254–2261.

Carafoli, E., Balcavage, W. X., Lehninger, A. L. and Mattoon, J. R. (1970). *Biochim. biophys. Acta* **205**, 18–26.

Carafoli, E., Hansford, R. G., Sacktor, B. and Lehninger, A. L. (1971). *J. biol. Chem.* **246**, 964–972.

Carafoli. E., Crovetti, F. and Ceccarelli, D. (1973a). *Archs. Biochem. Biophys.* **154**, 40–46.

Carafoli, E., Gazzotti, P., Saltini, C., Rossi, C. S., Sottocasa, G. L., Sandri, G., Panfili, E and De Bernard, B. (1973b). *In* "Mechanisms in Bioenergetics" (Eds G. F. Azzone, L. Ernster, S. Papa, E. Quagliariello and N. Siliprandi) 293–307. Academic Press, New York.

Carafoli, E., Prestipino, G. F. Ceccarelli, D and Conti, F. (1974a). *In* "Membrane Proteins in Transport and Phosphorylation" (Eds G. F. Azzone, M. Klingenberg, E. Quagliariello and N. Siliprandi) 85–90. North-Holland, Amsterdam.

Carafoli, E., Tiozzo, R., Lugli, G., Crovetti, F. and Kratzing, C. (1974b). *J. molec. cell. Cardiol.* **6**, 361–371.

Carafoli, E., Crompton, M., Malmström, K., Sigel, E., Salzmann, M., Chiesi, M. and Affolter, H. (1977). *In* "Biochemistry of Membrane Transport" (Eds G. Semenza and E. Carafoli) 535–551. Springer-Verlag, Heidelberg.

Carafoli, E., Caroni, P., Chiesi, M. and Famulski, K. (1982). *In* "Metabolic Compartmentation" (Ed. H. Sies). Academic Press. New York. In press.

Chance, B. (1956) *In* "Proc. 3rd Intern. Congr. Biochem". (1955) (Ed. C. Liébecq). 300–304. Academic Press, New York.

Chance, B. (1965). *J. biol. Chem.* **240**, 2729–2748.

Chance, B. and Mela, L. (1966a). *Proc. natn. Acad. Sci. U.S.A.* **55**, 1243–1251.

Chance, B. and Mela, L. (1966b). *J. biol. Chem.* **241**, 4588–4599.

Chappell, J. B., Cohn, M. and Greville, G. D. (1963). *In* "Energy-Linked Functions of Mitochondria" (Ed. B. Chance). 219–231. Academic Press, New York.

Chudapongse, P. and Haugaard, N. (1973). *Biochim. biophys. Acta* **307**, 599–606.

Clark, A. and Roman, I. (1980). *J. biol. Chem.* **255**, 6556–6558.

Cooper, R. H., Randle, P. J. and Denton, R. M. (1974). *Biochem. J.* **143**, 625–641.

Crompton, M. and Heid, I. (1978). *Eur. J. Biochem.* **91**, 599–608.

Crompton, M., Sigel, E., Salzmann, M. and Carafoli, E. (1976a). *Eur. J. Biochem.* **69**, 429–434.

Crompton, M., Capano, M. and Carafoli, E. (1976b). *Eur. J. Biochem.* **69**, 453–462.

Crompton, M., Kunzi, N. and Carafoli, E. (1977). *Eur. J. Biochem.* **79**, 549–558.

Crompton, M., Heid, I., Baschera, C. and Carafoli, E. (1978a). *FEBS Lett.* **104**, 352–354.

Crompton, M., Moser, R., Lüdi, H. and Carafoli, E. (1978b). *Eur. J. Biochem.* **82**, 25–31.

Crompton, M., Hediger, M. and Carafoli, E. (1978c). *Biochem. biophys. Res Commun.* **80**, 540–546.

Crompton, M., Heid, I. and Carafoli, E. (1980). *FEBS Lett.* **115**, 257–259.

De Luca, H. F. and Engstrom, G. (1961). *Proc. natn. Acad. Sci. U.S.A.* **47**, 1744–1750.

Denton, R. M. and McCormack. J. G. (1980). *Trans. Biochem. Soc.* **8**, 266–268.
Denton, R. M., Randle, P. J. and Martin, B. R. (1972). *Biochem. J.* **128**, 161–163.
Denton, B. R., Richards, D. A. and Chin, J. G. (1978). *Biochem. J.* **176**, 899–906.
Dorman, D. M., Barrit, G. J. and Bygrave, F. L. (1975). *Biochem. J.* **50**, 389–395.
Drahota, Z., Carafoli, E., Rossi, C. S., Gamble, R. L. and Lehninger, A. C. (1965). *J. biol. Chem.* **240**, 2712–2720.
Engstrom, G. W. and De Luca, H. F. (1963). *Biochemistry* **3**, 379–383.
Fiskum, G. and Lehninger, A. L. (1979). *J. biol. Chem.* **254**, 6236–6239.
Gear, A., Rossi, C. S., Reynafarje, B. and Lehninger, A. L. (1967). *J. biol. Chem.* **243**, 3403–3413.
Greenawalt, J. W., Rossi, C. S. and Lehninger, A. L. (1964). *J. Cell Biol.* **23**, 21–38.
Gunter, T. E. and Puskin, J. S. (1976). *Ann. N.Y. Acad. Sci.* **264**, 112–123.
Haworth, R. A., Hunter, D. R. and Berkoff, H. A. (1980). *FEBS Lett.* **110**, 216–218.
Heaton, G. M. and Nicholls, D. G. (1976). *Biochem. J.* **156**, 635–646.
Hughes, B. P. and Barritt, G. J. (1978). *Biochem. J.* **176**, 295–304.
Hunter, D. R. and Haworth, R. A. (1979). *Archs. Biochem. Biophys.* **195**, 468–477.
Hunter, D. R., Haworth, R. A. and Southard, J. H. (1976). *J. biol. Chem.* **251**, 5069–5077.
Hutson, S. M., Pfeiffer, D. R. and Lardy, H. A. (1976). *J. biol. Chem.* **251**, 5251–5258.
Jacobus, W. E., Tiozzo, R., Lugli, G., Lehninger, A. L. and Carafoli, E. (1975). *J. biol. Chem.* **250**, 7863–7870.
Jeng, A. Y. and Shamoo, A. (1980a). *J. biol. Chem.* **255**, 6897–6903.
Jeng, A. Y. and Shamoo, A. (1980b). *J. biol. Chem.* **255**, 6904–6912.
Kimberg, D. V. and Goldstein, S. A. (1966). *J. biol. Chem.* **241**, 95–103.
Kimberg, D. V. and Goldstein, S. A. (1967). *Endocrinology* **80**, 89–98.
Kimura, S. and Rasmussen, H. (1977). *J. biol. Chem.* **252**, 1217–1225.
Lehninger, A. L. (1970). *Biochem. J.* **119**, 129–138.
Lehninger, A. L. (1974). *Proc. natn. Acad. Sci. U.S.A.* **71**, 1520–1524.
Lehninger, A. L. (1981). *In* "Calcium and Phosphate Transport across Membranes" (Eds F. Bronner and M. Peterlik). Academic Press, New York, In press.
Lehninger, A. L. and Carafoli, E. (1969). *In* "Biochemistry of the Phagocytic Process" (Ed. J. Schultz) 922–931. North-Holland, Amsterdam.
Lehninger, A. L. and Carafoli, E. (1971). *Archs. Biochem. Biophys.* **146**, 506–515.
Lehninger, A. L., Rossi, C. S. and Greenawalt, J. W. (1963). *Biochem. biophys. Res. Commun.* **10**, 444–448.
Lehninger, A. L., Vercesi, A. and Bababunmi, E. A. (1978). *Proc. natn. Acad. Sci. U.S.A.* **75**, 1690–1694.
Lötscher, H. R., Schwerzmann, K. and Carafoli, E. (1979a). *FEBS Lett.* **99**, 194–198.
Lötscher, H. R., Winterhalter, K. H., Carafoli, E. and Richter, C. (1979b) *Proc. natn. Acad. Sci. U.S.A.* **76**, 4340–4344.

Lötscher, H. R., Winterhalter, K. H., Carafoli, E. and Richter, C. (1980a). *Eur. J. Biochem.* **110**, 211–216.
Lötscher, H. R., Winterhalter, K. H., Carafoli, E. and Ritcher, C. (1980b). *J. biol. Chem.* **255**, 9325–9330.
McCormack, J. G. and Denton, R. M. (1979). *Biochem. J.* **180**, 533–544.
Malmström, K. and Carafoli, E. (1975a). *Archs Biochem. Biophys.* **171**, 418–423.
Malmström, K. and Carafoli, E. (1975b). *Biochem. biophys. Res. Commun.* **69**, 658–664.
Malmström, K. and Carafoli, E. (1976). *Biochem. biophys. Res. Commun.* **69**, 658–665.
Matlib, A. and O'Brien, P. K. (1974). *Biochem. Soc. Trans.* **2**, 997–1000.
Mela, L. (1968). *Archs. Biochem. Biophys.* **123**, 286–293.
Mitchell, P. (1966). *In* "Chemiosmotic Coupling in Oxidative and Photosynthetic Phosphorylation". Glynn Research, Bodmin, UK.
Mitchell, P. and Moyle, J. (1967). *Biochem. J.* **109**, 1147–1162.
Moore, C. L. (1971). *Biochem. biophys. Res. Commun.* **42**, 405–418.
Moyle, J. and Mitchell, P. (1977a). *FEBS Lett.* **73**, 131–136.
Moyle, J. and Mitchell, P. (1977b). *FEBS Lett.* **77**, 136–145.
Nicholls, D. G. (1978a). *Biochem. J.* **170**, 511–522.
Nicholls, D. G. (1978b). *Biochem. J.* **176**, 463–474.
Nicholls, D. G. (1981). *In* "Calcium and Phosphate Transport across Membranes" (Eds F. Bronner and M. Peterlik). Academic Press, New York. In press
Nicholls, D. G. and Brand, M. D. (1980). *Biochem. J.* **188**, 113–118.
Nicholls, D. G. and Crompton, M. (1980). *FEBS Lett.* **111**, 261–268.
Niggli, V., Gazzotti, P. and Carafoli, E. (1978). *Experientia* **34**, 1136–1137.
Otto, D. A. and Ontko, J. A. (1978). *J. biol. Chem.* **253**, 789–795.
Panfili, E., Sandri, G., Sottocasa, G. L., Graziosi, G., Lunazzi, G. C. and Liut, G. F. (1976). *Nature, Lond.* **264**, 185–186.
Panfili, E., Sottocasa, G. L., Sandri, G. and Liut, G. F. (1980). *Eur. J. Biochem.* **105**, 205–210.
Peng, C. F., Price, D. W., Bhuvaneswaran, C and Wadkins, C. L. (1974). *Biochem. biophys. Res. Commun.* **56**, 134–141.
Prestipino, G., Ceccarelli, D., Conti, F. and Carafoli, E. (1974). *FEBS Lett.* **45**, 99–103.
Puskin, J. S., Gunter, T. E., Gunter, K. K. and Russell, P. R. (1976). *Biochemistry* **15**, 3839–3842.
Rasmussen, H., Chance, B. and Ogata, E. (1965). *Proc. natn. Acad. Sci. U.S.A.* **53**, 1069–1076.
Reed, K. C. and Bygrave, F. L. (1975). *Eur. J. Biochem.* **55**, 497–504.
Reynafarje, B and Lehninger, A. L. (1977). *Biochem. biophys. Res. Commun.* **77**, 1273–1279.
Roman, I., Gmaj, P., Nowicka, C. and Angielski, S. (1979). *Eur. J. Biochem.* **102**, 615–623.
Roos, I., Crompton, M. and Carafoli, E. (1978). *FEBS Lett.* **94**, 418–421.
Roos, I., Crompton, M. and Carafoli, E. (1980). *Eur. J. Biochem.* **110**, 319–325.
Rossi, C. S. and Lehninger, A. L. (1963). *Biochem. Z.* **338**, 698–713.
Rossi, C. S. and Lehninger, A. L. (1964). *J. biol. Chem.* **239**, 3971–3980.
Rossi, C. S., Bielawski, J. and Lehninger, A. L. (1966). *J. biol. Chem.* **241**, 1919–1921.
Rossi, C., Azzi, A. and Azzone, G. F. (1967). *Eur. J. Biochem.* **1**, 141–146.

Rossi, C. S., Vasington, F. D. and Carafoli, E. (1973). *Biochem. biophys. Res. Commun.* **50**, 846–852.

Rottenberg, H. and Scarpa, A. (1974). *Biochemistry* **13**, 4811–4819.

Sandri, G., Panfili, E., Liut, G. F. and Sottocasa, G. L. (1979). *Biochim. biophys. Acta* **558**, 214–220.

Saris, N. E. (1959). *Finska Kemistsamfundets Medd.* **68**, 98–107.

Saris, N. E. (1963). *Soc. Sci. Fennica, Comment. Physico-Mathem.* **28**, 3–59.

Scarpa, A. and Azzone, G. F. (1970). *Eur. J. Biochem.* **12**, 328–335.

Scarpa, A. and Graziotti, P. (1973). *J. gen. Physiol.* **62**, 756–772.

Scarpa, A., Malmström, K., Chiesi, M. and Carafoli, E. (1976). *J. Membr. Biol.* **29**, 205–208.

Scott. I. D. and Nicholls, D. G. (1980). *Biochem. J.* **186**, 21–33.

Selwyn, H. J., Dawson, A. P. and Dunnett, S. J. (1970). *FEBS Lett.* **10**, 1–5.

Siliprandi, N., Rugolo, M., Siliprandi, D., Toninello, F and Zoccarato, F. (1979). *In* "Function and Molecular Aspects of Biomembrane Transport" (Eds E. Quagliariello, F. Palmieri, S. Papa and M. Klingenberg) 147–156. Elsevier/North-Holland, Amsterdam.

Slater, E. C. and Cleland, K. W. (1953). *Biochem. J.* **55**, 566–580.

Sordahl, L. A. (1974). *Archs. biochem. Biophys.* **167**, 104–115.

Sottocasa, G. L., Sandri, G., Panfili, E., De Bernard, B., Gazzotti, P., Vasington, F. D. and Carafoli, E. (1972). *Biochem. biophys. Res. Commun.* **47**, 808–813.

Spencer, T. and Bygrave, F. L. (1973). *Bioenergetics* **4**, 347–362.

Stücki, J. W. and Ineichen, E. A. (1974). *Eur. J. Biochem.* **48**, 365–375.

Sul, H. S., Shrago, E. and Shug. A. L. (1976). *Archs. Biochem. Biophys.* **172**, 230–237.

Tew, W. P. (1977). *Biochem. biophys. Res. Commun.* **78**, 624–630.

Vasington, F. D. and Murphy, J. V. (1962). *J. biol. Chem.* **237**, 2670–2677.

Vasington, F. D. and Murphy, J. V. (1961). *Fed. Proc.* **20**, 146.

Vasington, F. D., Gazzotti, P, Tiozzo, R. and Carafoli, E. (1972). *Biochim. biophys. Acta.* **256**, 43–54.

Vercesi, A., Reynafarje, B. and Lehninger, A. L. (1978). *J. biol. Chem.* **253**, 6379–6385.

Wehrle, J. P. and Pedersen, P. L. (1979). *J. biol. Chem.* **254**, 7269–7275.

Will, H., Blanck, J., Smettan, G. and Wollenberger, A. (1976). *Biochim. biophys. Acta* **449**, 295–303.

Wolkowicz, P. E. and McMillin-Wood, J. (1980). *J. biol. Chem.* **255**, 10348–10353.

Yamazaki, R. K. (1975). *J. biol. Chem.* **250**, 7924–7930.

4

The Transport of Calcium by Sarcoplasmic Reticulum and Various Microsomal Preparations

LEOPOLDO DE MEIS

Instituto de Ciências Biomédicas, Departmento de Bioquímica Universidade Federal do Rio de Janeiro, Rio de Janeiro, Brasil

GIUSEPPE INESI

Department of Biochemistry, University of Maryland, Baltimore, Maryland, USA

INTRODUCTION

The cytoplasmic Ca^{2+} concentration is generally several orders of magnitude lower than that of extracellular fluids. This gradient must be maintained by energy-dependent transport mechanisms located in the cell membrane. In addition, intracellular membranous structures may form compartments which serve as Ca^{2+} sinks and reservoirs inside the cell. In some tissues, in which specific functional controls require rapid Ca^{2+} delivery to and sequestration from the cytoplasm, these intracellular systems are developed to a highly differentiated and prominent degree. A prime example is the sarcoplasmic reticulum (SR) of fast striated muscles.

Microsomal fractions obtained from various tissues demonstrate some calcium transport activity supported by ATP hydrolysis. This suggests that the calcium sequestering function of intracellular membranes is not limited to striated muscle, but may also be present to some extent in other tissues, such as smooth muscle, nervous tissue, secretory glands and liver. However, in most microsomal preparations it is very difficult to identify with certainty the source of the membrane vesicles sustaining Ca^{2+} transport. These vesicles may derive from cell membrane, endoplasmic reticulum or other specialized structures. For instance, brain microsomes include vesicles of various dimensions as well as synaptosomes which can be enriched to some degree by subfractionation. However the calcium transport activity is low and cannot be definitely attributed to a specific membrane structure contained in the microsomal preparation.

Owing to the prominence of SR in certain muscle tissues, it is possible to obtain from these tissues microsomal preparations which can be identified with pure SR membrane. This has permitted detailed studies on the mechanism of Ca^{2+} transport coupled to ATPase activity, which will be described below. SR may be considered a highly specialized form of endoplasmic reticulum. It is likely that fundamental features of transport and energy interconversion which are observed in this membrane, also apply to other systems involved in calcium transport and energy transduction.

THE SARCOTUBULAR SYSTEM OF MUSCLE FIBRES

The reticular formation of striated muscle fibres was demonstrated in early studies by Veratti (1902) using light microscopy. Its ultra-structure was subsequently defined by electron microscopy (Huxley, 1957; Porter and Palade, 1957; Anderson-Cedergreen, 1958; Franzini-Armstrong, 1964; Peachey, 1965).

Two distinct components may be identified within the sarco-tubular system: (a) transverse (T) tubules formed by invaginations of the external membrane, penetrating the fibre in correspondence to Z lines or A–I junctions of each sarcomere, depending on the animal species (the lumen of the T-tubules communicates freely with the extracellular environment as demonstrated by penetration of large molecules (Huxley, 1964; Peachey and Schild, 1968), but is sealed with respect to the myoplasm); (b) sarcoplasmic reticulum (SR) which is a membrane bound, intracellular compartment com-posed by cysternal formations adjacent to T-tubules. The cysternae communicate with each other by means of longitudinal tubules. The SR compartment is sealed with respect to the myoplasm.

Sections of T-tubules and two adjacent cisternal formations reveal an arrangement of these structures which is referred to as a "triad". In this regard, an ultrastructural feature which has received attention (Franzini-Armstrong, 1970, 1973; Eisemberg and Gilai, 1979; Somlyo, 1979) is the presence of foot-like processes which appear to be joining tubular membranes and adjacent segments of cisternal membranes ("junctional SR"). According to a model presented by

A. Somlyo (1979) these processes are bridging structures which provide a direct continuity of the cytoplasmic, but not the luminal leaflets, of these two membranes.

Development of SR is most prominent in fast skeletal muscle, and less developed in slow skeletal and cardiac muscle. However, a prominent development of T-tubules and other specific ultrastructural features are found in heart muscle (Sommer and Johnson, 1978). This is probably related to functional adaptations of the tissue.

ISOLATION OF SR AND T-TUBULAR MEMBRANES

Microsomal fractions were first obtained by differential centrifugations of muscle homogenates by Portzehl (1957), Ebashi (1958) and Nagai et al. (1960), and later identified with fragmented sarcotubular membranes (Muscatello et al., 1961). These microsomal fractions are usually obtained from muscles which are rich in SR; therefore, they contain mostly fragments of SR membrane. Further purification may be obtained by the use of sucrose gradient centrifugations (Meissner et al., 1973; Meissner, 1975).

More recently, considerable effort has been devoted to isolation of T-tubular membranes (Lau et al., 1977; Scales and Sabbadini, 1979). One of the problems encountered in these isolations is recognition of different membranes. In fact, while Ca^{2+}-activated ATPase may be considered a specific marker for SR, it is not yet clear which marker should be considered specific for T-tubule membranes. Comparison of ultrastructural and labeling patterns of the membranes in situ and following isolation may be helpful in the progression of this work.

STRUCTURE OF THE SR MEMBRANE

The Protein Component

Protein accounts for approximately 55% of the SR membrane, while the remaining 45% is attributed to lipid. Taking into account the different specific weight, it can be estimated that approximately half the membrane volume is necessary of accommodate the protein. The prominence of the protein component acquires a greater specificity considering that electrophoretic analysis (McFarland and Inesi, 1971; Martonosi and Halpin, 1971; MacLennan et al., 1978) indicates that the Ca^{2+}-activated ATPase accounts for 60–80% of the SR protein.

In addition to the Ca^{2+}-ATPase small amounts of other proteins are found in preparations of SR vesicles and can be demonstrated by gel electrophoresis. These minor components include "calsequestrin" (MacLennan and Wong, 1971). a "calcium binding protein" (Ostwald and MacLennan, 1974), a proteolipid (MacLennan et al., 1972) and glycoprotein (Michalak et al., 1980). Presently, no definitive information on the role of these minor protein components is available. It has been proposed that calsequestrin might serve as a Ca^{2+} sink in SR (Jorgensen et al., 1979). However, calsequestrin is not present in the SR of all animal species (Malan et al., 1975).

The Ca^{2+}-activated ATPase was first purified by MacLennan (1970) with a procedure including differential solubilization with deoxycholate, salt precipitation and gel chromatography. Other successful procedures for solubilization and partial purification of Ca^{2+}-ATPase were reported by Martonosi (1968), Ikemoto et al. (1971), McFarland and Inesi (1971), Meissner et al. (1973), Warren et al. (1974a, b), Dean and Tanford (1978) and Banarjee et al. (1979).

Comparative use of various solubilization procedures reveals that the MacLennan's procedure yields a lower amount of protein, but insures greater purity and removal of detergents which appear to modify some kinetic parameters of the enzyme (Takisawa and Tonomura, 1978, 1979), Following removal of detergent the ATPase protein reassembles in the form of membranous vesicles retaining the original lipid composition. With the use of selected detergents it is possible to obtain and maintain the enzyme in a more dissociated and still active form (Dean and Tanford, 1978).

Values for the molecular weight of single ATPase polypeptide chains are within the 100 000–115 000 range, as determined by electrophoresis and ultracentrifugation analysis (Inesi et al., 1970; MacLennan, 1970; Louis and Shooter, 1972; Thorley-Lawson and Green, 1975; Le Maire et al., 1976, Rizzolo et al., 1976).

Assembly of the ATPase Within the Membrane

The isolated fragments of SR membrane appear under the electron microscope as isolated vesicles of approximately 150 nm diameter (Fig. 1). A characteristic ultrastructural feature of negatively stained preparations is the presence of outer granules (3.5–4.0 nm diameter) densely packed on the outer surface of the vesicles (Ikemoto et al., 1968; Inesi and Asai, 1968). Another ultrastructural feature of the SR membrane is the presence of 9 nm particles noted on electron micrographs of freeze fracture preparations (Deamer and Baskin,

1969). These particles are almost exclusively found on the concave outer surface of the cytoplasmatic membrane leaflet of the fractured vesicles, while the convex faces (outer surface of the luminal leaflet) appear smooth. Therefore, the SR membrane presents a character of marked structural asymmetry, inasmuch as a significant portion of the protein component protrudes from the outer surface of the vesicles. This is consistent with early studies demonstrating that labeling of -SH groups results in attachment of electron dense molecules exclusively in the outer, but not on the inner surface of the vesicles (Hasselbach and Elfvin, 1967; Hasselbach and Beil, 1977).

Electron microscopic studies on trypsin treated vesicles (Inesi and Scales, 1974; Stewart and MacLennan, 1974) and on incorporation of purified ATPase in liposomes (Thorley-Lawson and Green, 1973; Packer et al., 1974), have demonstrated that both the outer granules noted in negatively stained preparations, and the 9 nm particles observed on freeze fracture faces, correspond to the ATPase protein. These structural features are limited to the cytoplasmic leaflet of native vesicles, suggesting that the ATPase protein has an anphiphilic character whereby the polar ends of the enzyme protrude from the outer surface into the aqueous medium, while the hydrophobic ends remain within the membrane bilayer. This model for protein disposition in the SR membrane is also suggested by low angle X-ray diffraction studies yielding asymmetric profiles for electron density distribution across the membrane (Dupont et al., 1973; Herbette et al., 1977). These profiles include the presence of an outer density peak, in addition to two peaks corresponding to the polar portions of the phospholipid bilayer. The outer peak is related to the outer granules observed by electron microscopy. The other two peaks and the trough in between them can be attributed to a bilayer profile deformed by protein penetrating the hydrophobic interior (Fig. 2). Contrary to the native membrane, reconstituted vesicles show protein particles which are incorporated randomly in both the cytoplasmic and luminal leaflets (Packer et al., 1974). Nevertheless, the reconstituted vesicles provide a useful system for studies on residual functions (Racker, 1972; Chiesi et al., 1978; Wang et al., 1979) and on the role of lipids (Warren et al., 1974a, b; Knowles et al., 1976; Peterson and Deamer, 1977).

An intriguing question is whether the 100 000 mol.wt polypeptide chains demonstrated by electrophoresis following solubilization in SDS, are contained in the native membrane as single units ("monomers"), or interact with each other to form "oligomeric" structures. The density of outer granules observed on replicas of deeply etched

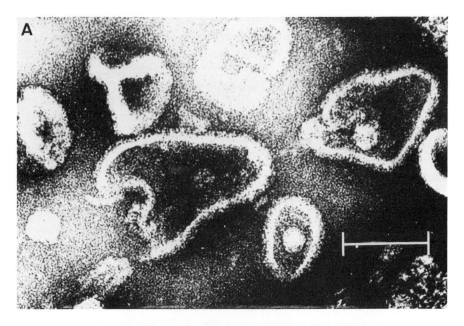

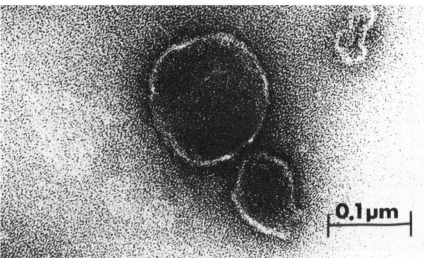

Fig. 1. *A, electronmicroscopic appearance of negatively stained SR vesicles before (top) and following (bottom) proteolytic digestion. Note the outer protein granules which disappear as a consequence of digestion. B, electronmicroscopic appearance of freeze fractured SR vesicles before (top) and following (bottom) proteolytic digestion. Note the protein particles which are exclusively on the concave faces of control vesicles and are markedly affected in their density and distribution as a consequence of proteolytic digestion. Scale bars = 0.1 μm (Inesi and Scales, 1974).*

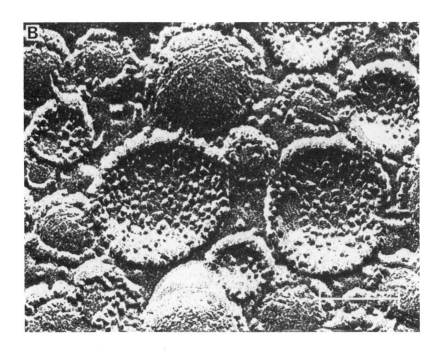

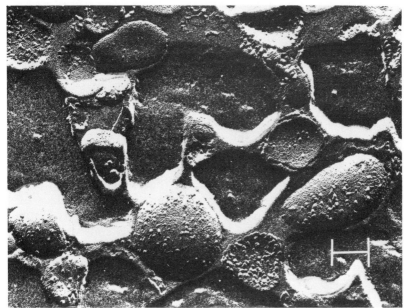

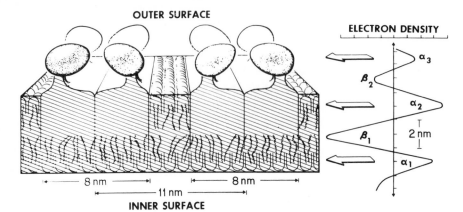

Fig. 2. *Diagramatic representation of structural features of the SR membrane revealed by biochemical determinations, electromicroscopic observations and X-ray diffraction. An attempt is made to match the structural model with the electron density profile given by Dupont et al., (1973). A striking structural feature is the prominence of the ATPase protein which accounts for more than half the membrane mass. The protein is exposed on the external surface of the vesicles in the form of outer projections corresponding to individual polypeptide chains. The nonpolar segments of these chains penetrate the membrane interior, thereby appearing as inner particles on the concave freeze fracture faces. The diagram suggests that the discrepancy between the number of visible outer projections and that of inner particles is due to a tendency of the chains to aggregate. Membrane thickness, as well as dimensions and densities of outer granules and inner particles are drawn to actual scale as found in rabbit SR vesicles (Inesi, 1979).*

vesicles appears to match the theoretical density of individual 100 000 mol.wt polypeptide chains calculated on the basis of the ATPase content and the surface area of the SR vesicles (Scales and Inesi, 1976). On the other hand, the density of 9 nm particles observed on replicas of freeze fracture faces appears lower than that of the outer granules (Jilka *et al.*, 1975; Scales and Inesi, 1976). This discrepancy between the densities of the polar and non polar portions of the ATPase chains may be due to a number of artifacts, including uneven effacement of nonpolar portions of the chains on the freeze fracture faces. On the other hand, it is possible that while the polar partions of the chains retain their individuality on the outer surface of the membrane, the nonpolar partions have a tendency to establish oligomeric interactions. The latter possibility is consistent with the fluorescence energy transfer experiments of Vanderkoi *et al.* (1977) who found that energy transfer among labels of the polypeptide chains was not reduced by a 10-fold change in the protein–lipid ratio. This was interpreted as excluding random collision of the chains, and suggesting a rather stable relationship of chains with each other.

A different line of evidence for interactions of the ATPase chains was obtained in experiments on solubilization with Triton-X-100, demonstrating by analytic centrifugation the presence of molecular aggregates of various size (McFarland and Inesi, 1970). More specifically, Le Maire *et al.* (1976) solubilized active ATP chains which retained approximately 30 molecules of phospholipid each, and found that this preparation consisted prevalently of trimers and tetramers. However, following phospholipid substitution with appropriate detergents, an active enzyme can be obtained in the monomeric form (Dean and Tanford, 1978). The monomeric form displays a simple dependence on substrate concentration, when compared to the complex substrate dependence of the enzyme in its native form (Dean and Tanford, 1978; Taylor and Hattan, 1979; Möller *et al.*, 1980). Presently it is still uncertain whether the native enzyme which is capable of sustaining ATP hydrolysis coupled to Ca^{2+} transport, is in a monomeric or an oligomeric form within the membrane.

When native SR vesicles are subjected to mild tryspin digestion, the 100 000 mol.wt polypeptide chains are cleaved in two complementary fragments of approximately 55 000 and 45 000 mol.wt (Thorley-Lawson and Green, 1973; Migala *et al.*, 1973; Stewart and MacLennan, 1974; Inesi and Scales, 1974). This initial cleavage is not accompanied by a significant loss of protein or reduction in Ca^{2+} transport activity, indicating that the two fragments are intimately associated and stabilized by multiple weak interactions (Rizzolo *et al.*, 1976). The two fragments dissociate upon solubilization in SDS and can be demonstrated by gel electrophoresis.

Subsequent tryspin digestion causes further cleavage of the 55 000 mol.wt fragment into 30 000 and 20 000 mol.wt subfragments. At this point the ability of the vesicles to accumulate calcium is lost, but ATPase activity is retained (Thorley-Lawson and Green, 1973; Stewart and MacLennan, 1974; Stewart *et al.*, 1976). It is apparent that amino acid residues located in different tryptic fragments participate in the structures involved in Ca^{2+} translocation and calalytic activity. In fact, a phosphoylated residue is found in the 30 000 mol.wt subfragment of the 55 000 mol.wt fragment, when digested vesicles are exposed to $[\gamma\text{-}^{32}P]$-ATP and the fragments are then dissociated with SDS. On the other hand ionophoric activity with respect to Ca^{2+} has been attributed to the 20 000 mol.wt subfragment (Stewart *et al.*, 1976; Shamoo *et al.*, 1976).

Isolation of various peptide fragments has led to sequencing work on the Ca^{2+} ATPase (Allen, 1977; Tong, 1977; Klip *et al.*, 1980).

The partial sequencing which is presently available permits assignment of the main tryptic fragments in for form $NH_2-AC-20$, 000–30, 000–45, 000–COOH (Klip et al., 1980). Each fragment contains segments of polar and nonpolar amino acids. The later segments may represent portions of the chain which are inbedded in the membrane bilayer.

The Lipid Component

Lipids accound for approximately 45% of the membrane dry weight and consist mostly (90%) of phosphlipids (Drabikowski et al., 1966; Martonosi et al., 1968; Fiehn and Hasselbach, 1970; Tagaki, 1971; Fiehn and Peter, 1971; Meissner and Fleischer, 1971; MacLennan et al., 1971; Waku et al., 1971; Owens et al., 1972; Marai and Kuksis, 1973). Phosphatidylcholine (50–70%) and phosphatidylethanolamine constitute the large portion of phospholipids, but sphyngomyelin, phosphatidylserine, phosphatidylinositol and plasmologen compounds (Waku et al., 1971; Owens et al., 1972; Marai and Kuksis, 1973) are also present. Neutral lipids include cholesterol, cholesterol esters, triacylglycerols and free fatty acids. It was recognized in early studies (Kielley and Meyerhof, 1950) that phospholipids are required for the enzymatic activity of muscle microsomes which were later identified with vesicular fragments of SR membrane (Ebashi and Lipmann, 1962). Subsequent studies demonstrated that phospholipase treatment and removal of the digestion products interfere with the hydrolytic activity of Ca^{2+}-ATPase (Martonosi et al., 1968; Fiehn and Hasselbach, 1970; MacLennan et al., 1971). Addition of phospholipid to the depleted vesicles restores the hydrolytic activity (Martonosi et al., 1971; Meissner and Fleischer, 1972).

Another approach to the study of lipid requirement for the SR ATPase is the technique introduced by Warren et al. (1974a, b), Hesketh et al. (1976), consisting of lipids exchange in the presence of cholate and subsequent gradient centrifugation of the vesicles subjected to exchange. It was found by this technique that the phospholipid requirement is not specific. This was also demonstrated by Knowles et al. (1976), and Le Maire et al. (1976). Quantitatively the requirement is limited to approximately 30 phospholipid molecules per ATPase chain. All or nearly all these phospholipids can be replaced by nonionic detergents with a preference for detergents with a polar moiety of limited size (Dean and Tanford, 1978).

It is apparent then that the Ca^{2+}-ATPase, an anphiphilic molecule itself, requires interaction with other nonspecific anphiphilic

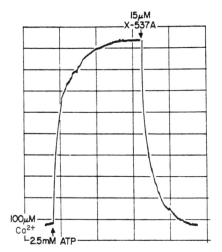

Fig. 3. *Ca²⁺ uptake and subsequent release of Ca²⁺ obtained by adding ATP and the iono-phore X-537 A to SR vesicles (Scarpa et al., 1972).*

molecules to acquire an active conformation. The affinity of the enzyme for native phospholipids is greater than for nonionic detergents.

THE CALCIUM PUMP OF SR

It was discovered independently by Hasselbach and Makinose (1961, 1963) and Ebashi and Lipmann (1962) that the main functional feature of SR vesicles is their ability to accumulate Ca^{2+} in the presence of ATP. When ATP is added to SR vesicles in a medium containing Mg^{2+} and a neutral pH buffer, the vesicles take up Ca^{2+} from the medium with a time curve reaching asymptotic levels in 60–90 s (Fig. 3). If the passive permeability of the SR membrane to Ca^{2+} is increased by the addition of divalent cations ionophores, the accumulated Ca^{2+} is rapidly released (Scarpa *et al.*, 1972).

The energy required for Ca^{2+} transport and establishment of a Ca^{2+} gradient across the SR membrane is derived from hydrolytic cleavage of the ATP terminal phosphate. This was first established by an experimental expedient utilizing the effect of oxalate or phosphate to prolong the brief transport activity normally obtained after the addition of ATP (Hasselbach and Makinose, 1961, 1963; Martonosi and Feretos, 1964; Weber, 1966). In the absence of precipitating anions, the ATP-driven Ca^{2+} influx is impared when the Ca^{2+} concentration in the vesicles lumen rises to the millimolar range

("Back inhibition"). Formation of calcium oxalate or phosphate complexes reduces the high Ca^{2+} concentration produced in the lumen of the vesicles as a consequence of active transport. Therefore, "back inhibition" of the Ca^{2+} pump is prevented, and both transport and hydrolytic activity proceed to an extent permiting P_i determination by usual colorimetric methods. Following accumulation of large amounts of Ca^{2+}, calcium oxalate and (or) calcium phosphate reach their limits of solubility inside the vesicles and calcium oxalate (Hasselbach, 1964; Deamer and Baskin, 1969; Agostini and Hasselbach, 1971; Carsten and Reedy, 1971) or calcium phosphate crystals (de Meis et al., 1974) may be visualized by electron microscopy. The catalytic and transport cycle of SR ATPase is initiated by the transfer of the ATP terminal phosphate of ATP into the protein, yielding a phosphorylated enzyme intermediate (Makinose, 1967, 1969; Yamamoto and Tonomura, 1967). Following phosphorylation, Ca^{2+} is translocated across the membrane and the intermediate phosphoprotein undergoes hydrolytic cleavage. In steady-state conditions Ca^{2+} transport and ATP hydrolysis are related with a molar ratio of $2:1$.

Under suitable conditions, the entire process of Ca^{2+} transport can be reversed and the very same ATPase involved in Ca^{2+} transport is able to catalyse the synthesis of ATP from ADP and P_i in a process coupled with the release of Ca^{2+}. When SR vesicles are previously loaded with Ca^{2+} and then incubated in a medium containing EGTA, Ca^{2+} flows out of the vesicles at a slow rate due to the low Ca^{2+} permieability of the membrane. Barlogie et al. (1971), Makinose and Hasselbach (1971) and Makinose (1972), first observed that the Ca^{2+} efflux is sharply increased when Mg^{2+}-ADP and P_i are added to the incubation medium. The increment of Ca^{2+} efflux is preceeded by phosphorylation of the enzyme by P_i (Makinose, 1972; Yamada et al., 1972) and coupled with ATP synthesis. For every two calcium ions released from the vesicles, one ATP molecule is synthesized. Another index of the reversal of the Ca^{2+} pump is the ATP $\rightleftharpoons P_i$ exchange reaction. In 1971, Makinose observed that when SR vesicles are incubated in a medium containing 0.2 mM $CaCl_2$, ATP, ADP, Mg^{2+} and $^{32}P_i$, during the initial incubation intervals, Ca^{2+} was accumulated by the vesicles at the expense of ATP hydrolysis and a Ca^{2+} concentration gradient was built up until a steady state was reached in which the Ca^{2+} efflux was balanced by the ATP-driven influx. When this condition was reached, a steady rate of exchange between $^{32}P_i$ and the γ-phosphate of ATP was observed. The ATP $\rightleftharpoons P_i$ exchange indicates that the ATPase operates simultaneously in the forward (ATP hydrolysis) and reverse (ATP synthesis) directions. From this finding it was inferred that when the steady state between Ca^{2+} influx and Ca^{2+} efflux was reached, the energy derived from the hydrolysis of ATP was used to maintain the gradient and the energy derived from Ca^{2+} gradient was used for the synthesis of $[\gamma\text{-}^{32}P]$ ATP from ADP and $^{32}P_i$.

The evidences available in 1972 led to the conclusion that the SR ATPase would be able to interconvert osmotic and chemical energy (Barlogie *et al.*, 1971; Makinose and Hasselbach, 1971; Makinose, 1971, 1972; Deamer and Baskin, 1972; Panet and Selinger, 1972; Yamada *et al.*, 1972; Yamada and Tonomura, 1973; de Meis and Carvahlo, 1974). During the process of Ca^{2+} accumulation, the chemical energy derived from the hydrolysis of ATP would be used to build up a Ca^{2+} concentration gradient across the vesicles membrane and in the reverse process, the osmotic energy derived from the Ca^{2+} gradient would be used for the synthesis of ATP. However, new evidence has led to a second interpretation which is also consistent with the earlier data, namely that the role of the Ca^{2+} gradient is less important than the energy provided by the binding of Ca^{2+} to two different sites of the enzyme located on the outer and inner surface of the SR membrane respectively (Masuda and de Meis, 1973; de Meis and Carvahlo, 1974; de Meis and Masuda, 1974; de Meis and Sorenson, 1975; Knowles and Racker, 1975; Post *et al.*, 1975; Taniguchi and Post, 1975; Carvahlo *et al.*, 1976; de Meis, 1976; Kuriki *et al.*, 1976; de Meis and Tume, 1977; de Meis *et al.*, 1980). According to this new view, transport ATPases would be able to interconvert binding energy into chemical energy.

The following sections of this chapter will describe the partial reactions of the catalytic and transport cycle in the forward and reverse directions, in the light of recent experimental findings.

CALCIUM BINDING AND ENZYME ACTIVATION

The Ca^{2+} pump is activated by very low Ca^{2+} concentrations in the reaction medium. Experimental determination of such concentrations is rendered difficult by the presence of Ca^{2+} contaminations of the reaction mixture, and by rapid variations of these low Ca^{2+} concentrations by the action of the pump. These difficulties have rendered necessary the use of EGTA (Ethylene glycol *bis* (β-aminoethyl ether) N, N', N', N'-tetraacetic acid) which is a chelating agent displaying a much greater affinity for Ca^{2+} than for Mg^{2+}. Although the values given in the literature for the Ca-EGTA dissociation constant (5×10^{-7}–4×10^{-6} M) are not identical (Schmid and Reilly, 1957; Schwartzenbach *et al.*, 1957; Holloway and Reilly, 1960; Ebashi, 1961; Ogawa, 1968; de Meis and Hasselbach, 1971), they are sufficiently low to reduce the Ca^{2+} concentration in the medium below levels producing activation of the pump. Addition of increasing concentrations of $CaCl_2$ then yields a Ca-EGTA buffer for control of the free Ca^{2+} concentration and stepwise activation of the SR vesicles, while total calcium is sufficiently high to support the activity of the pump for experimentally useful times. It is noteworthy that EGTA does not penetrate the SR membrane (Weber *et al.*, 1966) and its Ca^{2+} buffering effect is limited to the medium outside the vesicles.

Generally, it is reported in the literature that activation of the SR pump displays a sigmoidal dependence on the Ca^{2+} concentration in the medium, with half-maximal saturation at approximately 0.5 μM Ca^{2+} (de Meis, 1971; The and Hasselbach, 1972a, b; Gattass and de Meis, 1975; Vianna, 1975; Suko and Hasselbach, 1976; Neet and Green, 1977). Such a sigmoidal dependence, as well as the 2:1 molar ratio observed between calcium transport and ATP hydrolysis, suggest binding of two calcium ions to the same enzyme unit. The sigmoidal dependence of functional parameters on Ca^{2+} concentration, however, does not distinguish between a cooperative mechanism involving two interacting sites, and a mechanism requiring occupancy of a second binding site on the enzyme for activation of the pump. This question can be answered only by direct measurements of calcium binding.

The presence of various classes of calcium binding sites in SR vesicles was demonstrated by several investigators (Carvahlo, 1966; Chevalier and Butow, 1971; Fiehn and Migala, 1971; Meissner *et al.*, 1973; Ikemoto, 1974, 1975; Chin and Haynes, 1977). Of the various classes, the high affinity class ($K_d < 1$ μM) can be attributed to binding sites on the ATPase protein (Meissner *et al.*, 1973), which are exposed on the outer surface of the vesicles. This class of binding sites is involved in activation of the enzyme. Unfortunately, measurement of binding at such low concentrations presents serious difficulties and most of the early studies report few experimental points in the Ca^{2+} concentration range in which this class is saturated. Therefore the experimental data were fitted with linear Scatchard plots, without taking into consideration alternative mechanisms of binding. Some of the more recent studies (Ikemoto, 1975; Kalbitzer *et al.*, 1978; Inesi *et al.*, 1980a) indicate that calcium binding to high affinity sites occurs through a cooperative mechanism which is fully consitent

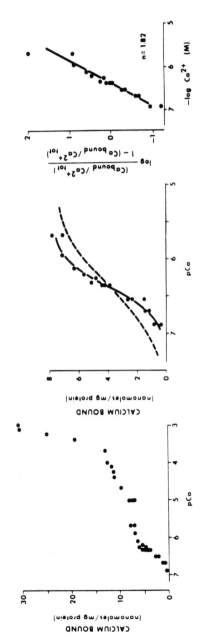

Fig. 4. Calcium binding to SR vesicles in the absence of ATP. The graph on the left side shows various classes of binding sites. The cooperative character of the high affinity binding is demonstrated in the other two graphs [Inesi et al., 1980a].

with the Ca^{2+} concentration dependence of functions parameters (Fig. 4).

The high affinity calcium binding sites correspond to 8–10 nmol/mg SR protein. This value is twice as high as the maximal level of phosphorylated enzyme intermediate found in steady-state conditions in the presence of ATP, and corresponds to two calcium sites per phosphorylated enzyme unit. The stochiometric account is less definite on whether one or two 100 000 mol.wt polypeptide chains should be attributed to such a unit, since 1 mg of SR protein contains approximately 8 nmol of 100 000 mol.st ATPase chains. The stochiometric relationship of high affinity calcium sites, phosphorylation sites and 100 000 polypeptide chains raises the possibility that in certain conditions the ATPase acquires the character of a dimeric function unit.

Ample experimental evidence based on EPR spectrosopy of spin labelled SR (Coan and Inesi, 1977; Champeil et al., 1978), kinetics of -SH reactivity (Ikemoto et al., 1978; Murphy, 1978), and measurements of intrinsic fluorescence (Dupont and Leigh, 1978) indicates that calcium binding to high affinity sites is accompanied by a protein conformational change. This change which is fully reversible, was attributed by Inesi et al. (1980a) to a mechanism of cooperative calcium binding and enzyme activation, as in:

$$E + Ca^{2+} \rightleftharpoons E.Ca \rightleftharpoons E'.Ca + Ca^{2+} \rightleftharpoons E''.Ca_2 \qquad (1)$$

The functional significance of other classes of calcium sites detected in the absence of ATP, is less clear. A class of intermediate ($K_d > 10^{-5}$ M) affinity has been detected at $0°$ C (Sumida and Tonomura, 1974; Ikemoto, 1975) and attributed to a temperature effect whereby half of the high affinity sites would undergo a reduction in affinity when the temperature is decreased from $22°$ C to $0°$ C. However, recent measurements by Dupont (1980) do not show any significant temperature effect on high affinity binding.

Finally, a low affinity ($K_d \approx 10^{-3}$ M) class of large binding capacity can be detected at millimolar Ca^{2+} concentrations. This class is likely to include nonspecific sites contributed by protein and lipid components of the SR membrane.

It is noteworthy that during the ATPase catalytic cycle the high affinity sites are converted to a low affinity state following enzyme phosphorylation with ATP. As it will be discussed in a later section of this review, specific functional effects are detected when the Ca^{2+} concentration is raised to the millimolar range, permitting titration of the sites converted to low affinity. In this connection, the nonphosphorylated form of the enzyme may be also considered to be in equilibrium between high affinity and low affinity states, as in:

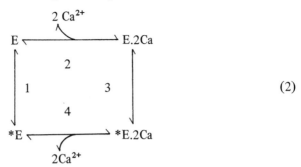

$$(2)$$

where $K_2 \approx 10^6$ M^{-1} and $K_4 \approx 10^3$ M^{-1}. In this case, the fraction of specific ATPase sites in a low affinity state (*E and/or *E.2Ca) depends on the values given to the equilibrium constants K_1 and K_2, and on the Ca^{2+} concentration.

SUBSTRATE SPECIFICITY AND BINDING

The physiological nucleotides ATP, ITP, GTP, CTP and UTP (Hasselbach, 1964; Makinose and The, 1965; Makinose, 1966), and other phosphorylated compounds such as acetylphosphate, p-nitrophenyl phosphate, carbamylphosphate, methylumbelliferulphosphate and furylacryloylphosphate (de Meis, 1969a, b; Friedman and Makinose, 1970; Inesi, 1971; Nakamura and Tonomura, 1978; Rossi et al., 1979; Kurzmack et al., 1980), can be utilized as substrates for the SR ATPase. Of these substrates, ATP is the most specific inasmuch as it is utilized at very low concentrations and high rates. The true substrate is the ATP.Mg complex (Weber et al., 1966; Yamamoto and Tonomura, 1967; Vianna, 1975.

The ATP concentration dependence for the ATPase reaction has been studied first by measuring the hydrolytic reaction (e.g. release of P_i) in steady state conditions. A complex dependence was found in this case, with a first activation obtained with ATP concentrations ranging between 0.5 and 50 μM, and a further rise of activity at higher concentrations (Inesi et al., 1967; Yamamoto and Tonomura, 1967; de Meis, 1971; The and Hasselbach, 1972a, b; de Meis and de Mello, 1973; Froehlich and Taylor, 1975; Vianna, 1975; Yates and Duance, 1976; Dupont, 1977; Ribeiro and Vianna, 1978; Shikegawa et al., 1978; Scofano et al., 1979; Taylor and Hattan, 1979).

The observed dependence cannot be fitted according to simple Michaelis-Menten kinetics. It is possible that the effect of high ATP is related to activation of partial reactions which follow the initial enzyme phosphorylation of ATP and contribute to rate limitation with respect to the enzyme turnover. Consistent with this suggestion, it is found that the ATP concentration dependence of the initial enzyme phosphorylation is limited to the lower concentration range (Kanazawa *et al.*, 1971; Froehlich and Taylor, 1975, 1976; Verjovski-Almeida *et al.*, 1978; Scofano *et al.*, 1979), but the hydrolytic cleavage of the phosphoenzyme formed in the initial part of the reaction, is activated by high ATP concentrations (de Meis and de Mello, 1973; Froehlich and Taylor, 1975; Verjovski-Almeida and Inesi, 1979).

It is clear that the ATP concentration dependence (K_m) for the SR ATPase which is a multistep reaction, is related but does not reflect directly the concentration dependence of ATP binding to the enzyme. This is due to contributions of rate constants for partial reactions preceeding and following that chosen as an experimental parameter. Direct measurements of ATP binding are quite difficult from the experimental point of view. Recent measurements by Pang and Briggs (1977) and Dupont (1977) have detected nucleotide binding sites in the 10^{-5} M range, and suggest the presence of lower affinity sites which can be titrated with some experimental scatter at high ATP concentrations.

ENZYME PHOSPHORYLATION WITH ATP

Early studies on the functional characterization of SR vesicles described an ATP $\rightleftharpoons$ ADP exchange consisting of (^{14}C)ATP formation following incubation of (^{14}C)ADP and ATP with the vesicles in the presence of Ca^{2+} and Mg^{2+} (Ebashi and Lipmann, 1962; Hasselbach and Makinose, 1962). This exchange of the ADP moiety is related to a phosphorylation reaction involving transfer of the ATP terminal phosphate onto the enzyme protein as in:

$$ATP + E \longleftrightarrow ADP + EP \tag{3}$$
$$(^{14}C)ATP \qquad (^{14}C)ADP$$

A direct demonstration of this reaction was obtained by the use of $[\gamma\text{-}^{32}P]$ATP and recovery of an acid stable $[^{32}P]$ phosphoenzyme

(Makinose, 1967, 1969; Yamamoto and Tonomura, 1967, 1968). It was later reported that the transferred phosphoryl group is covalently bound to the β-carboxyl group of an aspartyl residue of the acid denatured protein (Degani and Boyer, 1973). The phosphorylated aspartyl residue is found in a tripeptide sequence which is common to Ca^{2+} and Na^+-K^+ activated ATPases (Bastide *et al.*, 1973).

Both ADP $\rightleftharpoons$ ATP exchange and enzyme phosphorylation display a Ca^{2+} dependence identical fo that of the overall ATPase reaction, corresponding to the Ca^{2+} concentration range (< 1 μM) required for titration of the high affinity calcium sites of the ATPase protein. Furthermore, Mg^{2+} is required as part of the ATP.Mg complex which is the true substrate for the phosphorylation reaction. Finally, if the phosphorylation is quenched by Ca^{2+} chelation with EGTA and the protein is then denatured by addition of acid at sequential times, hydrolytic cleavage of the phosphoenzyme proceeds with a rate constant compatible with the over-all turnover of the ATPase reaction (Inesi *et al.*, 1970). Therefore, it is apparent that enzyme phosphorylation with ATP is an intermediate step in the mechanism of Ca^{2+}-dependent utilization of ATP by SR ATPase, as in:

$$E \underset{1}{\overset{2Ca^{2+}}{\rightleftharpoons}} E.Ca_2 \underset{2}{\overset{ATP}{\rightleftharpoons}} ATP.E.Ca_2 \underset{3}{\rightleftharpoons} ADP.EP.Ca_2 \underset{n}{\rightleftharpoons} ADP + 2\ Ca^{2+} + E + P_i \quad (4)$$

where 1 and 2 represent Ca^{2+} binding to activating sites and formation of the substrate–enzyme complex, 3 is the phosphorylation reaction, and *n* signifies a number of subsequent steps for completion of the ATPase catalytic cycle.

The amount of phosphoenzyme measured following addition of ATP represents the steady-state levels of different intermediate species in the enzyme cycle. The phosphoenzyme reaches significant levels owing to a favorable ratio between the rate constants of the reaction causing formation, and those causing disappearance of the intermediate.

In optimal conditions with respect to ATP and Ca^{2+} concentrations, temperature and pH, maximal steady state levels of phosphoenzyme range between 3 and 5 nmol/mg protein when SR vericles are used (Froehlich and Taylor, 1975; Verjovski-Almeida *et al.*, 1978). However, at variance with the constant phosphoenzyme levels obtained with saturating ATP, different steady state levels of phosphoenzyme are obtained with saturating ITP or GTP (Souza and de Meis, 1976; Verjovski-Almeida and de Meis, 1977; de Meis and Boyer, 1978). When calcium binding is limited to the sites in the high affinity state, the level of phosphoenzyme obtained with ITP or GTP

TABLE I.

Temperature	Substrate	Ca^{2+} (mM)	Phosphoenzyme (μmol/g of protein)
37° C	ATP	0.1	3.97–4.46
37° C	ATP	20.0	4.22–4.25
2° C	ATP	0.1	3.70–4.34
2° C	ATP	20.0	4.05–4.18
37° C	GTP	0.1	2.37–2.54
37° C	GTP	20.0	7.72–7.91
2° C	GTP	0.1	6.24–6.29
2° C	GTP	20.0	7.60–7.81

Phosphoenzyme levels obtained incubating purified ATPase (MacLennan *et al.*, 1971) with 1.0 mM [γ-^{32}P] ATP and [γ-^{32}P] GTP in the presence of 30 mM Tris-Maleate (pH 7.4) and 5 mM $MgCl_2$. Protein concentration: 0.15–0.30 mg/ml. Incubation was carried out for times required to obtain maximal steady state levels of phosphoenzyme in each set of conditions: 5 s (ATP) or 20 s (GTP) at 37° C, and 20 s (ATP) or 60 s (GTP) of 2° C.

The reaction was stopped by solution of 0.3 M perchloric acid, and the denatured enzyme was washed by repeated centrifugations and resuspensions. Phosphoenzyme was estimated by determination of radioactivity [^{32}P] and protein in the washed samples (de Meis and Inesi, unpublished observations).

is lower than that obtained with ATP. This indicates that in the steady state obtained with ITP or GTP, a significant portion of the enzyme is in a nonphosphorylated form. On the other hand, when saturation of the calcium binding sites is permitted even in their low affinity state ("back inhibition") the phosphoenzyme levels obtained with ITP or GTP are approximately twice as high as observed with ATP in the same conditions (Table I). Purified ATPase yields phosphoenzyme levels that are approximately 20% higher than those obtained with SR vesicles.

Considering that 1 mg protein contains 8–10 nmol of monomeric ATPase chains, the maximal levels of phosphoenzyme obtained with GTP are consistent with phosphorylation of single chains. This high value verifies the purity of the ATPase preparations since it demonstrates the presence of a potential phosphorylation site per 100 000 mol.wt chain. It is not clear why the maximal levels of phosphoenzyme obtained with ATP cannot be increased significantly when cleavage of the phosphoenzyme is markedly inhibited with high Ca^{2+} at low temperatures. These findings raise the possibility of oligomerization or alternate site reactivity in the presence of ATP, but not in the presence of GTP and high Ca^{2+}.

We have already mentioned that the enzyme state reacting with ATP is $E.Ca_2$. Therefore, the phosphorylation reaction is expected to be sensitive to the extent of enzyme saturation with Ca^{2+}, e.g. partition of the enzyme population in the E (or *E) and $E.Ca_2$ states.

It should be also pointed out that following the phosphorylation reaction, the phosphoenzyme acquires different states as a consequence of other sequential reactions of the catalytic and transport cycle. These various states are distinguished by specific features such as sensitivity to ADP. As it will be explained in later sections, enzyme phosphorylation is accompanied by change in orientation and reduction in affinity of the specific calcium sites, whereby Ca^{2+} can be released inside the vesicles. When the Ca^{2+} concentration in the lumen of the vesicles increases to and above the millimolar level, the transport sites remain occupied by calcium with consequent inhibition of the enzyme turnover ("back inhibition"). It is only the phosphoenzyme with transport sites occupied by calcium, which is sensitive to ADP, permitting reversal of the pump and formation of ATP (de Meis and Carvahlo, 1974; de Meis and Sorenson, 1975; Knowles and Racker, 1975; de Meis and Tume, 1977; Plank et al., 1979; Shigekawa and Dougherty, 1978a, b).

Owing to heterogeneity of the enzyme states, and to variable contributions of different rate constants of intermediate steps, kinetic experiments on phosphoenzyme formation with ATP may yield apparent inconsistencies. Both, steady state phosphoenzyme levels and transient time profiles of enzyme phosphorylation are sensitive to experimental variables such as K^+, temperature, pH and Ca^{2+} concentration inside the vesicles. These variable modify in different manners the rate constants of intermediate steps following phosphorylation of the enzyme by the nucleotide, including phosphoenzyme cleavage. Furthermore, the step limiting the enzyme turnover is changed by these variables.

An apparent inconsistency is obtained when K_m values for the phosphorylation reaction are derived from measurements of phosphoenzyme levels in "back inhibited" vesicles, yielding values as low as 10^{-8} M (de Meis and de Mello, 1973; Yates and Duance, 1976). In fact, "back inhibition" is produced in intact vesicles within a few catalytic cycle, and the phosphoenzyme builds up to levels which are not directly related to the nucleotide concentration and the initial saturation of the enzyme with substrate. Higher K_m values ($10^{-6}-10^{-5}$ M) are obtained if measurements of initial velocities of phosphorylation are used (Kanazawa et al., 1971; Froehlich and Taylor, 1975; Verjovski-Almeida et al., 1978; Scofano et al., 1979).

When ATP is added to the enzyme preincubated with Ca^{2+} ($E.Ca_2$), the phosphorylation reaction occurs quite rapidly and experimental resolution of its time profile requires rapid mixing techniques. A rate constant of 80–150/s has been obtained for this reaction at 25° C, in media containing optimal concentrations of cofactors and neutral pH buffers (Froehlich and Taylor, 1975; Verjovski-Almeida et al., 1978; Scofano et al., 1979). These values were derived from experimental data obtained with ATP concentrations between 1 and 10 μM. Since time resolution cannot be obtained at high ATP concentrations, it is not clear whether this rate constant is valid for all ATP concentrations. This is a serious limitation since other experimental parameters show a diphasic dependence on the ATP concentration.

The phosphorylation time profile changes depends on whether the enzyme is or is not allowed to equilibrate with Ca^{2+} before addition of the substrate. If the reaction is started by adding ATP and Ca^{2+} to the enzyme preincubated with EGTA (rather than ATP to the enzyme preincubated with Ca^{2+}) a slower time curve of phosphorylation is observed (Sumida et al., 1978; Scofano et al., 1979; Inesi et al., 1980a). This is due to a time limit imposed by the Ca^{2+}-induced activation of the enzyme. This activation is accompanied by a protein conformational change which can be shown by measurements of intrinsic fluorescence (Dupont and Leigh, 1978), EPR spectra of spin labelled SR (Coan and Inesi, 1977), and -SH reactivity (Ikemoto et al., 1978; Murphy, 1978). Accordingly, the reaction sequence may be considered to proceed as in:

$$*E \rightleftharpoons E \overset{2Ca^{2+}}{\rightleftharpoons} E.Ca_2 \overset{ATP}{\rightleftharpoons} ATP.E.Ca_2 \rightleftharpoons ADP.EP.Ca_2 \qquad (5)$$

or

$$E \overset{Ca^{2+}}{\rightleftharpoons} E.Ca \rightleftharpoons E'.Ca \overset{Ca^{2+}}{\rightleftharpoons} E''.Ca_2 \overset{ATP}{\rightleftharpoons} ATP.E''.Ca_2 \rightleftharpoons ADP.EP.Ca_2 \qquad (6)$$

where the asterisk and the prime signs signify specific conformational states. In (5) the transformation is related to a Ca^{2+}-induced displacement of the equilibrium between a low affinity (*E) and a high affinity form (E) of the enzyme for Ca^{2+} (Scofano et al., 1979). In (6) the transformation is considered to be an enzyme transition following calcium binding and yielding enzyme activation (Dupont and Leigh, 1978; Guillain et al., 1980; Inesi et al., 1980a, b).

It is of interest that fitting the phosphorylation time profiles requires attribution of a 100/s rate constant to the enzyme transformation (Inesi et al., 1980a, b). This constant is quite faster than

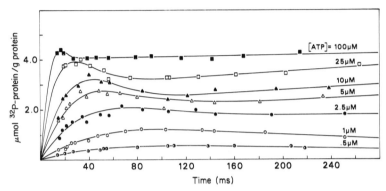

Fig. 5. *ATP concentration dependence of transient phosphorylation of sarcoplasmic reticulum vesicles. Reproduced from Froehlich and Taylor (1975).*

that observed (5/s) by fluorescence changes upon addition of Ca^{2+} in the absence of ATP (Dupont and Leigh, 1978). This indicates that the Ca^{2+}-induced transformation proceeds at faster rates in the presence of ATP (Carvalho *et al.*, 1976; Souza and de Meis, 1976; Vieyra *et al.*, 1979).

The specificity of the activation step is also revealed (Scofan *et al.*, 1979) by: (a) insensitivity of the phosphorylation curve to pH variations from 6.0 to 7.4 when ATP is added to $E.Ca_2$, as opposed to its sensitivity when ATP and Ca^{2+} are added to E; (b) different K_m values obtained with respect to ATP in the former (3 μM) as opposed to the later (50 μM) condition. The different dependence on ATP concentration suggests a secondary effect of ATP on the Ca^{2+}-induced enzyme activation (e.g. conformational transition), perhaps related to the diphasic ATP dependence of the enzyme turnover in steady-state conditions. In fact, EPR spectroscopy of spin labelled SR detects not only a Ca^{2+} effect, but also changes due to ATP binding (Landgraff and Inesi, 1969; Coan and Inesi, 1977). A less pronounced effect is produced by ITP, GTP and other substrates.

Another interesting feature of the time profiles of enzyme phosphorylation was first pointed out by Froehlich and Taylor (1975). These authors noted that the rise of phosphoenzyme following addition of ATP to $E.Ca_2$ soon undergoes a slight downward inflection (overshoot), followed by a trough and a secondary rise (Fig. 5). These fluctuations are very small and can hardly be detected at 25° C (Verjovski-Almeida *et al.*, 1978; Scofano *et al.*, 1979), while they can be obtained more consistently at low temperatures (Kanazawa *et al.*, 1971) owing to a much higher temperature dependence of

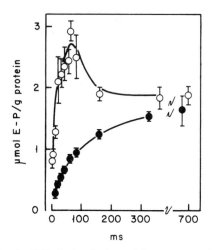

Fig. 6.   *Phosphorylation by ITP. Preincubation of the enzyme with Ca^{2+} or with EGTA. Syringe A contained 1.0 mg/ml of leaky vesicles, 30 mM Tris—maleate buffer (pH 7.4) and either 0.1 mM $CaCl_2$ (o) or 0.2 mM EGTA (•). Syringe B contained 30 mM Tris—maleate buffer (pH 7.4), 10 mM $MgCl_2$ and either 10 mM [γ-^{32}P] ITP (o) or 1.0 mM [γ-^{32}P] ITP plus 0.3 mM $CaCl_2$ (•). Equal volumes of solutions contained in two syringes were forced mixed together through a capillary tube connected by means of "γ" junction. The other end of the capillary was immersed in the quenching solution (4 mM P_i in a 250 mM perchloric acid solution). Different reaction times were obtained by varying bothe the length of the capillary tube and the flow rate of injection. The reactions were performed at room temperature (22—25° C). The values represent the average ± SE of 4 experiments. Reproduced from Scofano et al. (1979).*

phosphoenzyme cleavage as compared to phosphoenzyme formation. In this case, the initial downward inflection can be simply attributed to initiation of hydrolytic sctivity following the primary build up of phosphorylated enzyme intermediate. The secondary rise is then related to "back inhibition" of hydrolytic activity following the rise in the Ca^{2+} concentration inside the vesicles.

A more pronounced overshoot is observed at pH 7.4 when ITP, instead of ATP, is added to $E.Ca_2$ (Scofano *et al.*, 1979). It was suggested by these authors that in this case the downward inflection may be due to a slower rate of *E to E transition [see (5)] obtained in the presence of ITP (as compared with ATP), and a consequent time limit for recycling of the enzyme.

The "overshoot" phenomenon is not observed when Ca^{2+} and either ATP or ITP are added to the enzyme preincubated with EGTA (Fig. 6).

Reversal of enzyme phosphorylation with ATP, is ATP formation by phosphoryl transfer from the phosphoenzyme to ADP. Experiments on ADP $\rightleftharpoons$ ATP exchange suggest that the rate constant for this reaction is of the same order of magnitude as that of the forward reaction (Verjovski-Almeida *et al.*, 1978). Phosphorylation time profiles were fitted satisfactorely assuming values ranging from 50/s to 20/s for this reaction, while the values used for the forward directions were 150–80/s (Froehlich and Taylor, 1975; Inesi *et al.*, 1980a, b). These constants are also consistent with the observed time dependence of phosphoenzyme disappearance upon addition of ADP in the presence of "back inhibition" (high Ca^{2+} inside the vesicles).

CALCIUM SITE TRANSFORMATION FOLLOWING ENZYME PHOSPHORYLATION

Several lines of experimentation indicate that following enzyme phosphorylation, the high affinity ("activating") sites undergo transformations which represents a central step for the catalytic and transport mechanism.

One type of transformation is related to the sideness of the sites with respect to the SR membrane, as revealed by the kinetics of Ca^{2+} dissociation upon addition of EGTA.

In the absence of ATP, the apparent rate constant for Ca^{2+} dissociation from the high affinity sites is 15/s (Sumida *et al.*, 1978). On the other hand, when EGTA is added to vesicles loaded with calcium in the presence of ATP, calcium remains associated with the vesicles for several seconds (Chiesi and Inesi, 1979). In fact, not only calcium accumulated inside the vesicles in steady state conditions, but even calcium bound to high affinity ("activating") sites becomes non accessible to EGTA following enzyme phosphorylation with ATP (Kurzmack *et al.*, 1977; Verjovski-Almeida *et al.*, 1978). It was demonstrated by a variety of rapid quench experiments performed at 25° C that within the first enzyme cycle following addition of ATP, phosphorylation of one enzyme unit produces translocation (or occlusion) of two calcium sites, before the phosphoenzyme proceeds through a time limiting step to hydrolytic cleavage (Inesi *et al.*, 1978; Chiesi and Inesi, 1979). The sequence of events consisting of initial phosphorylation and change in calcium sites orientation, followed by steady state P_i production and calcium transport, is shown in Fig. 7.

In experiments performed at 0° C with Ca^{2+} permeable vesicles, Dupont (1980) found that calcium bound to phosphorylated ATPase

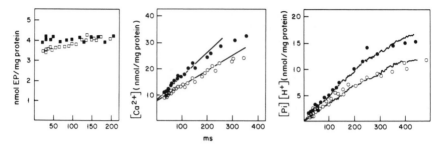

Fig. 7. *Transient kinetics of phosphoenzyme formation, Ca²⁺ uptake and hydrolytic activity measured in parallel experiments. Sarcoplasmic reticulum vesicles are incubated in the presence of calcium and the reaction is initiated by the addition of ATP at a final concentration of 0.1 mM (open symbols) or 1.0 mM (closed symbols). Left: the reaction is quenched with acid and the acid-stable phosphoenzyme is measured. Middle: the reaction is rapidly quenched with an excess of EGTA and then the intact vesicles are separated from the medium by filtration. The Ca²⁺ accumulated can be estimated by marsuring the calcium remaining in the filtrate. Right: the reaction is quenched with acid and the P_i concentration in the medium is measured (circles). Alternatively, hydrolytic activity is estimated by measuring H^+ production which accompanies phosphoenzyme cleavage. The continuous lines (right) represent oscilloscope tracings in stopped-flow measurements of absorbance changes undergone by the pH indicator phenol red. Reproduced from Verjovski-Almeida et al. (1978).*

is not accessible either from the outer or the inner surface of the membrane. At low temperature, nearly all the calcium sites of the phosphorylated enzyme are maintained in an occluded state pre-ceeding complete translocation and release of Ca²⁺ in the lumen of the vesicles. In this state the phosphoenzyme reacts readily with ADP leading to formation of ATP and returning the calcium sites to exposure on the outer surface of the vesicles (Sumida *et al.*, 1978; Dupont, 1980).

Another type of transformation undergone by the calcium sites following enzyme phosphorylation, is a loss of affinity which renders possible the release of Ca²⁺ against a concentration gradient inside the vesicles, as suggested by Hasselbach (1964). Initial experimental evidence for this transformation was obtained when it became apparent that the phosphoenzyme formed by reacting the enzyme with P_i (Masuda and de Meis, 1973; de Meis and Masuda, 1974; de Meis, 1976) acquires the ability to revert the ATPase cycle to ATP only in the presence of millimolar, rather than micromolar Ca²⁺ (de Meis and Carvalho, 1974; de Meis and Sorenson, 1975; Knowles and Racker, 1975; de Meis and Tume, 1977; Plank *et al.*, 1979).

Direct evidence for the change in affinity was obtained by Ikemoto (1974, 1975, 1976) with an ATPase preparation which, as opposed to intact vesicles, does not permit net accumulation of calcium.

In the presence of low (micromolar) Ca^{2+}, it was found that upon transient ATPase phosphorylation with limiting amounts of ATP, Ca^{2+} is released from the enzyme, to be bound again soon after hydrolytic cleavage of the phosphoenzyme. On the other hand, titration of the sites with high (millimolar) Ca^{2+} produces ATPase inhibition, e.g. the "back inhibition" observed in intact Ca^{2+} loaded vesicles.

A final piece of evidence for the change in affinity of the calcium sites was obtained in spin label experiments (Coan *et al.,* 1979) demonstrating that the Ca^{2+} concentration required to obtain the "calcium conformation" of the ATPase, shifts from the micromolar to the millimolar range upon phosphorylation of the enzyme with ATP.

At this point then, the ATPase reaction sequence may be written to include the following steps:

$$\overset{ATP}{E.Ca_2 \rightleftharpoons} ATP.E.Ca_2 \rightleftharpoons \overset{ADP}{ADP.E{\sim}P.Ca_2 \rightleftharpoons} {}^*E\text{-}P.Ca_2 \rightleftharpoons {}^*E\text{-}P + 2\ Ca_i^{2+} \quad (7)$$

where $E{\sim}P$ represents the ADP sensitive phosphoenzyme, and $^*E\text{-}P$ the phosphoenzyme state permitting Ca^{2+} release inside the vesicles.

HYDROLYTIC CLEAVAGE OF THE PHOSPHOENZYME

Completion of the catalytic and transport cycle requires hydrolytic cleavage of the phosphorylated enzyme intermediate and return of the enzyme to a state with high affinity and outward orientation of the calcium binding sites. In steady state conditions this step is manifested by production of P_i and H^+ at velocities corresponding to the enzyme turnover.

In rapid kinetic experiments initiated by addition of ATP to $E.Ca_2$, P_i and H^+ production can be shown to start following the initial burst of phosphoenzyme formation and calcium translocation (Inesi *et al.,* 1978). Changes in the "shape" of P_i production may be observed (Kanazawa *et al.,* 1971; Froehlich and Taylor, 1975, 1976; Takisawa and Tonomura, 1978) due to shifts in rate-limiting steps during the progression of the reaction. On the other hand when the reaction is carried out with solubilized ATPase (as opposed to SR vesicles) P_i production is linear, demonstrating a constant turnover of the phosphorylated enzyme intermediate (Takisawa and Tonomura, 1979).

The partial reactions related to hydrolytic cleavage of the phosphoenzyme can be studied in double-quench experiments consiting of a preliminary phosphorylation of E.Ca$_2$ with ATP, followed by a first-quench with EGTA in order to prevent further ATP utilization while hydrolytic cleavage of the enzyme is allowed to proceed. This is in turn followed by a second quench with acid at sequential times in order to stop further hydrolytic cleavage of the enzymes. In this manner the time profile of phosphoenzyme decay can be studied (Inesi *et al.*, 1970; Chiesi and Inesi, 1979).

This type of experimentation has permitted a clear demonstration of the phosphoenzyme distribution in two states: one state is sensitive to ADP for formation of ATP in the reverse direction of the cycle. The other state which is not sensitive to ADP, is the phosphoenzyme actually undergoing hydrolytic cleavage (Shigekawa and Dougherty, 1978a, b; Shigekawa and Akowitz, 1979; Takisawa and Tonomura, 1979). It is apparent that the two steps of the phosphoenzyme correspond to sequential steps of the catalytic cycle, as proposed by most laboratories involved in the study of this system.

While Ca^{2+} activates the enzyme for ATP utilization and formation of the phosphorylated intermediate, Mg^{2+} is an activator in the progression of the catalytic cycle to hydrolytic cleavage of the phosphoenzyme (Inesi *et al.*, 1970; de Meis and de Mello, 1974; Garrahan *et al.*, 1976; Shigekawa and Dougherty, 1978a, b). With respect to this activation, it was proposed by Kanazawa *et al.* (1971) that Mg^{2+} may exchange for Ca^{2+} and serve as a counter-ion in a Ca^{2+}–Mg^{2+} antiport system. However, in experiments in which Mg^{2+} was substituted with Mn^{2+} (due to the availability of a more convenient radioactive isotope of the latter), no Mn^{2+} fluxes in exchange for Ca^{2+} could be detected (Chiesi and Inesi, 1980). Presently, it is not clear whether Mg^{2+} accelerates specifically the transformation of the ADP sensitive to the ADP insensitive phosphoenzyme or the cleavage of the latter.

A "basic" ATPase which catalyse slow rates of ATP hydrolysis in the absence of Ca^{2+} and without formation of phosphorylated enzyme intermediate was noted in early studies on SR (Hasselbach, 1964). This activity is associated with native SR vesicles, but disappears upon solubilization with detergents. It is possible that the "basic" ATPase is a special form of the Ca-ATPase (Walter and Hasselbach, 1973; Froehlich and Taylor, 1976; Inesi *et al.*, 1976). In this regard, it was recently reported (Inesi *et al.*, 1980b) that modification of SR ATPase with dimethylsulphoxide renders the enzyme capable of hydrolysing the pseudosubstrate furylacryloylphosphate

(FAP) at high rates in the absence of Ca^{2+}. In this condition the enzyme is phosphorylated by FAP even in the absence of Ca^{2+}. Utilization of this substrate by the enzyme in the absence of organic solvent is Ca^{2+} dependent.

REVERSAL OF THE CATALYTIC AND TRANSPORT CYCLE

The phosphorylation and phosphohydrolase reactions are fully reversible as demonstrated by Barlogie et al. (1971). Makinose (1971, 1972) and Makinose (1971). Reversal of the Ca pump is initiated by enzyme phosphorylation by P_i. This was first demonstrated by incubating Ca^{2+} loaded vesicles with P_i, Mg^{2+} and EGTA (Makinose, 1972; Yamada et al., 1972; Yamada and Tonomura, 1973). Enzyme phosphorylation with P_i is impaired by calcium binding to the high affinity sites on the outer surface of the membrane (Masuda and de Meis, 1973; de Meis and Masuda, 1974; de Meis, 1976).

Addition of ADP to the phosphoenzyme formed in these conditions, leads to formation of ATP through phosphoryl transfer from the phosphoenzyme to ADP, and Ca^{2+} release into the outside medium (Barlogie et al., 1971; Hasselbach, 1978; Makinose, 1972; de Meis, 1976; Dupont, 1978; Hasselbach, 1978). The stochiometry of this reaction involves two calciums per ATP formed, and the reaction continues for repeated cycles as long as the Ca^{2+} concentration inside the vesicles remains high, and the Ca^{2+} concentration outside the vesicles is maintined low (Fig. 8). In addition to ADP (apparent K_m = 0.1–0.5 mM), IDP (apparent K_m = 1.0–2.0 mM) can also function as a phosphate acceptor (Hasselbach, 1978).

Another experimental evidence for reversal of the ATPase cycle is the occurence of ATP $\rightleftharpoons P_i$ exchange. This reaction is detected when the vesicles are filled with Ca^{2+} (Makinose, 1971; de Meis and Carvalho, 1974; Carvalho et al., 1976) and indicates that in the presence of ATP, low ($\sim 10^{-6}$ M) Ca^{2+} in the medium, and high (approximately millimolar) Ca^{2+} inside the vesicles, the catalytic and transport cycle proceeds in both directions. This can be illustrated using the reaction scheme proposed by Carvalho et al. (1976):

$$
\begin{array}{cccccc}
 & 2Ca^{2+} & ATP & [\gamma\text{-}^{32}P]\,ATP & ADP & \\
E & \rightleftharpoons E.Ca_2 & \rightleftharpoons & ATP.E.Ca_2 & \rightleftharpoons & E{\sim}P.Ca_2 \\
\updownarrow & & & & 2Ca_i^{2+} & \updownarrow \\
{}^*E & \longrightarrow {}^*E.P_i & \longrightarrow & {}^*E\text{-}P & \longleftarrow & {}^*E\text{-}P.Ca_2 \\
 & {}^{32}P_i & P_i & H_2O & &
\end{array}
\tag{8}
$$

170   L. DE MEIS AND G. INESI

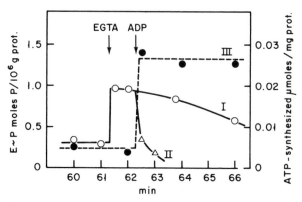

Fig. 8. *Phosphoprotein formation in SR membrane in the absence of energy rich phosphoryl substrate. The SR vesicles (0.5 mg/ml) are incubated for 1 h in a medium containing (mM): histedine 20, [³²P] orthiphosphate 5, KCl 100, CaCl₂ 3, glucose 100, MgCl₂ 7 and 0.02 mg HK/ml (ph 7). At 61 min 15 s, 10 mM EGTA was added. The EGTA solution is alkalized previously to neutralize the proton liberated from EGTA at neutral pH. Immediately after the addition of EGTA, phosphoprotein is formed in a significant amount and disappears slowly (curve I). Subsequent addition of 2 mM ADP causes a rapid drop of phosphoprotein level (curve II) leading to the ATP synthesis (curve III). Reproduced from Makinose (1972).*

The initial observations on enzyme reactivity to P_i in the presence of a Ca^{2+} gradient across the SR membrane, and the subsequent formation of ATP, suggested a specific thermodynamic significance of the gradient, in the belief that the related chemical and/or osmotic potential may be operative in energy transduction for ATP synthesis (Makinose, 1971, 1972; Yamada et al., 1972; Yamada and Tonomura, 1973) in analogy to the chemiosmotic mechanism proposed for mitochondrial function (Mitchell, 1966, 1969). However, subsequent findings revealed that both, enzyme phosphorylation with P_i and formation of ATP, can be obtained in the absence of a Ca^{2+} concentration gradient across the SR membrane.

ENZYME PHOSPHORYLATION WITH P_i IN THE ABSENCE OF A CALCIUM GRADIENT

In experiments preluding demonstration of enzyme reactivity to P_i in the absence of a transmembrane Ca^{2+} gradient, Kanazawa and Boyer (1973) found an activity of $P_i \rightleftharpoons HOH$ oxygen exchange when non-loaded (e.g. not subjected to preincubation with Ca^{2+}) vesicles were exposed at pH 7.0 to EGTA, P_i and Mg^{2+}. The oxygen exchange indicated the occurrence of steady-state enzyme phosphorylation

and hydrolytic cleavage. In fact a very small amount of P_i incorporation was demonstrated directly. In presence of EGTA the $P_i \rightleftharpoons HOH$
exchange was inhibited by the addition of small concentrations of
Triton-X-100 which render the vesicles leaky without damaging the
ATPase. Therefore, it could not be excluded that a minimal gradient
established across the SR membrane by contaminating Ca^{2+} inside
the vesicles and EGTA in the outside medium was required both for
the $P_i \rightleftharpoons HOH$ exchange and for the small phosphorylation by P_i
detected.

Reaction of the enzyme with P_i in the absence of a gradient was
demonstrated unambigously (Masuda and de Meis, 1973; de Meis
and Masuda, 1974) by the use of vesicles rendered leaky by treatment with diethylether and mechanical distruption.

In these studies the conditions for the reaction were optimized
and highest levels of phosphoenzyme were found at pH 6.0, in the
presence of millimolar P_i and EGTA, and in the absence of either
K^+ or Na^+. The maximal levels (2–4 nmol/mg of protein) of phosphoenzyme obtained in these conditions are equal to those obtained
with in the presence of a gradient, or with ATP in the presence of
Ca^{2+} in the ouside medium. Monovalent cations and pH above 6.5
increase markedly the dependence of the reaction on P_i concentration. This effect is not as pronounced in the presence of a transmembrane Ca^{2+} gradient (Masuda and de Meis, 1973; de Meis, 1976;
Beil et al., 1977; Chaloub and de Meis, 1980).

Phosphorylation with P_i in the absence of a transmembrane Ca^{2+}
gradient has been repeated with intact vesicles, vesicles rendered
leaky by various procedures, and purified ATPase (Masuda and de
Meis, 1973; de Meis and Masuda, 1974; Kanazawa, 1975; Knowles
and Racker, 1975; de Meis, 1976; Beil et al., 1977; Boyer et al.,
1977; Rauch et al., 1977; Hasselbach, 1978; Punzengruber et al.,
1978; Verjovski-Almeida et al., 1978; Kolassa et al., 1979; Prager
et al., 1979).

Both in the absence and in the presence of a Ca^{2+} gradient, the
P_i reaction requires Mg^{2+}. This requirement has been the subject of
detailed analysis (Punzengruber et al., 1978; Kolassa et al., 1979;
Prager et al., 1979) indicating that Mg^{2+} binds to the enzyme as an
independent species, rather than a Mg-phosphate complex.

Another requirement is that the Ca^{2+} concentration in the outside
medium be lower than that producing occupancy of the high affinity
sites involved in activation of ATP utilization. In fact, by varying
the Ca^{2+} concentration in the medium it can be demonstrated that
in both leaky and loaded vesicles, an equal number of phosphorylation

sites react either with ATP or with P_i depending on whether the high affinity calcium sites are occupied by Ca^{2+} or not (Masuda and de Meis, 1973; de Meis, 1976; Beil et al., 1977; Prager et al., 1979).

While enzyme phosphorylation with ATP yields steady state levels of phosphoenzyme, the reaction of the enzyme with P_i in the absence of Ca^{2+} and ADP yields equilibrium levels of phosphoenzyme, as in:

$$*E + P_i \rightleftharpoons *E.P_i \rightleftharpoons *E\text{-}P + H_2O \qquad (9)$$

where the overall equilibrium constant includes the constants for complex formation, and the equilibrium between the complex and the covalently reacted species.

The P_i concentration producing half-maximal levels of phospho-enzyme varies with the pH of the medium and is higher in the absence, as compared with the presence of Ca^{2+} gradient (de Meis, 1976; Beil et al., 1977; de Meis and Vianna, 1979).

The effect of pH on the P_i concentration dependence of the phos-phorylation reaction was attributed by Beil et al. (1977) to the extent of P_i ionization. In the absence of a Ca^{2+} gradient, the effect of pH is abolished when organic solvents such as glycerol, dimethyl-formamide and dimethylsulphoxide are included in the medium. In this case, a marked reduction of the P_i concentration required for enzyme phosphorylation is observed, independent of the pH (6.0–8.0) of the medium. This suggests that pH effects cannot be attribu-ted exclusively to P_i ionization as suggested by Beil et al. (1977), but may also include solvation effects (de Meis et al., 1980).

Further insight of the P_i reactions was obtained in studies of rapid kinetics (Boyer et al., 1977; Rauch et al., 1977; Chaloub et al., 1979; Guimarães-Motta and de Meis, 1979). It was found that when P_i and EGTA are added simultaneously to loaded or nonloaded vesicles in media containing micromolar Ca^{2+}, the phosphorylation reaction occurs with identical kinetics showing an approximately 100 ms lag period and then proceeding with a 9–10/s apparent rate constant (22–25° C). Considering that the enzyme is not reactive to P_i when th high affinity sites are occupied by Ca^{2+}, the lag period indicates a slow step associated with removal of Ca^{2+} from the high affinity sites, as in:

$$E.Ca_2 \underset{1}{\overset{2Ca^{2+}}{\rightleftharpoons}} E \underset{2}{\rightleftharpoons} *E \underset{3}{\overset{P_i}{\rightleftharpoons}} *E.P_i \underset{4}{\overset{HOH}{\rightleftharpoons}} *E\text{-}P \qquad (10)$$

where *E is the reactive species, and either 1 or 2 is the slow step.

Accordingly, when P_i is added to nonloaded vesicles preincubated with EGTA, the phosphorylation proceeds without a lag, and a

35–43/s rate constant. In this case, the enzyme state which is reactive to P_i (*E) is formed during the preincubation, and the phosphorylation reaction begins immediately upon addition of P_i.

Occupancy of the low affinity sites by Ca^{2+} modifies the kinetics of phosphorylation with P_i. Addition of P_i to vesicles previously loaded with calcium and resuspended in EGTA, yields a very slow time profile of phosphorylation ($k_{app} \approx 1/s$). In these loaded vesicles the enzyme is likely to reside in a state with low affinity and inward oriented binding sites which are occupied by Ca^{2+} due to its high concentration in the vesicular lumen. Therefore, the slow phosphorylation suggests that this enzyme state (*E.Ca$_2$) reacts with P_i at very slow rates. Alternatively, it is possible that the reacting species is *E rather than *E.Ca$_2$, and the reaction is limited by the *E.Ca$_2$ equilibration.

The rate constant for the hydrolytic cleavage of the phosphoenzyme formed by reacting P_i with nonloaded vesicles, ($k_{app} \approx 10/s$) are compatible with the turnover of the ATPase catalytic cycle in the absence of "back inhibition". This can be shown by diluting the [^{32}P]-phosphoenzyme with media containing nonradioactive P_i, so as to prevent further incorporation of [^{32}P]-P_i, while catalytic decay of [^{32}P]-phosphoenzyme is followed by serial quenching with acid.

Since no incorporation of P_i is obtained in the presence of Ca^{2+}, decay of the phosphoenzyme formed with P_i can be also initiated by the addition of Ca^{2+}. It is of interest that in this case decay of the phosphoenzyme level is much slower ($t_{1/2} \approx 1/s$) than observed by dilution of radioactive phosphate (de Meis and Tume, 1977; Rauch et al., 1977; Chaloub et al., 1979; Guimarães-Motta and de Meis, 1980). This indicates that transformation of the enzyme species reacting with P_i to a state with high affinity sites occupied by Ca^{2+}, is rate-limiting as in:

$$*E \xrightleftharpoons{\hspace{0.8cm}} E \xrightleftharpoons{2\,Ca^{2+}} E.Ca_2 \qquad (11)$$

or

$$E \xrightleftharpoons{Ca^{2+}} E.Ca \xrightleftharpoons{\hspace{0.8cm}} E'Ca \xrightleftharpoons{Ca^{2+}} E''.Ca_2 \qquad (12)$$

It will be recalled that this is the same transformation delaying enzyme phosphorylation with ATP, when ATP is added with Ca^{2+} to the enzyme preincubated with EGTA, as opposed to the enzyme preincubated with Ca^{2+}.

$P_i \rightleftharpoons$ ATP EXCHANGE AND ATP FORMATION IN THE ABSENCE OF A TRANSMEMBRANE CALCIUM GRADIENT

Several experiments on the reversal of the Ca^{2+} pump (Makinose, 1971, 1972; Yamada *et al.*, 1972; de Meis, 1976) indicate that the phosphoenzyme formed through the P_i reaction in the presence of a Ca^{2+} gradient is able to phosphorylate ADP and form ATP. On the contrary, a distinctive feature of the phosphoenzyme obtained through the P_i reaction in the absence of a Ca^{2+} gradient is its inability to form ATP. This difference demonstrates clearly the existence of two forms of phosphoenzyme: a form which is reactive, and another form which is not reactive with ADP (de Meis and Sorenson, 1975; Knowles and Racker, 1975; de Meis, 1976). It will be recalled that a similar conclusion was reached on the basis of kinetic experimentation on the hydrolytic cleavage of the phosphoenzyme formed with ATP in the forward direction of the ATPase cycle (Shigekawa and Dougherty, 1978a, b; Shigekawa and Akowitz, 1979; Takisawa and Tonomura, 1979).

Even though the reaction of the enzyme with P_i in the absence of a Ca^{2+} gradient produces a phosphoenzyme which does not react with ADP, it can be demonstrated that this reaction is an integral part of the ATPase catalytic cycle. In fact, while the enzyme does not react with P_i in most reaction mixtures containing micromolar Ca^{2+}, this Ca^{2+} inhibition can be by-passed if ITP or low ATP concentration are added to sustain steady state ATPase activity of leaky vesicles (e.g. unable to form a Ca^{2+} gradient and to reduce the Ca^{2+} concentration in the medium). In these conditions it is found that a sizable portion of the phosphoenzyme steady state level is formed by incorporation of $^{32}P_i$ (de Meis and Masuda, 1974; Carvalho *et al.*, 1976). This indicates that during progression of the ATPase cycle, an enzyme species which is reactive to P_i (*E) and is able to sustain HOH $\rightleftharpoons P_i$ exchange (de Meis and Boyer, 1978) is present at significant concentrations. Its presence can be better detected in conditions in which the Ca^{2+} induced transition is slow as in the presence of acetyl phosphate, ITP or low concentrations of ATP (Fig. 9). The P_i reaction during progression of the cycle can be illustrated in the following diagram:

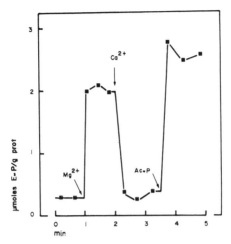

Fig. 9. *Effect of Ca^{2+} and acetyl phosphate on E-P formation. The assay medium composition was 10 mM Tris—maleate buffer (pH 6.5), 4 mM [^{32}P] P$_i$, 0.05 mM EGTA, and 0.6 mg of SRV protein/ml. The reaction was started by the addition of SRV. After different incubation intervals at 37° C, aliquots of 2.5 ml of this medium were removed for E—P determination. Arrows point to the addition of MgCl$_2$, final concentration in the assay medium of 10 mM, CaCl$_2$ 9.25 mM; AcP, 2 mM. Essentially the same results were observed in three different SRV preparations tested. Reproduced from de Meis and Masuda (1974).*

where 1 and 2 contribute to the slow transition with a consequent rise of the steady state level of *E.

In similar experiments with leaky vesicles or purified ATPase (de Meis and Carvalho, 1974) it is possible to obtain complete P$_i$ ⇌ ATP exchange if the Ca^{2+} concentration in the medium is raised to levels (millimolar) permitting occupancy of the calcium binding sites in the low affinity state and rise of the *E-P.Ca$_2$ steady state levels.

The occurance of P$_i$ ⇌ ATP exchange in these conditions (Fig. 10) demonstrates that complete reversal of the catalytic and formation of ATP can be obtained even *in the absence of a transmembrane Ca^{2+} gradient.* ADP sensitivity is confered to the phosphoenzyme by occupancy of the low affinity calcium sites.

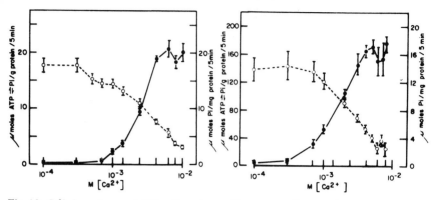

Fig. 10. Ca^{2+} dependence of $ATP \rightleftharpoons P_i$ exchange. The assay medium was 30 mM Tris–maleate buffer (pH 6.8), 10 mM ATP, 6 mM $^{32}P_i$, 20 mM $MgCl_2$ and 0.3 mg/ml of SRV protein. For the ATPase activity (o---o), $[\gamma\text{-}^{32}P]$ ATP was used. For $ATP \rightleftharpoons P_i$ exchange (●—●), $^{32}P_i$ was used. The reaction was performed at $37°C$. (left) solubilized SRV; (right) leaky SRV. The values shown in the figure represent the average ± SE of five experiments. Reproduced from de Meis and Carvahlo (1974).

Net synthesis of ATP in the absence of a Ca^{2+} gradient can be demonstrated directly (knowles and Racker, 1975; de Meis and Tume, 1977) by adding high (millimolar) Ca^{2+} and ADP to the phosphoenzyme formed with P_i and leaky vesicles or purified ATPase (Fig. 11). This reaction is limited to a single cycle consisting of equilibrium displacement from

$$*E + P_i \rightleftharpoons *E\text{-}P \qquad\qquad (14)$$

to

$$*E\text{-}P + Ca^{2+} + ADP \rightleftharpoons E.Ca_2 + ATP \qquad\qquad (15)$$

It should be pointed out that the concentration of ATP which is formed in these conditions is three or four orders of magnitude higher than predicted from simple equilibration of the ADP and P_i concentrations in the reaction medium. Therefore, phosphoenzyme and Ca^{2+} must also contribute to the equilibrium. The experimental findings indicate that calcium and ADP binding to the enzyme renders the equilibrium possible.

A role of calcium binding is also indicated by the influence of pH, as it was shown that the affinity of the binding sites as well as the Ca^{2+} concentration dependence of ATP formation are changed by pH shifts (de Meis and Tume, 1977; Verjovski-Almeida and de Meis, 1977). In fact, the Ca^{2+} concentration in the reaction mixture can be adjusted to suitable levels to permit phosphorylation with

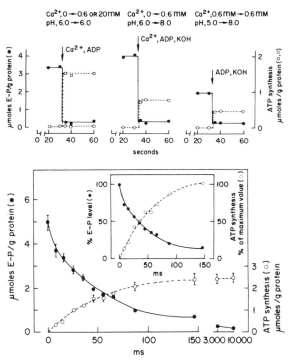

Fig. 11. *Top: pH and Ca²⁺ jump. Left, the reaction medium (0.5 ml) contained 20 mM MgCl₂, 8 mM ³²Pᵢ (2.5 × 10⁷ cpm/μmol), 2 units of hexokinase and 1 mg of Ca²⁺-dependent ATPase protein. The pH of the medium was 6.0. The reaction was performed at 25° C, it was started by the addition of the enzyme, and after 30 s of incubation (arrow) 0.5 ml of a solution containing 4 mM ADP, 10 mM glucose, and either 1.2 (○) or 40 mM (□) CaCl₂ were added. Note that the concentrations of these reagents were halved after mixing. The pH of the assay medium remained constant (6.0) after this addition. The reaction was terminated at different incubation intervals as shown in the figure by the addition of 0.1 ml of 100% Cl₃AcOH (w/v). Aliquots of the deproteinized supernatant and the protein precipitates were analysed for [γ-³²P] ATP and E~³²P formation, respectively. Centre, the experimental conditions and assay medium were the same as above, except that the second solution, in addition to 4 mM ADP, 10 mM glucose, and 1.2 mM CaCl₂ also contained 23 mM KOH. Thus, the pH of the medium increased from 6.0 to 8.0 after mixing. Right, the reaction medium (0.8 ml) contained 9 mM maleic acid, 20 mM MgCl₂, 8 mM ³²Pᵢ, 0.6 mM CaCl₂, 2 units of hexokinase, and 1 mg of Ca²⁺-dependent ATPase protein. The pH was 5.0. The reaction was started by the addition of the enzyme and 30 s later 0.2 ml of a mixture of ADP, glucose, CaCl₂ and KOH was added to give final concentrations of 2 mM ADP, 5 mM glucose, and 40 mM KOH. The CaCl₂ concentration remained constant (0.6 mM). After this addition, the pH of the medium increased from 5.0 to 8.0. Other experimental conditions were as described for left: (○) ATP synthesis; (●) E~P. Bottom: dephosphorylation of E~P and ATP formation. The reaction medium and experimental technique were as described above except that mixing and quenching were attained with a fast kinetics equipment as described in Fig. 6. Each point represents the mean ± SE of three experiments. The inset shows the same data but expressed either as a per cent of the E~P level or as a per cent of maximum ATP synthesized. Reproduced from de Meis and Tume (1977).*

P_i at pH 5.0, and then obtain ATP formation simply by the addition of ADP and a pH jump to 8.0, without changing the Ca^{2+} concentration (de Meis and Tume, 1977; Ratkje and Shamoo, 1980). In this case, the pH jump increases the affinity of the calcium sites to obtain binding at the Ca^{2+} concentrations already present in the medium, thereby permitting ATP formation (Fig. 11).

It is of interest that changes in temperature can also affect the Ca^{2+} dependence of the enzyme reaction with P_i, and the reaction of the phosphoenzyme with ADP to form ATP. The former reaction is inhibited at low temperature, and no phosphoenzyme is formed with P_i at $0°$ C. On the other hand, the Ca^{2+} concentration required to render the phosphoenzyme reactive to ADP at pH 8.0, is three orders of magnitude lower at $0°$ C than at $30°$ C (de Meis *et al.*, 1980). It is likely that this effect is due to a shift of the equilibrium between different forms of the enzyme in favour of the form reactive to ATP, which according to (15) would yield higher concentrations of $ADP.E{\sim}P.Ca_2$ and $ATP.E.Ca_2$. In agreement with this conclusion is the fact that the levels of phosphoenzyme formed with ATP in the presence of Ca^{2+} are not reduced by lowering the temperature.

Finally, phosphorylation of the enzyme with P_i and formation of ATP can be affected by the presnce of organic solvents in the reaction medium. In a comparative study of several organic solvents, de Meis *et al.* (1980) found that the P_i concentration dependence of enzyme phosphorylation with P_i is very much reduced in the presence of solvents such as dimethylsulphoxide, while subsequent phosphoryl transfer from the phosphoenzyme to ADP is inhibited. However, the ADP sensitivity is acquired again by the phosphoenzyme if the concentration of dimethylsulphoxide is reduced by dilution of the reaction mixture with aqueous media. Based on these experiments, de Meis *et al.* (1980) have suggested that solvation of substrates and appropriate residues at the active site may be involved in determining free energy changes in phosphorylation reactions within the ATPase cycle.

CALCIUM TRANSPORT AND ATPase ACTIVITY IN VARIOUS MICROSOMAL PREPARATIONS

In addition to SR vesicles, microsomes obtained from various tissue display some degree of calcium transport and ATPase activity. Although this activity is low and it is difficult to characterize the

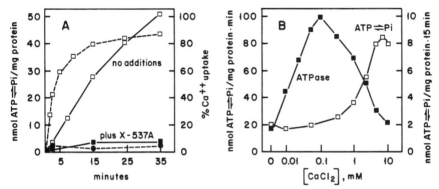

Fig. 12. *Left: effect of the ionophore X-537 A on ATP ⇌ P_i exchange and calcium uptake by brain microscomes; time course of Ca^{2+} uptake (dashed line) and $ATP \rightleftharpoons P_i$ exchange (solid lines) in the absence (□) and presence (●, □) of 25 μM/ml of X-537 A. Assay medium contained 40 mM Tris–maleate buffer (pH 7.0), 50 mM KCl, 0.1 mM ouabain, 10 mM $MgCl_2$, 2.5 mM ATP, 8 mM phosphate buffer (pH 7.0), 0.4 mg/ml of microsomal protein, and 10 μM $CaCl_2$. Right: calcium dependence for $ATP \rightleftharpoons P_i$ exchange and ATPase activity of soluble enzyme. The $CaCl_2$ concentrations are plotted on a logarithmic scale. ■, ATPase activity; $ATP \rightleftharpoons P_i$ exchange. The assay medium was as for left. Reproduced from Trotta and de Meis (1978).*

microsomes thoroughly, it is possible that the observed transport is related to a function of intracellular membrane systems such as SR and endoplasmic reticulum. Active microsomes have been obtained from smooth muscle (Fitzpatrick *et al.,* 1972; Hess and Ford, 1974; d'Auriac *et al.,* 1972; Hurwitz *et al.,* 1973), kidney (Moore *et al.,* 1974), pancreatic islets (Howell and Montague, 1975; Sehlin, 1976), salivary glands (Alonso *et al.,* 1971), liver (Moore *et al.,* 1975), crustacean periferal nerves (Lieberman *et al.,* 1967) and brain (Otsuka *et al.,* 1965; Yoshida *et al.,* 1966; Robinson and Lust, 1968; de Meis *et al.,* 1970; Diamong and Goldberg, 1971; Nakamaru and Schwartz, 1971; Nakamura and Konishi, 1974; Satomi, 1974; Trotta and de Meis, 1975, 1978).

Characterization of the calcium transport machinery in these microsomal preparations has not been carried out to the same depth as for sarcoplasmic reticulum vesicles. This is due to serious technical difficulties related to heterogeneity of the microsomal fractions and low specific activity. At any rate, it is apparent that the transport mechanism resembles that of SR vesicles. For instance, calcium accumulated by brain microsomes is abolished by divalent cation ionophores, enhanced by high concentrations of oxalate or phosphate (Trotta and de Meis, 1978) produced "back inhibition" and sustains $ATP \rightleftharpoons P_i$ exchange (Fig. 12) both in the presence and in the absence of a Ca^{2+} gradient (Trotta and de Meis, 1978).

CONCLUDING REMARKS

Calcium is involved in regulation of a variety of physiological events such as the contraction–relaxation cycle of muscle, the release of transmitters at the nerve terminals, cell-to-cell communication through gap junctions, and the release of secretory granules. These events are modulated by discrete variations of the Ca^{2+} concentration in the cytosol. It is apparent that in addition to the outer membrane, intracellular membranes such as SR and endoplasmic reticulum are also involved in regulation of the cytoplasmic Ca^{2+} concentration. Among various intracellular membranes, SR represents a most specific and highly differenciated system for Ca^{2+} transport. The possibility of purifying SR vesicles and the relative simplicity of the enzyme, provides a very convenient experimental system for the study of energy interconversion at the level of biological membranes.

In SR, the free energy of hydrolysis of the ATP terminal phosphate is utilized stepwise for a shift in orientation and binding affinity of the calcium sites in the enzyme. This shift is mediated by protein conformational changes and, as a result of repeated cycles, a transmembrane Ca^{2+} gradient is formed. The dimension of the gradient increases as long as calcium can be bound on the outer surface, and released on the inner surface of the vesicles. In the reverse process, the same conformational changes associated with Ca^{2+} binding on the inner surface, and calcium dissociation on the outer surface of the vesicles, lead to ATP synthesis from ADP and P_i. Therefore, a large difference of Ca^{2+} concentration on the two sides of the membrane is only needed to meet the difference in affinity of the calcium sites in the inward and outward orientation. This conclusion is derived from the finding that the protein conformational changes associated with calcium binding, and synthesis of ATP within a single catalytic cycle, can be obtained in the absence of a Ca^{2+} gradient. Thus the basic mechanism of Sr ATPase is the interconversion of chemical (ATP) into conformation and binding energy.

REFERENCES

Agostini, B. and Hasselbach, W. (1971). *Histochemie* **27**, 303.

Allen, G. (1977). *Proc. FEBS 11th Meeting, Copenhagen* **45**, Symposium A4, 159–168.

Alonso, G. L., Bazerque, P. M., Arrigo, D. M. and Tumilasci, O. R. (1971). *J. gen. Physiol* **59**, 340.

Andersen-Cedergreen, E. (1958). *J. Ultrastruct. Res.* Suppl. 1, 126.

Banerjee, R., Eptein, M., Kandrach, M., Simniak, P. and Racker, E. (1979). *Membrane Biochem.* 2, 283.

Barlogie, B., Hasselbach, W. and Makinose, M. (1971). *FEBS Lett.* 12, 267.

Bastide, F., Meissner, G., Fleischer, S. and Post, R. L. (1973). *J. biol. Chem.* 248, 8485.

Beil, F. U., Chak, D. and Hasselbach, W. (1977). *Eur. J. Biochem.* 81, 151.

Boyer, P. D., de Meis, L., Carvalho, M. G. C. and Hackney (1977). *Biochemistry* 16, 136.

Carsten, M. E. and Reedy, M. K. (1971). *J. Ultrastruct. Res.* 38, 554.

Carvalho, A. P. (1966). *J. cell Physiol.* 67, 73.

Carvalho, M. G. C., Souza, D. O. and de Meis, L. (1976). *J. biol. Chem.* 253, 1179.

Chaloub, R. M. and de Meis, L. (1980). *J. biol. Chem.* 255, 6168.

Chaloub, R. M., Guimarães-Motta, H., Verjovski-Almeida, S., de Meis, L. and Inesi, G. (1979). *J. biol. Chem.* 254, 9464.

Champeil, P., Büschlen-Boucly, S., Bastide, F. and Gary-Bobo, C. (1978). *J. biol. Chem.* 253, 1179.

Chevalier, G. and Butow, R. A. (1971). *Biochemistry* 10, 2733.

Chiesi, M. and Inesi, G. (1979). *J. biol. Chem.* 254, 10 370.

Chiesi, M. and Inesi, G. (1980). *Biochemistry* 19, 2912.

Chiesi, M., Peterson, S. and Acuto, O. (1978). *Archs. Biochem. Biophys.* 189, 132.

Chin, V. C. K. and Haynes, D. H. (1977). *Biophys. J.* 18, 3.

Coan, C. and Inesi, G. (1977). *J. biol. Chem.* 252, 3044.

Coan, C., Verjovski-Almeida, S. and Inesi, G. (1979). *J. biol. Chem.* 254, 2968.

d'Auriac, G. A., Baudoim, M. and Meyre, P. (1972). *Circ. Res.* 30, Suppl. II, 151.

Deamer, D. W. and Baskin, R. J. (1969). *J. cell. Biol.* 42, 296.

Deamer, D. W. and Baskin, R. J. (1972). *Archs. Biochem. Biophys.* 153, 47.

Dean, W. L. and Tanford, C. (1978). *Biochemistry* 17, 1683.

Degani, C. and Boyer, P. D. (1973). *J. biol Chem.* 248, 8222.

de Meis, L. (1969a). *Biochim. biophys. Acta* 172, 343.

de Meis, L. (1969b). *J. biol. Chem.* 244, 3733.

de Meis, L. (1971). *J. biol. Chem.* 246, 4767.

de Meis, L. (1976). *J. biol. Chem.* 251, 2055.

de Meis, L. and Boyer, P. D. (1978). *J. biol. Chem.* 253, 1556.

de Meis, L. and Carvalho, M. G. C. (1974). *Biochemistry* 13, 5032.

de Meis, L. and de Mello, M. C. F. (1973). *J. biol. Chem.* 248, 3691.

de Meis, L. and Hasselbach, W. (1971). *J. biol. Chem.* 246, 4759.

de Meis, L. and Masuda, H. (1974). *Biochemistry* 13, 2057.

de Meis, L. and Sorenson, M. M. (1975). *Biochemistry* 14, 2739.

de Meis, L. and Tume, R. K. (1977). *Biochemistry* 16, 4455.

de Meis, L. and Vianna, A. L. (1979). *A. Rev. Biochem.* 48, 275.

de Meis, L., Rubin-Altschul, B. L. and Machado, R. D. (1970). *J. biol. Chem.* 245, 1883.

de Meis, L., Hasselbach, W. and Machado, R. D. (1974). *J. cell. Biol.* 62, 505.

de Meis, L., Martins, O. B. and Alves, E. W. (1980). *Biochemistry* 19, 4252.

Diamond, I. and Goldberg, A. L. (1971). *J. Neurochem.* 18, 1419.

Drabikowski, W., Dominas, H. and Dabrowska, M. (1966). *Acta. Biochim. Pol.* 13, 11.

Dupont, Y. (1967). *Eur. J. Biochem.* **72**, 185.
Dupont, Y. (1977). *Biochem. biophys. Res. Commun.* **82**, 893.
Dupont. Y. (1980). *Eur. J. Biochem.* **109**, 231.
Dupont, Y. and Leigh, J. B. (1978). *Nature, Lond.* **273**, 396.
Dupont, Y., Harrison, S. C. and Hasselbach, W. (1973). *Nature, Lond.* **244**, 555.
Ebashi, E. and Lipmann, F. (1962). *J. cell Biol.* **14**, 389.
Ebashi, S. (1958). *Archs. Biochem. Biophys.* **76**, 410.
Ebashi, S. (1961). *J. Biochem, Tokyo* **50**, 236.
Eisemberg, B. and Gilai, A. (1979). *J. gen. Physiol.* **74**, 1.
Fiehn, W. and Hasselbach, W. (1970). *Eur. J. Biochem.* **20**, 245.
Fiehn, W. and Migala, W. (1971). *Eur. J. Biochem.* **20**, 245.
Fiehn, W. and Peter, J. B. (1971). *J. biol. Chem.* **246**, 5616.
Fitzpatrick, D. F., Landon, E. J., Debbas, G. and Hurwitz, L. (1972). *Science N.Y.* **176**, 305.
Franzini-Armstrong, C. (1964). *Fedn. Proc. Fedn. Am. Socs. exp. Biol.* **23**, 887.
Franzini-Armstrong, C. (1970). *J. cell Biol.* **47**, 448.
Franzini-Armstrong, C. (1973). *J. cell Biol.* **56**, 120.
Friedman, Z. and Makinose, M. (1970). *FEBS Lett.* **11**, 69.
Froehlich, J. P. and Taylor, E. W. (1975). *J. biol. Chem.* **250**, 2013.
Froehlich, J. P. and Taylor, E. W. (1976). *J. biol. Chem.* **251**, 2037.
Garrahan, P. J., Rega, A. F. and Alonso, G. L. (1976). *Biochim. biophys. Acta* **448**, 121.
Gattass, C. R. and de Meis, L. (1975). *Biochim. biophys. Acta* **389**, 506.
Guillain, F., Gingold, M. P., Büschlen, S. and Champeil, P. (1980). *J. biol. Chem.* **255**, 2072.
Guimarães-Motta, H. and de Meis, L. (1980). *Archs. Biochem. Biophys.* **203**, 395.
Hasselbach, W. (1964). *Prog. biophys. molec. Biol.* **14**, 169.
Hasselbach, W. (1978). *Biochim. Biophys. Acta* **515**, 23.
Hasselbach, W. and Beil, F. U. (1977). *In* "Biochemistry of Membrane Transport" (Eds G. Semenza and E. Carafoli) 416. Springer-Verlag, Berlin.
Hasselbach, W. and Elfvin, L. G. (1967). *J. Ultrastruct. Res.* **17**, 598.
Hasselbach, W. and Makinose, M. (1961). *Biochem. Z.* **333**, 518.
Hasselbach, W. and Makinose, M. (1962). *Biochem. biophys. Res. Commun.* **7**, 132.
Hasselbach, W. and Makinose, M. (1963). *Biochem. Z.* **339**, 94.
Herbette, L., Marquardt, J., Scarpa, A. and Blassie, J. (1977). *Biophys. J.* **20**, 245.
Hesketh, T. R., Smith, G. A., Houslay, M. D., McGill, K. A., Birdsall, N. J. M., Metcalfe, J. C. and Warren, G. B. (1976). *Biochemistry* **15**, 4145.
Hess, M. L. and Ford, G. D. (1974). *J. mol. cell. Cardiol.* **6**, 275.
Holloway, J. H. and Reilly, C. M. (1960). *Analyt. Chem.* **32**, 249.
Howell, S. L. and Montague, W. (1975). *FEBS Lett.* **52**, 48.
Hurwitz, L., Fitzpatrick, D. F., Debbas, G. and Landon, E. J. (1973). *Science, N.Y.* **179**, 384.
Huxley, H. E. (1957). *J. biophys. Biochem. Cytol.* **3**, 631.
Huxley, H. E. (1964). *Nature, Lond.* **202**, 1067.
Ikemoto, N. (1974). *J. biol. Chem.* **249**, 649.
Ikemoto, N. (1975). *J. biol. Chem.* **250**, 7219.
Ikemoto, N. (1976). *J. biol. Chem.* **251**, 7275.

Ikemoto, N., Streter, F., Nakamura, A. and Gergely, J. (1968). *J. Ultrastruct. Res.* **23**, 216.

Ikemoto, N., Bathnagar, G. M. and Gergely, J. (1971). *Biochem biophys. Res. Commun.* **44**, 1510.

Ikemoto, N., Morgan, J. F. and Yamada, S. (1978). *J. biol. Chem.* **253**, 8027.

Inesi, G. (1971). *Science, N.Y.* **171**, 901.

Inesi, G. (1979). *In* "Membrane Transport in Biology" (Eds G. Giebisch, D. C. Tosteson and H. H. Hussing) Vol. 2, 357. Springer-Verlag, Berlin, Heidelberg, New York.

Inesi, G. and Asai, H. (1968). *Archs. Biochem. Biophys.* **126**, 439.

Inesi, G. and Scales, D. (1974). *Biochemistry* **13**, 3298

Inesi, G., Goodman, J. J. and Watanabe, S. (1967). *J. biol. Chem.* **242**, 4637.

Inesi, G., Maring, E., Murphy, A. J. and MacFarland, B. (1970). *Archs. Biochem. Biophys.* **138**, 285.

Inesi, G., Cohen, J. A. and Coan, C. R. (1976). *Biochemistry* **15**, 5293.

Inesi, G., Kurzmack, M. and Verjovski-Almeida, S. (1978). *Ann. N.Y. Acad. Sci.* **307**, 224.

Inesi, G., Kurzmack, M., Coan, C. and Lewis, D. E. (1980a). *J. biol. Chem.* **255**, 3025.

Inesi, G., Kurzmack, M., Nakamoto, R., de Meis, L. and Bernhard, S. A. (1980b). *J. biol. Chem.* **255**, 6040.

Jilka, R. L., Martonosi, A. and Tillack, T. W. (1975). *J. biol. Chem* **250**, 7511.

Jorgensen, A. O., Kalnins, V. and MacLennan, D. H. (1979). *J. cell Biol.* **80**, 372.

Kalbitzer, H. H., Stehlik, D. and Hasselbach, W. (1978). *Eur. J. Biochem.* **82**, 245.

Kanazawa, T. (1975). *J. biol. Chem.* **250**, 113.

Kanazawa, T. and Boyer, P. D. (1973). *J. biol. Chem.* **248**, 3163.

Kanazawa, T., Yamada, S., Yamamoto, T. and Tonomura, Y. (1971). *J. Biochem. Tokyo* **70**, 95.

Keilley, W. and Meyerhof, O. (1948). *J. biol. Chem.* **174**, 387.

Klip, A., Reithmeier, R. A. F. and MacLennan, D. H. (1980). *J. biol. Chem.* **255**, 6562.

Knowles, A. F. and Racker, E. (1975). *J. biol. Chem.* **250**, 1949.

Knowles, A. F., Eytan, E. and Racker, E. (1976). *J. biol. Chem.* **251**, 5161.

Kolassa, N., Punzengruber, C., Suko, J. and Makinose, M. (1976). *FEBS Lett.* **108**, 495.

Kuriki, Y., Halsey, J., Biltomen, R. and Racker, E. (1976). *Biochemistry* **15**, 4956.

Kurzmack, M., Verjovski-Almeida, S. V. and Inesi, G. (1977). *Biochem. biophys. Res. Commun.* **78**, 772.

Kurzmack, M. Inesi, G., Tal. H. and Bernhard, S. (1980). *Biochemistry* in press.

Landgraf, W. C. and Inisi, G. (1969). *Archs. Biochem. Biophys.* **130**, 11.

Lau, Y., Caswell, A. and Brunschurg, J. (1977). *J. biol. Chem.* **252**, 5565.

Le Maire, M., Moller, J. and Tanford, C. (1976). *Biochemistry* **15**, 2336.

Lieberman, E. M., Palmer, R. F. and Collins, G. H. (1967). *Exp. Cell Res.* **46**, 419.

Louis, C. and Shooter, E. (1972). *Archs. Biochem. Biophys.* **153**, 641.

McFarland, B. H. and Inesi, G. (1970). *Biochem. biophys. Res. Commun.* **41**, 239.

McFarland, B. H. and Inesi, G. (1971). *Archs. Biochem. Biophys.* **145**, 456.

MacLennan, D. H. (1970). *J. biol. Chem.* **245**, 4508.

MacLennan, D. H. and Wong. P. T. S. (1971). *Proc. natn. Acad. Sci. U.S.A.* **68**, 1231.

MacLennan, D. H., Seeman, P., Iles, G. H. and Yip. C. C. (1971). *J. biol. Chem.* **246**, 2702.

MacLennan, D. H., Yip. C. C., Iles, G. H. and Seeman, P. (1972). *Cold Spring Harb. Symp. quant. Biol.* **37**, 469.

MacLennan, D. H., Subrzycka, E., Jorgensen, A. O. and Kalnins, V. I. (1978). *In* "Molecular Biology of Membranes" (Eds S. Fleischer, Y. Hatefy and D. H. MacLennan) 309–320. Plenum, New York.

Makinose, M. (1966). *Biochem. Z.* **345**, 80.

Makinose, M. (1967). *Pflügers Arch. ges. Physiol.* **294**, 8.

Makinose, M. (1969). *Eur. J. Biochem.* **10**, 74.

Makinose, M. (1971). *FEBS Lett.* **12**, 269.

Makinose, M. (1972). *FEBS Lett.* **25**, 113.

Makinose, M. and Hasselbach, W. (1971). *FEBS Lett.* **12**, 271.

Makinose, M. and The, R. (1965). *Biochem. Z.* **343**, 383.

Malan, N., Sabbadini, R., Scales, D. and Inesi, G. (1975). *FEBS Lett.* **60**, 122.

Marai, L. and Kuksis, A. (1973). *Can. J. Biochem.* **51**, 1365.

Martonosi, A. (1968). *J. biol. Chem.* **243**, 71.

Martonosi, A. and Feretos, R. (1964). *J. biol. Chem.* **239**, 648.

Martonosi, A. and Halpin, R. A. (1971). *Archs. Biochem. Biophys.* **144**, 66.

Martonosi, A., Donley, J. and Halpin, R. A. (1968). *J. biol. Chem.* **243**, 61.

Martonosi, A., Donley, J. R., Pucell, A. G. and Halpin, R. A. (1971). *Archs. Biochem. Biophys.* **144**, 529.

Masuda, H. and de Meis, L. (1973). *Biochemistry* **12**, 4581.

Meissner, G. (1975). *Biochim. biophys. Acta* **389**, 51.

Meissner, G. and Fleischer, S. (1971). *Biochim. biophys. Acta.* **241**, 356.

Meissner, G. and Fleischer, S. (1972). *Biochim. biophys. Acta.* **255**, 19.

Meissner, G., Connor, G. E. and Fleischer, S. (1973). *Biochim. biophys. Acta.* **298**, 246.

Michalak, M., Campbell, K. and MacLennan, D. H. (1980). *J. biol. Chem.* **255**, 1317.

Migala, A., Agostini, B. and Hasselbach, W. (1973). *Z. Naturf.* **28**, 178.

Mitchell, P. (1966). *In* "Chemiosmotic coupling in Oxidative and Photosynthetic Phosphorylation." Glynn Research, Bodmin, UK.

Mitchell, P. (1969). *Eur. J. Biochem.* **95**, 1.

Möller, J. V., Lind, K. E. and Andersen, J. B. (1980). *J. biol. Chem.* **255**, 1912.

Moore, L., Fitzpatrick, D. F., Chen. T. S. and Landon, E. J. (1974). *Biochim. biophys. Acta.* **345**. 405.

Moore, L., Chen. T., Knapp, H. R. and Landon, E. J. (1975). *J. biol. Chem.* **250**, 4562.

Murphy, A. J. (1978). *B. biol. Chem.* **253**, 385.

Muscatello, V., Anderson-Cedergreen, E., Azzone, G. F. and Van Der Decken, A. (1961). *J. Biochem. biophys. Cytol.* **10**, 201.

Nagai, T., Makinose, M. and Hasselbach, W. (1960). *Biochim. biophys. Acta* **43**, 223.

Nakamura, K. and Konishi, K. (1974). *J. Biochem. Tokyo* **75**, 1129.

Nakamura, Y. and Schwartz, A. (1971). *Archs. Biochem. Biophys.* **144**, 16.

Nakamura, Y. and Tonomura Y. (1978). *J. Biochem. Tokyo* **83**, 571.

Neet, K. E. and Green. N. M. (1977). *Archs. Biochem. Biophys.* 178, 588.

Ogawa, Y. (1968). *J. Biochem. Tokyo* 64, 255.

Ostwald, T. and MacLennan, D. H. (1974). *J. biol. Chem.* 249, 974.

Otzuka, M., Ohtsuki, I. and Ebashi, S. (1965). *J. Biochem. Tokyo* 58, 188.

Owens, K., Ruth, R. C. and Weglecki, W. B. (1972). *Biochim. biophys. Acta* 288, 479.

Packer, L., Melherd, C. W., Meissner, G., Zahler, W. L. and Fleischer, S. (1974). *Biochim. biophys. Acta* 363, 159.

Panet, R. and Selinger, Z. (1972). *Biochim. biophys. Acta* 255, 34.

Pang, D. C. and Briggs, F. N. (1977). *J. biol. Chem.* 252, 3262.

Peachey, L. D. (1965). *J. cell Biol.* 25, 209.

Peachey, L. D. and Schild, R. F. (1968). *J. Physiol.* 194, 249.

Peterson, S. and Deamer, D. (1977). *Archs. Biochem. Biophys.* 179, 218.

Plank, B., Hellmann, G., Punzengruber, C. and Suko, J. (1979). *Biochim. biophys. Acta* 550, 259.

Porter, K. and Palade, G. E. (1957). *J. biophys. biochem. Cytol.* 3, 269.

Portzehl, H. (1957). *Biochim. biophys. Acta* 26, 373.

Post, R. L., Tada, G. and Rogers, F. N. (1975) *J. biol. Chem.* 250, 3010.

Prager, R., Punzengruber, C., Kolassa, N., Winkler, F. and Suko, J. (1979). *Eur. J. Biochem.* 97, 239.

Punzengruber, C., Prager, R., Kolassa, N., Winkler, F. and Suko, J. (1978). *Eur. J. Biochem.* 92, 349.

Racker, E. (1972). *J. biol. Chem.* 247, 8198.

Ratkje, S. K. and Shamoo, A. (1980). *Biophys. J.* 30, 523.

Rauch, G., Chak, D. and Hasselbach, W. (1977). *Z. Naturf.* 32c, 828.

Ribeiro, J. M. C. and Vianna, A. L. (1978) *J. biol. Chem.* 253, 153.

Rizzolo, L., Le Maire, M., Reynolds, J. and Tanford, C. (1977). *Biochemistry* 15, 3433.

Robinson, J. D. and Lust, W. D. (1968). *Archs. Biochem. Biophys.* 125, 286.

Rossi, B., Leone, F. A., Gache, C. and Lazdunski, M. (1979). *J. biol. Chem.* 254, 2302.

Satomi, D. (1974). *J. Biochem. Tokyo* 76, 391.

Scales, D. and Inesi, G. (1976). *Biophys. J.* 16, 735.

Scales, D. and Sabbadini, R. (1979). *J. cell Biol.* 83, 33.

Scarpa, A., Baldassare, G. and Inesi, G. (1972). *J. gen. Physiol.* 60, 735.

Schmid, R. W. and Reilly, C. N. (1957). *Analyt. Chem.* 29, 264.

Schwartzenbach, G., Senn, H. and Anderegg, G. (1957). *Helv. chim. Acta* 40, 1886.

Scofano, H. M., Vieyra, A. and de Meis, L. (1979). *J. biol. Chem.* 254, 10227.

Sehlin, J. (1976). *Biochem. J.* 156, 63.

Shamoo, A. E., Ryan, T. E., Stewart, P. S. and MacLennan, D. H. (1976). *J. biol. Chem.* 251, 4147.

Shigekawa, M. and Akowitz, A. A. (1979). *J. biol. Chem.* 254, 4726.

Shigekawa, M. and Dougherty, J. B. (1978a). *J. biol. Chem.* 253, 1451.

Shigekawa, M. and Dougherty, J. B. (1978b). *J. biol. Chem.* 253, 1458.

Shigekawa, M., Dougherty, J. B. and Katz, A. (1978). *J, biol. Chem,* 253, 1442.

Somloy, A. V. (1979). *J. Cell Biol.* 80, 743.

Soomer, J. and Johnson, E. (1978). *In* "Handbook of Physiology". (Eds R. Berne, N. Speralakis and S. Geiger) Vol. 1, 113. American Physiological Society, Bethesda, Md.

Souza, D. O. and de Meis, L. (1976). *J. biol. Chem.* **251**, 6355.
Stewart, P. S. and MacLennan, D. H. (1974). *J. biol. Chem.* **249**, 985.
Stewart, P. S., MacLennan, D. H. and Shamoo, A. E. (1976). *J. biol. Chem.* **251**, 712.
Suko, J. and Hasselbach, W. (1976). *Eur. J. Biochem.* **64**, 123.
Sumida, M. and Tonomura, Y. (1974). *J. Biochem. Tokyo* **75**, 283.
Sumida, M., Wang, T. Mandel, F., Froehlich, J. and Schwartz, A. (1978). *J. biol. Chem.* **253**, 8772.
Tagaki, A. (1971). *Biochim. biophys. Acta* **248**, 12.
Takisawa, H. and Tonomura, Y. (1978). *J. Biochem. Tokyo* **83**, 1275.
Takisawa, H. and Tonomura, Y. (1979). *J. Biochem. Tokyo* **86**, 425.
Taniguchi, D. and Post, R. L. (1975). *J. biol. Chem.* **250**, 3010.
Taylor, J. S. and Hattan, D. (1979). *J. biol. Chem.* **254**, 4402.
The, R. and Hasselbach, W. (1972a). *Eur. J. Biochem.* **28**, 357.
The, R. and Hasselbach, W. (1972b). *Eur. J. Biochem.* **30**, 318.
Thorley-Lawson, D. A. and Green, N. M. (1973). *Eur. J. Biochem.* **40**, 403.
Thorley-Lawson, D. A. and Green, N. M. (1975). *Eur. J. Biochem.* **59**, 193.
Tong. W. W. (1977). *Biochem. biophys. Res. Commun.* **74**, 1242.
Trotta, E. E. and de Meis, L. (1975). *Biochim. biophys. Acta* **394**, 239.
Trotta, E. E. and de Meis, L. (1978). *J. biol. Chem.* **253**, 7821.
Vanderkooi, J., Ierokamas, A., Nakamura, H. and Martonosi, A. (1977). *Biochemistry* **16**, 1262.
Veratti, E. (1902). *Memorie Instituto Lombardo di Scienze e Lettere* **19**, 87. Reprinted in *J. Biophys. Biochem. Cytol.* **10**, 1 (1961).
Verjovski-Almeida, S. and de Meis, L. (1977). *Biochemistry* **16**, 329.
Verjovski-Almeida, S. and Inesi, G. (1979). *J. biol. Chem.* **254**, 18.
Verjovski-Almeida, S., Kurzmack, M. and Inesi, G. (1978). *Biochemistry* **17**, 5006.
Vianna, A. L. (1975). *Biochim. biophys. Acta* **410**, 389.
Vieyra, A., Scofano, H. M. Guimarães-Motta, H., Tume, R. K. and de Meis, L. (1979). *Biochim. biophys. Acta* **568**, 437.
Walter, H. and Hasselbach, W. (1973). *Eur. J. Biochem.* **36**, 110.
Wang, C., Saito, A. and Fleischer, S. (1979). *J. biol. Chem.* **254**, 9209.
Waku, K. Uda, Y and Nakazawa, Y. (1971). *J. Biochem. Tokyo* **69**, 483.
Warren, G. B., Toon, P. A., Birdsall, N. J. M., Lee, A. G. and Metcalfe, J. C. (1974a). *Proc. Wash. Acad. Sci.* **71**, 622.
Warren, G. B., Toon, P. A., Birdsall, N. J. M., Lee, A. G. and Metcalfe, J. C. (1974b). *Biochemistry* **13**, 5501.
Weber, A. (1966). *In* "Current Topics in Bioenergetics" (Ed. D. Sanadi) Vol. 1, 203. Academic Press, New York.
Weber. A., Herz, R. and Reiss, I. (1966). *Biochem. Z.* **345**, 329.
Yamada, S., Sumida, M. and Tonomura, Y. (1972). *J. Biochem Tokyo* **72**, 1537.
Yamada, S. and Tonomura, Y. (1973). *J. Biochem. Tokyo* **74**, 1091.
Yamamoto, T. and Tonomura, Y. (1967). *J. Biochem. Tokyo* **62**, 558.
Yamamoto, T. and Tonomura, Y. (1968). *J. Biochem. Tokyo* **64**, 137.
Yares, D. W. and Duance, V. C. (1976). *Biochem. J.* **159**, 719.
Yoshida, H., Kadota, K. and Fijisawa, H. (1966). *Nature, Lond.* **212**, 291.

5
Calcium Transport in Microorganisms

BARRY P. ROSEN

Department of Biological Chemistry, University of Maryland,
Baltimore, Maryland, USA

INTRODUCTION

An introduction into the importance of calcium in biological systems should not be necessary for those who have progressed as far in this book as this chapter. However, the ubiquity of calcium transport systems both in eukaryotes and prokaryotes may not be generally recognized. In every organism thus far examined calcium transport systems have been found to exist, and the cytosolic concentration of calcium is maintained lower than environmental calcium (Silver, 1977).

In this chapter we will examine first the growth requirements for calcium among microorganisms, followed by a detailed description of the transport systems for this ion both in prokaryotes and in unicellular eukaryotes. Finally, in those few cases where information is available, the physiological function of calcium in the physiology of the organism will be reviewed. Several reviews on similar material have appeared within recent years (Silver, 1977, 1978), but the only one specifically concerning microbial calcium transport is now several years old (Silver, 1977).

MICROBIAL CALCIUM REQUIREMENTS

Whether or not a substance is required for growth depends in part on the definition of "required". Is calcium absolutely required in that no growth occurs in its absence? One must further define "absence", since contamination of media by calcium is unavoidable. If calcium is not "absolutely" required, does the presence of calcium enhance growth? Do only certain functions require calcium, functions which might not be necessary for growth, for example, sporulation?

In only a few cases have the lower limits for calcium been defined. In determining the ionic requirements for growth of *Escherichia coli* Young *et al.* (1944) were able to demonstrate that the growth rate of that organism was unaffected by the absence of calcium where the upper limit of calcium contamination in the medium was about 1.5 μM. The addition of as much as 0.15 mM $CaCl_2$ had no effect on growth rate. Brey and Rosen (unpublished results) found that *E. coli* could grow in 10 mM EGTA. If the concentration of calcium in the growth medium were 10^{-5} M, a reasonable guess for calcium contamination, the free calcium concentration would be 10^{-10} M. At the other extreme, Brey and Rosen (1979a) showed that *E. coli* could grow normally in medium containing 50 mM $CaCl_2$, although mutants sensitive to this high concentration of calcium could be selected. *Escherichia coli*, then, does not seem to require calcium for growth, setting the lower limit for calcium at 10^{-10} M.

A number of strains of *Azotobacter*, on the other hand, will not grow in the absence of exogenously added calcium, although other strains of *Azotobacter* do not share the calcium requirement (Norris and Jensen, 1957). Although the concentration of calcium present in the original medium was not reported, the medium was formulated to contain as little calcium as possible. The lowest amount of exogenously added calcium was 10 μM, and maximal growth required approximately 250 μM. Strontium could replace calcium; it was unlikely that this reflected a general divalent cation requirement since $MgSO_4$ was a component of the growth medium. Little information is available on the function of calcium in *Azotobacter*.

Several other organisms exhibit calcium requirements for growth, including *Leptospira pomona* (Johnson and Gary, 1963), *Flavobacterium* B-9 (MacLeod and Onofrey, 1957), and at high temperatures, *Bacillus stearothermophilus* (Stahl and Ljunger, 1976). One of the more interesting examples of calcium dependency is that of *Yersinia pestis*, the causative organism of bubonic plague. Virulent strains of *Y. pestis* do not require calcium when grown at room temperature, but a shift to 37°C produces an immediate need for calcium (Brubaker, 1972). Nonvirulent mutants, on the other hand, do not require calcium for growth at the higher temperature, suggesting a relationship between calcium and virulence (Higuchi *et al.*, 1959). If virulent cells are shifted to 37°C in the absence of calcium, growth ceases (Zahorchak *et al.*, 1979). Addition of calcium within a few hours allows for resumed growth, but after longer periods addition of calcium cannot reverse the growth stasis. Interestingly, the growth stasis can be reversed at any time by shifting back to the lower temperature. Strontium can replace calcium; neither magnesium nor manganese can. The biochemical events leading to stasis are not known. Zahorchak *et al.* (1979) have shown that early events following shift from 26°C to 37°C in the absence of calcium include leakage of nucleotide pools and a decrease in RNA synthesis. Their preliminary results suggest that addition of ATP to the growth medium can reverse the growth stasis, suggesting that a decrease in intracellular ATP is responsible for cessation of growth. The fact that the cells leak metabolites in the absence of calcium may mean that the role of calcium is in stabilization of the membrane rather than a metabolic role, and that there is no direct connection between calcium and virulence. Still, it is possible that calcium-mediated reactions are involved in virulence, and this possibility should be explored.

Several other bacteria exhibit altered growth patterns in the absence of calcium. Cell wall formation in *Lactobacillus bifidus* is defective in the absence of calcium, leading to a change from a bacilloid shape to a bifid shape (Kojima *et al.*, 1970). *Sphaerotilus natans* grows as free cells in the absence of calcium, but in the presence of calcium forms filaments which become covered with a slime sheath (Dias *et al.*, 1968). This is commercially important since sheathed filaments of *Sphaerotilus* clog recreational waterways. *Sphaerotilus natans* may, in fact, have an actual calcium requirement for growth. EDTA causes growth stasis which is overcome by addition of calcium, strontium or barium. *Leucothrix mucor*, a colourless derivative of blue-green bacteria, similarly undergoes morphological changes when calcium is omitted from the growth medium (Snellen and Raj, 1970). No attempts were made to reduce the endogenous calcium levels, so it is not known whether *L. mucor* has a more general calcium requirement. In addition to the general calcium requirement of *Azotobacter* (Norris and Jensen, 1957), *A. vinelandii* also requires calcium for encystment (Page and Sadoff, 1975). The formation of the slime layer is related to synthesis of uronic acids, which also has a calcium requirement. Thus calcium may play a role in both synthesis and integrity of the slime layer.

None of the above studies sheds any light on the question of whether calcium has an intracellular role in bacteria. Indeed, many of the effects can be explained by an extracellular role in stabilization of membrane and/or wall components or in activation of extracellular enzymes. Divalent cations, often either calcium or magnesium, are essential for the integrity of the outer membrane of Gram negative bacteria (Asbell and Eagon, 1966; Leive, 1968; van Alphen, 1978).

It would not be surprising if, in a few organisms, calcium alone fulfilled that role.

MICROBIAL CALCIUM TRANSPORT

Transport systems can couple to chemical energy directly, e.g. primary ion pumps, or secondarily, e.g. uniports, symports or antiports (Mitchell, 1973; Rosen and Kashket, 1978). Bacteria and unicellular eukaryotes have evolved a number of mechanisms for calcium transport. In many organisms both influx and efflux systems exist, with different mechanisms of energy coupling for each. There are similarities between the various bacteria, between prokaryotes and eukaryotic organelles, and even between prokaryotes and the eukaryotic plasma membrane systems. But there are differences as well. In this section the properties of the calcium transport systems in individual organisms will be discussed. Based on little evidence, we would propose a generalized scheme in which most if not all bacteria have both influx and efflux systems for calcium, and the intracellular concentration of calcium is controlled by a balance between the two types of systems. This is seen diagrammatically in Fig. 1. The properties of the various efflux systems are summarized in Table I.

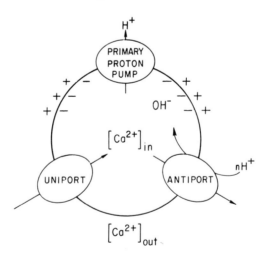

Fig. 1. *Calcium circulation in bacteria. Primary proton pumps establish electrochemical proton gradients, acid and positive, outside. Calcium enters the cell through a uniporter where the membrane potential provides the driving force. Calcium is extruded by exchange with protons via an antiporter, coupling to the ΔpH. The concentration of calcium in the cytosol depends on the ratio of activities of the two systems.*

TABLE I. Bacterial calcium extrusion systems.

Organism	Porter mechanism	K_m (mM)	V_{max} (nmol/min/mg of membrane protein)	Specificity	Reference
Escherichia coli		0.34	85	$Ca^{2+} = Mn^{2+} > Sr^{2+} > Ba^{2+}$	Tsuchiya and Rosen (1975a) Brey and Rosen (1979a)
Bacillus subtilis[a]		0.36	330[b]	$Ca^{2+} > Sr^{2+} > Mn^{2+}$	Silver et al. (1975)
Bacillus megaterium		0.088	0.047	—	Golub and Bronner (1974)
Azotobacter vinelandii	Ca^{2+}/H^+ antiporter	0.048	45	Ca^{2+} specific	Bhattacharyya and Barnes (1976)
Mycobacterium phlei		0.08	16	—	Kumar et al. (1979)
Rhodopseudomonas capsulata		—	—	—	Jasper and Silver (1978)
Clostridium perfringens		—	—	—	Hasan and Rosen (1979)
Halobacterium halobium	Ca^{2+}/Na^+ antiporter	0.071	2	—	Belliveau and Lanyi (1978)
Streptococcus faecalis	Ca^{2+}-ATPase	Varies with pH		—	Kobayashi et al. (1978)

[a] Data from intact, energy-poisoned cells.
[b] nmol/min/g dry weight of cells.

Escherichia coli

During the late 1960s several groups studied divalent cation uptake into *E. coli*. Lusk and Kennedy (1969), Nelson and Kennedy (1971) and Silver and Clark (1971) described transport systems for magnesium; while it appeared that manganese and cobalt were substrates, calcium did not appear to be. Silver and Kralovic (1969) compared uptake of $^{54}Mg^{2+}$ and $^{45}Ca^{2+}$ into intact cells. While manganese was accumulated by *E. coli*, there was an inverse relationship between manganese and calcium: at high temperatures the cells contained considerable manganese but little calcium; at $0°C$ calcium was taken up but not manganese. They attributed that to the presence of an active extrusion system for pumping calcium out of the cells.

The next step in the elucidation of calcium extrusion systems in bacteria relied on the use of membrane vesicle systems. Membrane vesicles can be prepared by several methods from the cytoplasmic membrane of *E. coli*. One method, devised by Kaback, produces vesicles which are in many ways right-side-out. A second method, which involves lysis of intact cells by passage through a French press, produces an entirely different sort of vesicle. Futai (1974) showed that the F_1 portion of the H^+-translocating ATPase (BF_0F_1) was located on the cytoplasmic surface of the cytoplasmic membrane of cells, but in French press vesicles all of the F_1 was located on the outer surface. Hertzberg and Hinkle (1974) showed that the entire BF_0F_1 was oriented functionally outwards: while intact cells would pump protons outwards during respiration or ATP hydrolysis via the BF_0F_1, these vesicles would pump protons inwards, creating a protonmotive force acid and positive inside. This protonmotive force of orientation opposite to that in intact cells would energize ATP synthesis if ADP and P_i were added to the outside of the vesicles. Rosen and McClees (1974), reasoning that extrusion of calcium from intact cells should be equivalent to uptake by everted vesicles, demonstrated $^{45}Ca^{2+}$ accumulation by vesicles prepared by French press lysis. They showed that the calcium transport system was energy dependent and was capable of using a protonmotive force as the direct energy donor. Although ATP could energize transport, in the absence of an F_1 no calcium transport was observed (Tsuchiya and Rosen, 1975b). Moreover, ATP-driven transport was sensitive to uncouplers, inhibitors of the BF_0F_1, and antiserum to F_1 (Tsuchiya and Rosen, 1975a). Thus, the calcium system is almost certainly a secondary porter rather than an ATP-driven calcium pump. Since the direction of calcium movement is into a region of acid and positive character, the most likely coupling mechanism is a calcium/proton antiporter. In direct confirmation of this postulate, Tsuchiya and Rosen (1976a) observed calcium uptake driven by an artificially imposed ΔpH, where the only possible mechanism would be exchange of calcium for protons.

Rosen and McClees (1974) found that the accumulation of $^{45}Ca^{2+}$ was greatly stimulated by the addition of phosphate to the reaction mix. Although the accumulated $^{45}Ca^{2+}$ could be removed from the vesicles by addition of EGTA (Rosen and Brey, 1979), uncouplers would not cause efflux nor would unlabled calcium cause exchange (Tsuchiya and Rosen, 1975a). Tsuchiya and Rosen (1975a) also showed that, although phosphate was necessary for gross accumulation of $^{45}Ca^{2+}$ oxalate could substitute. These observations suggested that most of the $^{45}Ca^{2+}$ accumulated was not in the free state, but, more likely, was deposited as the phosphate or oxalate salt. Double label experiments gave ratios of $^{45}Ca : {}^{32}P_i$ of approximately 2–3 at early times, but at later times a constant ratio of 1.4–1.5 was attained, suggesting calcium concentration at early times until the solubility of calcium phosphate was exceeded, after which the bulk of the $^{45}Ca^{2+}$ (and $^{32}P_i$) would be in the form of insoluble $Ca_3(PO_4)_2$. Phosphate could be added equally effectively from either side of the membrane, ruling out a calcium–phosphate symport. Interesting, though, no $^{32}P_i$ accumulated in the absence of calcium, and the rate of phosphate uptake increased with time, as if calcium accumulation were necessary for phosphate accumulation. It may be that the route of phosphate uptake by everted vesicles is via a phosphate porter which functions as an uptake system in intact cells.

Due to the fact that appreciable calcium uptake could be observed only in the presence of a precipitating co-ion, it was not possible to determine accurately such properties of the system as kinetics or specificity. A K_m of about 0.1 mM had been determined in the absence of phosphate (Rosen and McClees, 1974), but the velocities were low, so that the accuracy of the method was in doubt. In the presence of phosphate, it was difficult to decide whether the measured kinetics reflected an intrinsic property of the carrier or the formation of a calcium–phosphate complex. In order to circumvent this difficulty, Brey and Rosen (1979a) and Tsuchiya and Takeda (1979) examined the effect of calcium on the steady-state ΔpH. If the orientation of the proton motive force is acid interior in everted vesicles, then a weak lipophilic base will be accumulated in response to the ΔpH portion of the proton motive force. Since antiport activity involves exchange of protons for cation, uptake of the cation should be reflected in a transient decrease in ΔpH. Brey *et al.* (1978) measured the effect of cations on the fluoresence of quinacrine, a weak lipophilic base whose fluorescence is quenched in response to the formation of a ΔpH, acid interior. They found that *E. coli* membranes have three separate antiport systems: two for monovalent cations and one for divalent cations. The divalent cation/proton antiporter measured by this assay is most likely the same calcium/proton system measured by uptake of $^{45}Ca^{2+}$. Using this assay, Brey and Rosen (1979b) determined that depolarization of the membrane eliminated calcium/proton exchange, suggesting a role for the membrane potential in the activity of the antiporter. The simplest explanation of the result is that the stoichiometry of the antiporter is $H^+ : Ca^{2+} > 2$, so that the potential is a component of the driving force, and elimination of that component reduces the net force. It may also be that the potential has another function; otherwise it is difficult to explain why elimination of only part of the driving force results in total loss of activity. Perhaps there is a minimum voltage necessary for activity of the porter. Brey and Rosen (1979b) also found that the specificity of the antiporter is, in order of affinities, $Ca^{2+} = Mn^{2+} > Sr^{2+} > Ba^{2+}$. Mg^{2+} is not a substrate, but appears to be an allosteric regulator. Sigmoidal kinetics were observed with all of the substrates, with Hill coefficients of approximately 2. The concentration of calcium producing half-maximal rates was 0.1–0.5 mM in these assays. In the presence of Mg^{2+} the Hill coefficient declined to about 1. Thus, at low substrate concentrations, magnesium can greatly increase the reaction rate. Since the *in vivo* cytoplasmic calcium concentrations would always be subsaturating, small changes in the intracellular magnesium concentration could act as a switching mechanism to turn on or off the calcium/proton antiporter. Tsuchiya and Takeda (1979) also observed a K_m of 0.24 mM using a similar assay. Their experiments were performed in the presence of magnesium, and no cooperativity was observed. Calcium also has an effect on magnesium transport in intact cells of *E. coli*. Park *et al.* (1976) found that *cor* mutants, which have only the *mgt* transport system for magnesium, could not grow in the presence of calcium, suggesting that exogenous calcium may repress that magnesium transport system. The complex interrelationship between the transport systems for calcium and magnesium awaits further clarification.

While *E. coli* does not require calcium for growth, their ability to grow in high medium concentrations calcium allows for the selection of calcium-sensitive mutants. Brey and Rosen (1979b) isolated four distinct classes of such mutants, each of which affects the ability of everted vesicles to accumulate $^{45}Ca^{2+}$. The original calcium-sensitive mutant turned out to be a double mutant, with lesions in *corA* and another gene which they called *calA*. *CorA* mutants lack one of the magnesium transport systems and, as mentioned above, become sensitive to calcium, perhaps through repression of the remaining magnesium transport by medium calcium (Park *et al.*, 1976). *CorA* mutants are sensitive to 50 mM $CaCl_2$, but the double *corA calA* mutant is sensitive to 25 mM or less (Brey and Rosen, 1979b). The *calA* locus maps very close to *purA* at min 97 of the *E. coli* chromosome. Everted membrane vesicles from the double mutant unexpectedly exhibited elevated calcium transport, presumably reflecting increased calcium efflux from intact cells. Why this should result in calcium sensitivity is not clear. When the *calA* locus was separated from the *corA* gene by transfer of the gene into a *purA* strain, the resulting *corA*$^+$ *calA* strain was no longer calcium sensitive but still exhibited elevated calcium transport in everted vesicles. It appears that a mutation in the *calA* gene potentiates the calcium sensitivity of *corA* mutants but does not in itself cause calcium sensitivity. The double mutant was sensitive to calcium only at pH 8.0; when the medium pH was lowered to 5.6, the double mutant was resistant to calcium. Mutants sensitive to calcium at pH 5.6 could be selected using the double mutant as a parent strain. Everted membrane vesicles from the resulting triple mutant took up little $^{45}Ca^{2+}$, presumably reflecting decreased calcium extrusion from intact cells. Again, *cor*$^+$ derivatives were calcium resistant but still exhibited the transport defect. The new mutation was mapped near *calA*, but F′ complementation tests demonstrated that the two loci were different, suggesting a separate gene, termed *calB*, in the triple mutant. Selection for sensitivity to higher levels of calcium in a *cor*$^+$ strain resulted in the isolation of two other mutants termed *calC* and *calD*. Both are still *cor*$^+$, and everted membrane vesicles from either have reduced $^{45}Ca^{2+}$ uptake. In vesicles from these mutants addition of A23187, a calcium/proton exchanging ionophore, could restore calcium uptake to levels higher than the parent. In fact, A23187 stimulated uptake in the parent several-fold (Brey and Rosen, 1979a). The map positions of *calC* and *calD* are min 15.5 and 9, respectively.

While uptake of $^{45}Ca^{2+}$ was altered in vesicles from each of the mutants, no difference between the various mutants and wild-type strains were observed using the quinacrine assay (Brey and Rosen, 1979b). This disturbing fact calls into question the relationship of the mutations and the genes for the calcium/proton antiporter or, at the least, the relationship of the radioisotope assay to the quinacrine assay. However, a mutant in the potassium/proton antiporter exhibited reduced activity by both assays (Plack and Rosen, 1980). Perhaps, then, the assays may reflect more than one transport process. The fluoresence of quinacrine should be proportional to the steady-state ΔpH. Perturbation of that steady state by a cation through cation/proton exchange would be expected to decrease ΔpH. When the cation reaches equilibrium, no further energy expenditure should be required to maintain the gradient, and the pH gradient should increase again to near its original level. That this does not occur suggests that calcium never attains equilibrium, perhaps due to leak pathways for calcium. Such leak pathways could also explain the lack of significant calcium accumulation in the absence of a precipitating co-ion. The two assays may measure different aspects of the overall cation fluxes, so that some mutations may show with one assay but not the other or may show up in both. It is possible that the calcium-sensitive mutations are in the genes for systems which catalyse calcium uptake into cells, and affect only indirectly the antiporter.

Silver *et al.* (1975) had examined calcium uptake into intact cells and Kaback-type vesicles of *E. coli*. They observed uptake only at low temperature or in the presence of an uncoupler, suggesting that only net efflux of calcium occurs. The amounts were small compared to *B. subtilis*, and the amount of binding of calcium to the cell surface was high. Ueyama *et al.* (unpublished results) attempted to eliminate these problems by using flow dialysis and a calcium-selective electrode, assays which measure changes in medium calcium rather than cellular uptake. With both assays they observed calcium efflux from cells at pH 8.2, where the antiporter is known to be active, and calcium uptake into cells at pH 6.5, where the antiporter is inactive. This suggests that both calcium influx and efflux occur, perhaps simultaneously, but that the direction of net flux is variable. Examination of the calcium-sensitive mutants by these assays is clearly indicated.

Bacillus subtilis, megaterium and stearothermophilus

Bacillus subtilis and *megaterium* will be discussed together, since properties of the calcium transport systems appear nearly identical. *Bacillus subtilis* and *megaterium* show marked changes in calcium transport during the transition from vegatative growth to sporulation. In vegatative growth *Bacillus* is very much like *E. coli*; under energized conditions net efflux of calcium occurs, and facilitated transport of calcium occurs in the presence of uncoupler or at low temperature (Silver *et al.*, 1975). Silver *et al.* (1975) found that membrane vesicles prepared by osmotic lysis of *B. subtilis* protoplasts are similar to intact cells, but Golub and Bronner (1974) found that similar vesicles made from *B. megaterium* actively accumulated calcium. They later discovered that the vesicles were a mixed population of right-side-out and everted membranes, and that under certain conditions vesicles capable of calcium transport, but not amino acid transport, could be prepared (Bronner *et al.*, 1975). These results suggest that the two strains of *Bacillus* have a calcium/proton antiporter similar in some respects to the *E. coli* porter.

During the transition from vegetative growth to sporulation these strains of *Bacillus* begin to accumulate calcium (Halvorson, 1962; Murrell, 1967; Ellar, 1978). Eisenstadt and Silver (1972) have identified three stages in the calcium metabolism of sporulating *B. subtilis*. In the first stage calcium enters the cytoplasm of the sporulating cell, as shown by the fact that either toluene or lysozyme can effect release of the calcium. As the calcium is accumulated by the forespore, it becomes resistant to toluene. Forespores are sensitive to lysozyme, and lysozyme still causes release of the calcium. As the mature spore forms, the calcium becomes inaccessible except by the most drastic treatments such as incubation in hot trichloroacetic acid or autoclaving. Inside of the spore calcium is sequestered in a 1:1 complex with dipicolinic acid.

It is not at all certain what the individual transport reactions are that catalyze the movement of calcium into and out of the two compartments (cytosol and forespore). The works of Silver *et al.* (1975) and Golub and Bronner (1974) have clearly shown that the cytoplasmic membrane of vegetative cells extrudes calcium. Indeed, Golub and Bronner (1974) demonstrated that what are mostly everted membrane vesicles from the cytoplasmic membrane of sporulating cells of *B. megaterium* (cells which are capable of net calcium uptake) still exhibit calcium proton antiport activity. Bronner *et al.* (1975) went on to show that sporulating cells equilibrated with calcium in the absence of exogeneous energy sources extruded calcium when an energy donor, reduced phenazine methosulphate, was added. These results suggest that the antiport system for calcium efflux is functional even in sporulating cells. To account for the net accumulation of calcium, they postulated that sporulating cells are energy depleted, allowing calcium entry by the antiporter. Accumulation would be the result of complexation with a cytoplasmic factor.

Later the forespores would abstract the calcium from the cytosol. In support of this model they have isolated a cytoplasmic calcium binding factor which, although uncharacterized, is not dipicolinic acid (DPA). On the other hand, Seto (1979) found that the ATP pool of sporulating cells does not decrease, suggesting that they are not energy depleted.

Hogarth and Ellar (1978, 1979) have studied this process in detail in *B. megaterium* and have come to a somewhat different conclusion, namely that the uptake of calcium into the sporulating cell is an active, energy-dependent process, but that later transfer into the forespore is a non-energy-dependent, facilitated reaction. Uptake by sporulating cells is inhibited by inhibitors of the respiratory chain such as cyanide, antimycin A, and 2-heptyl-4-hydroxyquinoline-*N*-oxide, and by uncouplers such as FCCP. On the other hand, none of these inhibited uptake into isolated forespores. Moreover, the K_m for uptake by the sporulating cell is approximately 0.03 mM, while the K_m of the forespore is 2 mM. Cells incubated in submillimolar calcium accumulate calcium to cytosolic levels of 6–9 mM. Yet, forespores incubated in the same concentration of calcium do not accumulate calcium, as would be expected from the high K_m value. Only when incubated in 6–9 mM calcium do the forespores take up calcium in amounts consistent with the *in vivo* studies. Thus, forespore uptake is dependent on the ability of the cytoplasmic membrane calcium transport system to maintain a high intracellular concentration of calcium. The question arises as to whether that cytoplasmic membrane system could be the antiporter running in reverse. Since energy poisons inhibit uptake, it seems unlikely that the sporulating cell is energy depleted. So, the normal proton-pumping systems of the cell would produce a ΔpH which should couple to the antiporter for extrusion of calcium. More likely an entirely separate system is responsible for uptake. Seto (1979) has shown that an artificially imposed membrane potential can drive uptake of calcium into sporulating cells of *B. megaterium*. Since the process is saturable, an electrophoretic calcium uniporter seems reasonable. Golub and Bronner (1974) showed, though, that the antiporter was still present in the cytoplasmic membrane of sporulating cells. How do the two systems interact to prevent futile cycling of calcium? Seto (1979) demonstrated that nigericin, which dissipates ΔpH through potassium/proton exchange, greatly enhances calcium uptake by sporulating cells. Since the antiporter is driven by ΔpH, stimulation of net calcium uptake by nigericin is consistent with inhibition of calcium extrusion. Interestingly, Seto (1979) also found that vegetative cells likewise accumulate calcium in the presence of nigericin. These results suggest that two calcium transport mechanisms are active both in vegetative and sporulating cells, the uniporter for uptake and antiporter for extrusion. It might be expected that in vegetative cells the uniporter is relatively inactive, either because the number of carriers is low or because of regulation of activity. During the onset of sporulation the balance shifts in favour of the uniporter, so that net direction of calcium movement changes from efflux to influx. Again, this could be the result of synthesis of new uniporter, activation of pre-existing uniporter, inhibition of synthesis or activity of the antiporter, or any combination of effects. This is not unlike the situation in other organisms; as discussed above, the net direction of calcium movement in *E. coli* similarly seems to be a balance between influx and efflux systems (K. Ueyama, T. Tsuchiya and B. P. Rosen, unpublished results). In *Streptococcus faecalis* calcium fluxes in both directions have also been observed (Kobayashi *et al.*, 1978). The difference is that bacilli apparently have adapted the systems for use in the developmental process of sporulation.

The subsequent transfer of calcium from cytosol to forespore is less well understood. The work of Hogarth and Ellar (1978, 1979) definitely support the existence of carrier-mediated energy-independent transport. The forespore membrane is actually a two-membrane system. The outer forespore membrane is derived from invagination of the cytoplasmic membrane, enveloping the inner forespore membrane, which also arises from the cytoplasmic membrane; but the two membranes have opposite orientations (Ellar, 1978). If both were active in calcium transport, one would expect them to work in opposite directions. Perhaps this could result in the observed facilitated mode of transport, or perhaps an entirely different system is used for forespore calcium transport. Also, *in vivo* calcium accumulation into the forespore is apparently dependent on the synthesis of DPA (Ellar, 1978). The site of DPA synthesis is not known. If synthesized in the cytosol, is the calcium–DPA complex transported together into the forespore? Hoarth and Ellar (1978) found no effect of DPA on calcium uptake by isolated forespores, which seems to rule out cotransport of the two. Calcium uptake into the cytosol of sporulating cells occurs prior to DPA synthesis, so that the first step in deposition of calcium in spores is independent of DPA (Ellar, 1978). If, on the other hand, DPA is synthesized inside of the forespore, then the formation of the calcium–DPA chelate may provide the driving force for calcium accumulation inside the forespore.

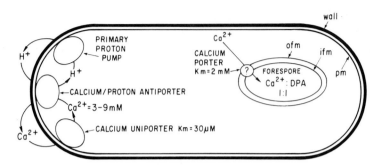

Fig. 2. *Calcium circulation in sporulating bacillus. The level of cytosolic calcium is a function of the ratio of the uniporter and antiporter both of which couple to components of the electrochemical proton gradient, as described in Fig. 1. During the onset of sporulation, the net direction of calcium movement is inwards, creating cytosolic calcium concentrations in the range of 3–9 mM. A calcium porter of unknown properties catalyses entry of calcium into the forespore, where it forms a complex with DPA, trapping the calcium in the forespore. imf, inner forespore membrane; omf, outer forespore membrane; pm, plasma membrane. Modified by Ellar (1978), with permission.*

In summary, we would favour the model shown in Fig. 2. Bacilli have a calcium circulation composed of calcium influx by an electrophoretic uniporter driven by the membrane potential and calcium extrusion by an antiporter coupled to the pH gradient. Both of these couple to proton circulation of the primary proton pumps of the cell such as the respiratory chain or BF_0F_1. The net direction of flux is dictated by the activity of the two calcium porters, either through a different number of carriers or through a differential regulation. In vegetative cells the antiporter is predominant, and the cytosolic calcium concentration is maintained low. During the onset of sporulation the uniporter attains dominance, and calcium begins to be accumulated in the cytosol. The developing forespore has a low affinity energy-dependent calcium porter. Assuming that DPA is synthesized inside of the forespore, calcium moves down its concentration gradient from cytosol to forespore and is deposited as the calcium–DPA complex. Admittedly, this model is derived from an outsider's view of this fascinating developmental process.

Although less well documented, adaptation to growth at high temperatures may also involve a shift in the balance between calcium influx and efflux. As mentioned previously, *Bacillus stearothermophilus* requires calcium for growth at temperatures over about 45°C. Stahl and Ljunger (1976) showed that at 55°C *B. stearothermophilus* accumulated calcium 50-fold, assuming that all of the calcium ions were free, and that the calcium could be released by toluene treatment of the cells. Uptake was sensitive to CCCP, suggesting active transport (Stahl, 1978a). Stahl (1978b) subsequently demonstrated that strains of *B. megaterium* selected for ability to grow at high temperature likewise accumulated calcium. In these strains, as in *B. stearothermophilus*, cells grown at low temperature neither required nor accumulated calcium, but required calcium for growth at higher temperatures. Although the evidence that actual accumulation of calcium is a prerequisite for thermophilic growth is not entirely convincing, the fact that both obligative and facultative thermophiles derived from a mesophile require calcium is indicative of a role for calcium in thermophily. Additionally, that thermophiles can be obtained from mesophiles by simple selection probably means that only a small number of genetic alterations can lead to temperature tolerance. It would be interesting, indeed, if one of those changes had to be induction of a calcium influx system.

Azotobacter vinelandii

Bhattacharyya and Barnes (1976) reported that membrane vesicles from *A. vinlandii* accumulate calcium with energy supplied from respiration or ATP hydrolysis via the BF_0F_1. In this organism the BF_0F_1 is latent and must be treated with trypsin before ATP hydrolysis can be coupled to proton pumping; only after trypsinization could ATP hydrolysis be coupled to calcium transport. The ability of the vesicles to transport calcium depends on the buffer in which the vesicles are prepared. Although the reason for the difference is unknown, vesicles prepared by osmotic lysis of spheroplasts in a phosphate buffer result in vesicles which are mainly right-side-out and incapable of calcium transport, whereas those made in Tris buffer are apparently everted and accumulate calcium. It is also not known whether the vesicles which accumulate calcium are derived from the cytoplasic membrane or from the internal membranes involved in nitrogen fixation.

Calcium transport by these vesicles occurs with a K_m of 48 μM and a V_{max} of 45 nmol/min/mg of membrane protein (Barnes *et al.*, 1978). Sensitivity to uncouplers and BF_0F_1 inhibitors demonstrate that the porter is coupled to the protonmotive force. An artificially imposed gradient of protons drives calcium uptake. Thus, this system is most likely a calcium/proton antiporter. In this system no precipitating co-ion is needed for extensive calcium uptake; it may be that the vesicles have less nonspecific calcium permeability. One disturbing property is the sensitivity to certain ionophores capable of catalysing calcium/proton exchange, such as A23187. A23187 completely inhibited calcium transport at $10 \mu M$, and was half-maximally inhibitory at $1 \mu M$. Given the orientation of the pH gradient (acid interior), one might expect A23187 to act similarly to the natural antiporter, stimulating uptake, as has been found for *E. coli* (Brey and Rosen, 1979a). One possible explanation is that A23187 catalyses exchange of intravesicular calcium for extra-vesicular magnesium.

Mycobacterium phlei

Kumar *et al.* (1979) found that membrane vesicles produced by sonication of *M. phlei* accumulate calcium with energy supplied by respiration. As in the case of *A. vinelandii*, the BF_0F_1 of *M. phlei* is latent, but trypsinized vesicles can couple ATP hydrolysis to calcium transport. Again, sensitivity to uncouplers suggests a calcium/proton antiporter mechanism, and a direct effect of calcium in decreasing the magnitude of the pH gradient was found, consistent with calcium/proton exchange. The K_m of $80 \mu M$ and V_{max} of $16 \, nmol/min/mg$ of membrane protein is similar to the kinetic properties of antiporters from other bacteria. In *M. phlei* A23187 was found to produce rapid efflux of calcium from the vesicles, although the reason for this apparently paradoxical effect is not known. Although a precipitating co-ion was not required for net accumulation of calcium, the presence of phosphate greatly stimulated uptake.

Rhodopseudomonas capsulata

Rhodoseudomonas capsulata has two sets of membranes: a cytoplasmic membrane and an internal membrane system of chromotophores involved in photosynthesis. Using chilled cells and cells poisoned with CCCP Jasper and Silver (1978) demonstrated uptake of calcium into *R. capsulata*. No uptake was observed in energized cells, leading to the conclusion that influx in chilled or poisoned cells is due to the presence of an efflux system. Chromatophores, which have an orientation similar to everted vesicles, actively accumulated calcium with energy supplied by respiration. Presumably, uptake into chromatophores would be the equivalent of removal of calcium from the cytosol. Since chromatophores produce a proton motive force which is positive and acid interior, a calcium/proton antiporter is the most likely mechanism of calcium accumulation.

Halobacterium halobium

While most bacteria use H^+ as the major coupling ion for bioenergetics, *Halobacterium halobium* uses protons for primary pumps but Na^+ as the coupling ion for transport. As might be expected, therefore, Belliveau and Lanyi (1978) showed that calcium/sodium exchange occurs in membrane vesicles of *H. halobium*. In this study right-side-out membrane vesicles were used rather than everted vesicles. To provide a driving force, the vesicles were loaded with 3 M NaCl. Calcium transport was sensitive to monesin, which exchanges sodium for protons, dissipating the sodium gradient, but was insensitive to FCCP. Formation of a potential by illumination had no effect on efflux of calcium from vesicles, suggesting that the carrier catalyses an electroneutral exchange, where $Na^+ : Ca^{2+} = 2$.

Clostridium perfringens

While investigating the properties of the BF_0F_1 of *Clostridium perfringens*, Hasan and Rosen (1979) found that a membrane vesicle preparation from this strict anaerobe was capable of calcium accumulation when energy was supplied in the form of ATP. Uptake was sensitive to FCCP, which dissipates the proton motive force, and to DCCD, which inhibits the BF_0F_1, suggesting that calcium uptake was coupled to the proton motive force generated by ATP hydrolysis via the BF_0F_1.

Streptococcus faecalis

While all of the bacterial species discussed above apparently maintain cytosolic calcium levels below that of the external medium by coupling secondary calcium/proton antiporters to proton fluxes derived from the proton motive force, *Streptococcus faecalis* uniquely couples calcium extrusion directly to ATP. In one of the more elegant studies of calcium fluxes in bacteria, Kobayashi *et al.* (1978) showed that this Gram positive anaerobe has net calcium extrusion when the cytosolic pH is acidic. The direction of calcium movement is a balance between electrophoretic uptake and active extrusion. Expecting the extrusion system to behave similarly to the calcium/proton antiporters of other bacteria, they were surprised to discover that it was not dependent on a protonmotive force. Using a vesicle preparation which contained a significant fraction of everted membranes, they observed calcium uptake in the presence of ATP. Since

S. *faecalis* is an anaerobe which lacks a complete respiratory chain, a protonmotive force is generated by the BF_0F_1 through ATP hydrolysis. Thus, dependence on ATP hydrolysis was expected. What was not expected was that uptake proved to be insensitive to uncouplers or to inhibitors of the BF_0F_1. Furthermore, the exchange reaction also required ATP but was uncoupler resistant. Thus, calcium uptake into everted vesicles, which should correspond to extrusion from intact cells, does not require a protonmotive force. Although ATP hydrolysis concomitant with calcium uptake could not be demonstrated, the most logical conclusion is that S. *faecalis* uses an ATP-driven calcium pump for calcium extrusion. In confirmation of this, Kobayashi *et al.* (1978) found that calcium-loaded cells extruded calcium when ATP was generated by glycolysis even when the proton motive force was clamped at zero with gramacidin. In the absence of glycolysis, though, neither efflux nor exchange occured; thus even downhill movement of calcium requires ATP, as was observed in vesicles.

Under certain conditions calcium uptake into cells was observed. Addition of nigericin, which caused acidification of the cytosol through K^+/H^+ exchange, stimulated calcium uptake. Imposition of a membrane potential to starved cells, which lack ATP, similarly caused calcium uptake. Sodium-loaded cells, which do generate a large $\Delta\psi$ but not a ΔpH, accumulated calcium as well. The results are consistent with an electrophoretic calcium influx driven by $\Delta\psi$. Only when the rate of uptake exceeds the rate of efflux is net uptake observed. Since acidification of the cytosol results in net calcium uptake, the authors concluded that the ATP-driven calcium pump is inhibited by an acidic cytosol. Although it is difficult to imagine a cell which is grossly leaky to calcium, the authors were unable to find evidence for a carrier-mediated calcium uptake system: uptake appeared to be nonsaturable, and at high external concentrations of calcium assumed massive proportions. It is possible, though, that at more physiological concentrations of calcium, an electrophoretic, carrier-mediated system is responsible for calcium uptake, but that at the higher levels of calcium this uptake is masked by a more general leak.

The existence of a primary calcium pump is remarkable. In general, bacteria utilize secondary porters for cation fluxes (Rosen and Kashket, 1978). Aside from the proton pumps such as the BF_0F_1, the only other cation-translocation ATPase thus far found in bacteria is the K^+-ATPase of E. *coli* (Laimins *et al.*, 1978). Since E. *coli* is a Gram negative facultative aerobe, and S. *faecalis* is a Gram positive anaerobe, the two are not very closely related. Other Gram positive bacteria such as the bacilli and clostridia were found to have secondary calcium/proton antiporters. Cation-translocating ATPases are commonly found in the plasma membrane of eukaryotes, so the evolutionary relationship between those pumps and the few found in bacteria is of considerable interest. Perhaps such pumps evolved very early, were retained to a large degree by the ancestor of the eukaryotic plasma membrane, but were largely replaced in the bacterial membrane by secondary porters.

ISOLATION OF CALCIUM CARRIER PROTEINS

The biochemistry of bacterial calcium transport proteins has shown a remarkable lack of progress. The only documented report of solubilization and partial purification of a calcium carrier is by Lee *et al.* (1979), who extracted membranes of *Mycobacterium phlei* with cholate and found that the extract could stimulate calcium uptake in reconstituted systems. After chromatography of the extract on Sepharose CL-6B, which is a hydrophobic molecular sieve, and SE-Sephadex, an ion exchange resin, an eight-fold purification of activity was effected. Activity is defined as the ability of the solubilized extract to stimulate calcium transport in the vesicles which remain after extraction. Reconstitution into liposomes made of *M. phlei* lipids was also attempted. While some uptake was noted, there was little time dependency. The ionophore A23187 inhibited uptake by reconstituted systems, as observed also for everted vesicles. Again, this is a rather unexpected phenomenon which might be explained by Ca^{2+}/Mg^{2+} exchange. In each case, the energy source for calcium uptake was an artificially imposed ΔpH, suggesting that a protein capable of calcium/proton exchange was present in the extract.

Two other preliminary claims of solubilization and reconstitution of calcium/proton antiport activity into liposomes have been reported. Zimniak and Barnes (1980) reported reconstitution of calcium carrier proteins from *Azotobacter vinelandii* into egg lecithin liposomes. Rosen *et al.* (1981) found that everted membrane vesicles from *E. coli* could be solubilized with octylglucoside. If the membrane protein were reconstituted into asolectin liposomes, calcium uptake was observed upon imposition of a ΔpH, acid exterior, formed by dilution of the vesicles into acidic buffer. Pronase-treated extracts exhibited no calcium transport activity. Addition of A23187 stimulated calcium uptake 10-fold and was effective by itself in liposomes, as would be expected from the known calcium/proton exchange activity of the ionophore.

Finally, Solioz and Carafoli (1981) have preliminary evidence that the Ca^{2+}-translocating ATPase of *Streptococcus faecalis* can be solubilized and reconstituted into liposomes capable of ATP-driven calcium transport. Activity was inhibited by vanadate, an inhibitor of eukaryotic plasma membrane cation pumps. However, calcium-dependent ATPase activity corresponding to the proposed calcium pump was not observed, most likely because of the presence of other ATP hydrolases such as the BF_0F_1 in the crude extract.

FUNCTIONS OF CALCIUM IN BACTERIA

It would be fair to say that at present no function has been found for intracellular calcium in bacteria. Calcium plays a role in a number of extracellular processes, as discussed below. The question arises as to whether passive entry of calcium into cells is unavoidable, and calcium extrusion systems function solely as bilge pumps to prevent its accumulation, to paraphrase Kobayashi *et al.* (1978). Calcium phosphate salts are rather insoluble, and intracellular phosphate levels are high in all bacteria. Calcium accumulation might lead to calcification, as is found after cell death in high calcium medium (Ennever *et al.*, 1974). A few enzymes are inhibited by calcium. *Escherichia coli* ribonucleases are inhibited *in vitro*, and *in vitro* protein synthesis is affected consequently (Cremer and Schlessinger, 1974). Although the *in vitro* ATP hydrolytic activity of most F_0F_1's is stimulated by calcium or magnesium, Tsuchiya and Rosen (1976b) demonstrated that in *E. coli* vesicles ATP synthesis by the BF_0F_1 required magnesium but was inhibited by calcium. These findings do not exclude a role for calcium in intracellular events. First of all, the inhibition of cellular enzymes by calcium is most likely a consequence of calcium extrusion rather than cause of it. If calcium extrusion systems had evolved for other purposes, intracellular enzymes would have had no selective pressure to evolve a specificity for or against calcium. Second, the levels of calcium which might be expected to occur in the cytosol are much lower than those used in most of the studies quoted above. It is possible that small changes in calcium concentrations around a very low cytosolic concentration are regulatory signals. Alternatively, it may be the flux itself which is important rather than the absolute concentration. By analogy with the known functions of calcium fluxes in behaviour and development of eukaryotic microbes the use of calcium fluxes as regulators in bacteria would not be unreasonable. Examples of such systems will be discussed below. Still, these arguments for a nontrivial role of calcium extrusion systems are weak. Direct experimental evidence is essential.

Extracellular Roles for Calcium in Bacteria

Exoenzymes

Aside from the generalized growth requirement for calcium shown by certain bacteria, as discussed above, there are a number of more specific functions. Bacteria secrete many proteins, some of which are released into the medium, some of which remain in periplasmic spaces, and others which remain associated with membranes. Most are enzymes involved in bacterial nutrition such as proteases, saccharidases, or nucleases. These break down complex polymers allowing for uptake of the resulting metabolic building blocks. In other cases the exoenzymes are protective devises capable of inactivating antibiotics such as penicillin. While not a universal property of exoenzymes, one striking property of many is their activation and/or stabilization by calcium (Pollock, 1962). In some cases, such as the thermolysin of B. thermoproteolyticus or the α-amylase of B. subtilis, calcium is a required cofactor, converting the apoenzyme into functional enzyme (Feder et al., 1971; Smolka et al., 1971). In other cases, such as the E. coli phospholipase A1, it is not clear whether the ion is a cofactor or part of the substrate, e.g. a calcium–phospholipid complex (Scandella and Kornberg, 1971). Calcium could have an even less specific role such as altering the packing of phospholipids in the membrane, making them more accessible to phospholipases. In most cases, however, magnesium cannot substitute for calcium. Considering that exoenzymes are synthesized in the cytosol, the lack of magnesium stimulation is logical. Activation of those catabolic enzymes intracellularly would lead to breakdown of cytosolic polymers. This is analogous to the compartmentation of catabolic enzymes in the lysozomes of eukaryotic cells. Secretion of the protein as an apoenzyme with extracellular activation by calcium presupposes a calcium-free cytosol: the evolution of calcium extrusion systems must have preceded that of exoenzymes, which evolved to take advantage of the low intracellular calcium, rather than calcium efflux systems evolving to prevent activation of pre-existing exoenzymes.

DNA transport

Calcium plays an essential role in processes which require uptake of exogenous DNA such as transformation and transfection. Pneumococcal transformation has long been known to require calcium (Fox and Hotchkiss, 1957). Seto and Tomasz (1976) recently investigated the role of calcium in this process. They found that the step in which calcium is required is the actual uptake step, rather than binding of the DNA to a membrane receptor. Magnesium not only cannot replace calcium, but appears to antagonize the effect of calcium on uptake by activating nucleases. On the other hand, activation of the cells to competence required magnesium but not calcium. Thus, calcium is not necessary for the initial steps of transformation, but only for the DNA transport step. What the molecular role of calcium is in this process has not been determined. One possibility is that the actual substrate for transport is calcium–DNA. Another equally plausible possibility is that calcium alters the conformation of the membrane by interaction with phospholipids, making the membrane more capable of DNA uptake. Finally, a direct role of calcium as a cofactor of an enzyme or even a DNA transport protein cannot be ruled out.

Escherichia coli differ from pneumococci in that competence for DNA uptake is not a genetic factor, but *E. coli* can be made competent for transfection by exogenous viral (Mandel and Higa, 1970) or plasmid (Cohen *et al.*, 1972) DNA by treatment with calcium. The process is an odd one, where the order of the steps are crucial. The cells must be incubated with calcium at $0°C$ for a considerable period, followed by a heat pulse at $42°C$. Without the heat pulse, the cells do not become competent. Similarly, calcium is necessary in the chilling stage. But once the cells are made competent, they remain competent for weeks if stored frozen or chilled. Taketo (1975), on the other hand, found no requirement for the heat pulse, but did find an absolute dependence on chilling the cells in the presence of calcium. Using the replicative form of ϕA phage, he found that the most effective cation for transfection is barium, with the order of effectiveness being $Ba^{2+} > Ca^{2+} > Sr^{2+} > Mg^{2+}$. Rather than calcium playing a role in the actual uptake of DNA, Taketo has postulated that the effect of calcium or barium is to promote crystallization of the surface phospholipids of the membrane, putting them into a form which allows for DNA uptake. Grinius (1980) has expanded on this idea, suggesting that calcium also enters the cell with the DNA. The calcium/proton antiport couples to the proton motive force to establish an electrochemical calcium gradient which provides the driving force for DNA uptake. Again, these are only hypotheses, and further work is necessary for determination of the molecular mechanism of DNA uptake and the role of calcium in that process.

Intracellular Roles for Calcium in Bacteria

Sporulation
The movement of calcium into sporulating bacilli has been described above. That calcium accumulation occurs concomitantly with sporulation is clear. That calcium is a major component of the spore is clear. What is less clear is whether calcium plays an essential role in sporulation, and if so, what the role is. Speculation about this phenomenon has been summarized by Silver (1977) without a satisfactory conclusion. Spores formed in calcium-free medium lack heat resistance, suggesting a role of calcium in heat resistance. Yet spores from DPA negative mutants contain little calcium but are still heat resistant.

In the somewhat related process of encystment of *A. vinelandii* calcium is required for synthesis of the polysaccharides of the slime layer (Page and Sadoff, 1975). In this case calcium fluxes are not involved, and the process is more an activation of exoenzymes involved in slime synthesis.

Thermophilicity
As discussed above, Stahl and Ljunger (1976) have postulated that calcium uptake into *B. stearothermophilus* is essential for growth at high temperature. Thermophilic bacilli selected from mesophilic strains simultaneously develop a requirement for calcium and calcium transport systems (Stahl, 1978b). However, the need for intracellular calcium has not been proven, nor has the exact role of calcium in thermal stability been elucidated.

The Roles of Calcium Fluxes in Eukaryotic Microbes

Calcium transport systems in eukaryotic microbes
Eukaryotic microbes have both influx and efflux systems for calcium, and in some cases the functions of these systems are known. Uptake of calcium by various fungi has been studied in several laboratories (Cuppoletti and Segel, 1975; Boutry *et al.*, 1977; Roomans *et al.*, 1979). At least one fungus *Neurospora crassa*, has been shown to have a calcium/proton antiporter in the cytoplasmic membrane

(Stroobant and Scarborough, 1979). This antiporter couples to the proton motive force established by the plasma membrane H^+-translocating ATPase (not an F_0F_1!) to lower the cytosolic calcium levels.

The fucoid algae *Fucus* and *Pelvetia* have transport systems for influx and efflux of calcium (Robinson and Jaffe, 1975). These related organisms are unique in that the location of the two systems are on the opposite poles of the developing egg. As the egg differentiates into rhizoid on the one end and thallus on the other, it passes an electrical current through itself. The current-carrying ion is calcium, which is pumped actively out of the tentative thallus end and enters through the tentative rhizoid. How the fertilized egg redistributes the calcium transport systems is not known. The total flux of calcium remains constant during development, which suggests that it is an actual redistribution rather than activation or synthesis of carriers. Influx appears to be passive, but whether it is mediated by an electrophoretic porter is not known. Likewise, the efflux system at the thallus end appears to be a pump, but whether it is a calcium-translocating ATPase is not known. Robinson and Jaffe (1975) hypothesize that the calcium current polarizes the cytosol of the egg by producing an electric field across it. The field could then cause partitioning of cytosolic (and membrane) components, resulting in differentiation.

Calcium fluxes and behavioural responses in Paramecium

The ciliated protozoan *Paramecium* exhibits a primitive type of behaviour characterized by the avoidance of high concentrations of ions such as sodium or potassium or of solid objects into which it bumps. The avoidance reaction occurs by means of a reversal of the direction of ciliary beating, so that when the animal hits an object or swims into a region of salt, the cilia reverse, and the animal swims backwards. After a time the cilia reverse again to their normal direction, and the animal swims forward again, but preferably in a direction which will avoid whatever repelled it in the first place. This phenomenon has been studied in detail using the tools of different disciplines: electrophysiology, genetics and biochemistry. Several reviews have appeared on this topic (Eckert, 1972; Kung *et al.*, 1975), so that only the outlines of the process will be repeated here. The initial event following stimulation is an activation of a gated calcium channel, which allows for electrophoretic calcium entry, depolarizing the membrane. The increased cytosolic calcium directly affects the cilia, producing reversal and avoidance. The calcium gate closes, and the membrane repolarizes. Finally, the calcium is pumped out of the cell, and the lowered internal calcium results in reversal of the cilia to the original orientation.

The electrophysiological evidence for the calcium channel is convincing (Satow and Kung, 1979). Uptake measurements using $^{45}Ca^{2+}$ and $^{133}Ba^{2+}$ confirm its existence (Browning and Nelson, 1976; Ling and Kung, 1979). "Pawn" mutants, which do not show an avoidance reaction, lack the calcium depolarization step (Kung and Eckert, 1972) and similarly do not show $^{45}Ca^{2+}$ uptake (Browning et al., 1976). Pawn mutants fall into three complementation groups: pwA, pwB and pwC. It is not known which, if any, code for the structural gene of the calcium channel. At least one, pwA, appears to be a regulatory gene affecting the number of channels (Satow and Kung, 1980).

The calcium efflux system is less well characterized. Cells preloaded with calcium at $0°C$ could be used for efflux studies after a temperature shift to $23°C$ (Browning and Nelson, 1976). Neither NaN_3 nor NaCN inhibited efflux, even though cellular ATP levels were reduced by 95%. Thus, although the calcium which entered through the channel must be pumped out of the cell again to allow for continued behaviour, information on the nature of the pump is lacking.

Calcium and bacterial chemotaxis
It is tempting to speculate that bacterial chemotaxis, by analogy with the behavorial responses of *Paramecium*, also involves calcium. The study of chemotaxis has proven to be both exciting and biochemically complicated (Macnab, 1978, Springer et al. 1979). Briefly, attractants or repellants bind to chemoreceptors on the cell membrane. This triggers a series of biochemical reactions (as opposed to electrical) which have the net result of altering the frequency of reversal of the rotation of the flagellum. In peritricous bacteria the flagella form a bundle which propels the cell in one direction. When the flagella reverse their direction of rotation, the bundle unwraps, and the cell tumbles. At some point the flagella spontaneously reverse again the direction of rotation, the bundle reforms, and the cell again swims. The tumbling ensures a different direction of swimming. Attractants depress the tumbling frequency, so that the "random walk" is biased in the direction of the attractant. Repellants increase the tumbling frequency (which is equivalent to increasing the frequency of flagellar reversals), so that swimming into a region of increasing repellant concentration causes increased tumbling. The net effect is a bias in swimming away from the repellant. The link between chemoreception and flagellar reversal is not yet known.

The extent of methylation of membrane proteins using S-adenosyl-methionine changes following filling of the membrane receptor. Still, there is a large gap in knowledge between the methylation step and the events surrounding flagellar reversal. By analogy with *Paramecium*, we and others (Rosen and McClees, 1974; Ordal, 1977; Brey and Rosen, 1979b; Mato, 1979) have speculated that a calcium channel opens, allowing calcium influx, which in some way causes the flagella to reverse their rotation; finally, the calcium/proton anti-porter would extrude calcium, allowing the cycle to continue. This is an attractive model for which, unfortunately, there is no evidence. The only suggestion that there might be a calcium involvement comes from the work of Ordal (1977), who showed that depletion of medium calcium with EGTA caused smooth swimming of *B. subtilis*. Smooth swimming is indicative of supression of flagellar reversal. In the presence of the ionophore A23187 calcium–EGTA added to give a free concentration of calcium of 10^{-7} M or higher caused incessant tumbling. These results are interpreted to mean that when cytosolic calcium reaches 10^{-7} M, the flagella reverse and do not recover. However, the studies do not directly demonstrate an involvement of intra-cellular calcium. For example, uncouplers act as repellants, causing tumbling. In the presence of calcium, A23187 might dissipate the pH gradient and thus act like an uncoupler. In itself cytoplasmic pH appears to be a regulator of the flagellar motor (R. M. Macnab, personal communication). Alternatively, calcium/magnesium exchange would occur, and Ordal (1976) had shown previously that removal of intracellular magnesium promoted tumbling. Other groups have been unable to repeat these observations, either in *E. coli* using mutants permeable to A23187 (R. N. Brey and B. P. Rosen, unpublished results) or in *B. subtilis* (R. M. Macnab, personal communication). Khan and Macnab (1980) were unable to find an ionic requirement for flagellar reversal in *Salmonella*.

Mato (1979) reported that S-adenosylmethionine inhibits calcium uptake into everted vesicles in the presence of a cytosolic factor. Since methylation with S-adenosylmethionine is one of the steps in chemotaxis, he postulated that the methylation reaction of chemo-taxis is for regulation of calcium fluxes. We have attempted to reproduce this work and numerous variations of the procedure without success (Brey and Rosen, unpublished results). Again, the idea is appealing; the evidence is lacking.

The recent results of Black *et al.* (1980) may be of relevance. They found that the levels of cytosolic cyclic GMP varied during the chemotactic process in *E. coli*. It is tempting to speculate that changes in intracellular calcium are involved, perhaps through the intermediary of a calmodulin-like protein. Even though no requirement for exogenous calcium has been demonstrated, the levels of calcium contained in most media might be sufficient for filling a site on a regulatory protein. Intracellular calcium concentrations could, in turn, be controlled by intracellular magnesium (Brey and Rosen, 1979a). In conclusion, the lack of supporting correlations between calcium metabolism and chemotaxis need not be interpreted negatively. A continuation of the quest for positive data may prove fruitful.

SUMMARY

Calcium is a ubiquitous ion which is necessary for the growth of some bacteria. Most, if not all, bacteria maintain cytosolic calcium levels considerably below that of the medium. In most cases calcium/proton antiporters couple to the proton motive force to extrude calcium, although in at least one case, that of *S. faecalis*, a calcium-translocating ATPase performs that function. Calcium influx into cells similarly appears to be universal, although the mechanisms are unclear. For the most part the calcium performs extracellular functions such as stabilization of cell walls and membranes, activation of extracellular enzymes, or participation in DNA uptake. Calcium may be involved in some specialized intracellular functions such as sporulation of Gram positive bacteria. The biochemistry of calcium carriers is in its infancy but is probably the direction in which the field will head in coming years.

ACKNOWLEDGEMENTS

I am grateful to R. M. Macnab, C. W. Slayman and E. N. Sorenson for their discussions and suggestions, and to G. A. Scarborough for supplying a preprint of his work. Support for the cited unpublished work from this laboratory was from grants by the National Science Foundation (PCM 77-17652) and the National Institute of General Medical Sciences of the National Institutes of Health (GM 21648).

REFERENCES

Asbell, M. A. and Eagon, R. G. (1966). *J. Bacteriol.* **92**, 380-387.
Barnes, E. M., Roberts, R. R. and Bhattacharyya, P. (1978). *Memb. Biochem.* **1**, 73-88.
Belliveau, J. W. and Lanyi, J. K. (1978). *Arch. Biochem. Biophys.* **186**, 98-105.
Bhattacharyya, P. and Barnes, E. M. (1976). *J. biol. Chem.* **251**, 5614-5619.
Black, R. A., Hobson, A. C. and Adler, J. (1980). *Proc. natn Acad. Sci. U.S.A.* **77**, 3879-3883.
Boutry, M., Foury and Goffeau, A. (1977). *Biochim. biophys. Acta* **464**, 602-612.
Brey, R. N. and Rosen, B. P. (1979a). *J. biol. Chem.* **254**, 1957-1963.
Brey, R. N. and Rosen, B. P. (1979b). *J. Bacteriol.* **139**, 824-834.
Brey, R. N., Rosen, B. P. and Beck, J. C., (1978). *Biochem. biophys. Res. Commun.* **83**, 1588-1594.
Bronner, F., Nash, W. C. and Golub, E. E. (1975). *In* "Spores VI" (Eds P. Gerhardt, R. N. Costilow and H. L. Sadoff) 356-361. American Society for Microbiology, Washington, DC.
Browning, J. L. and Nelson, D. L. (1976). *Biochim. biophys. Acta* **448**, 338-351.
Browning, J. L., Nelson, D. L. and Hansma, H. G. (1976). *Nature, Lond.* **259**, 491-494.
Brubaker, R. R. (1972). *Curr. Topics Microbiol. Immun.* **57**, 111-158.
Cohen, S. N., Chang, A. C. Y. and Hsu, L. (1972). *Proc. natn Acad. Sci. U.S.A.* **69**, 2110-2114.
Cremer, K. and Schlessinger, D. (1974). *J. biol. Chem.* **249**, 4730-4736.
Cuppoletti, J. and Segel, I. H. (1975). *Biochemistry* **14**, 4712-4718.
Dias, F. F., Okrend, H. and Dondero, N. C. (1968). *App. Microbiol.* **16**, 1364-1369.
Eckert, R. (1972). *Science, N.Y.* **176**, 473-481.
Eisenstadt, E. and Silver, S. (1972). *In* "Spores V" (Eds H. O. Halvorson, R. Hanson and L. L. Campbell) 180-186. American Society for Microbiology, Washington, DC.
Ellar, D. J. (1978). *Symp. Soc. gen. Microbiol.* **28**, 295-325.
Ennever, J., Vogel, J. J. and Streckfuss, J. J. (1974). *J. Bacteriol.* **119**, 1061-1062.
Feder, J., Garrett, L. R. and Wildi, B. S. (1971). *Biochemistry* **10**, 4552-4555.
Fox, M. S. and Hotchkiss, R. (1957). *Nature, N.Y.* **179**, 1322-1325.
Futai, M. (1974). *J. Membr. Biol.* **15**, 15-28.
Golub, E. E. and Bronner, F. (1974). *J. Bacteriol.* **119**, 840-843.
Grinius, L. (1980). *FEBS Lett.* **113**, 1-10.
Halvorson, H. (1962). *In* "The Bacteria" (Eds I. C. Gunsalus and R. Y. Stanier) Vol. 4, 223-264. Academic Press, New York.
Hasan, S. M. and Rosen, B. P. (1979). *J. Bacteriol.* **140**, 745-747.
Hertzberg, E. L. and Hinkle, P. C. (1974). *Biochem. biophys. Res. Commun.* **58**, 178-184.
Higuchi, K., Kupferberg, L. L. and Smith, J. L. (1959). *J. Bacteriol.* **77**, 317-321.
Hogarth, C. and Ellar, D. J. (1978). *Biochem. J.* **176**, 197-203.
Hogarth, C. and Ellar, D. J. (1979). *Biochem. J.* **178**, 627-630.
Jasper, P. and Silver, S. (1978). *J. Bacteriol.* **133**, 1323-1328.
Johnson, R. C. and Gary, N. D. (1963). *J. Bacteriol.* **85**, 983-985.

Khan, S. and Macnab, R. M. (1980). *J. molec. Biol.* **138**, 563-597.
Kobayashi, H., Van Brunt, J. and Harold, F. M. (1978). *J. biol. Chem.* **253**, 2085-2092.
Kojima, M., Suda, S., Hamada, K. and Suganuma, A. (1970). *J. Bacteriol.* **104**, 1010-1013.
Kumar, G., Deves, R. and Brodie, A. F. (1979). *Eur. J. Biochem.* **100**, 365-375.
Kung, C. and Eckert, R. (1972). *Proc. natn Acad. Sci. U.S.A.* **69**, 93-97.
Kung, C., C. Chang, S-Y., Satow, Y., Van Houten, J. and Hansma, H. (1975). *Science, N.Y.* **188**, 898-904.
Laimins, L. A., Rhoads, D. B., Altendorf, K. and Epstein, W. (1978). *Proc. natn Acad. Sci. U.S.A.* **75**, 3216-3219.
Lee, S-H., Kalra, V. K. and Brodie, A. F. (1979). *J. Biol. Chem.* **254**, 6861-6864.
Leive, L. (1968). *J. biol. Chem.* **243**, 2373-2380.
Ling, K-Y. and King, C. (1979). *J. exp. Biol.* **63**, 1-15.
Lusk, J. E. and Kennedy, E. P. (1969). *J. biol. Chem.* **244**, 1653-1655.
MacLeod, R. A. and Onofrey, E. (1957). *Can J. Microbiol.* **3**, 753-759.
Macnab, R. M. (1978). *Crit. Rev. Microbiol.* **5**, 291-341.
Mandel, M. and Higa, A. (1970). *J. molec. Biol.* **53**, 159-162.
Mato, J. M. (1979). *FEBS Lett.* **102**, 241-243.
Mitchell, P. (1973). *J. Bioenergetics* **4**, 63-91.
Murrell, W. G. (1967). *Adv. Microbiol. Physiol.* **1**, 133-251.
Nelson, D. L. and Kennedy, E. P. (1971). *J. biol. Chem.* **246**, 3042-3049.
Norris, J. R. and Jensen, H. L. (1957). *Nature, Lond.* **180**, 1493-1494.
Ordal, G. W. (1976). *J. Bacteriol.* **126**, 706-711.
Ordal, G. W. (1977). *Nature, Lond.* **270**, 66-67.
Page, W. J. and Sadoff, H. L. (1975). *J. Bacteriol.* **122**, 145-151.
Park, M. H., Wong, B. B. and Lusk, J. E. (1976). *J. Bacteriol.* **126**, 1096-1103.
Plack, R. H. and Rosen, B. P. (1980). *J. biol. Chem.* **255**, 3824-3825.
Pollock, M. R. (1962). *In* "The Bacteria" (Eds I. C. Gunsalus and R. Y. Stanier) Vol. 4, 121-178. Academic Press, New York.
Robinson, K. R. and Jaffe, L. F. (1975). *Science, N.Y.* **187**, 70-71.
Roomans, G. M., Thevenent, A. P. R., Van Den Berg, Th. P. R. and Borst-Pauwels, G. W. F. H. (1979). *Biochim. biophys. Acta* **551**, 187-196.
Rosen, B. P. and Brey, R. N. (1979). *In* "Microbiology 1979" (Ed. D. Schlessinger) 62-66. American Society for Microbiology, Washington, DC.
Rosen, B. P. and Kashket, E. R. (1978). *In* "Bacterial Transport" (Ed. B. Rosen) 559-620. Marcel Dekker, New York.
Rosen, B. P. and McClees, J. S. (1974). *Proc. natn Acad. Sci. U.S.A.* **71**, 5042-5046.
Rosen, B. P., Brey, R. N., Sorensen, E. N. and Tsuchiya, T. (1980). *In* "Calcium-Binding Proteins and Calcium Function" (Eds F. L. Siegel, E. Carafoli, R. H. Kretsinger, D. H. MacLennan and R. H. Wasserman) pp. 67-72. Elsevier/North-Holland, New York. In press.
Satow, Y. and Kung, C. (1979). *J. exp. Biol.* **78**, 149-161.
Satow, Y. and Kung, C. (1980). *J. exp. Biol.*, in press.
Scandella, C. J. and Kornberg, A. (1971). *Biochemistry* **10**, 4447-4456.
Seto, D. L. T. (1979). PhD thesis, Darwin College, Cambridge University.
Seto, H. and Tomasz, A. (1976). *J. Bacteriol.* **126**, 1113-1118.
Silver, S. (1977). *In* "Microorganisms and Minerals" (Ed. E. D. Weinberg) 49-103. Marcel Dekker, New York.

Silver, S. (1978). *In* "Bacterial Transport" (Ed. B. Rosen, ed.) 221-324. Marcel Dekker, New York.

Silver, S. and Clark, D. (1971). *J. biol. Chem.* **246**, 569-576.

Silver, S. and Kralovic, M. L. (1969). *Biochem. biophys. Res. Commun.* **34**, 640-645.

Silver, S., Toth, K. and Scribner, H. (1975). *J. Bacteriol.* **122**, 880-885.

Smolka, G. E., Birnbaum, E. R. and Darneall, D. W. (1971). *Biochemistry* **10**, 4556-4561.

Snellen, J. E. and Raj, H. D. (1970). *J. Bacteriol.* **101**, 240-249.

Solioz, M. and Carafoli, E. (1980). *In* "Calcium-Binding Proteins and Calcium Function" (Eds F. L. Siegel, E. Carafoli, R. H. Kretsinger, D. H. MacLennan and R. H. Wasserman) pp. 101-102. Elsevier/North-Holland, New York. In press.

Springer, M. S., Goy, M. F. and Adler, J. (1979). *Nature, Lond.* **280**, 279-285.

Stahl, S. (1978a). *Archs Microbiol.* **119**, 17-24.

Stahl, S. (1978b). *FEMS Microbiol. Lett.* **4**, 77-81.

Stahl, S. and Ljunger, C. (1976). *FEBS Lett.* **63**, 184-187.

Stroobant, P. and Scarborough, G. A. (1979). *Proc. natn Acad. Sci. U.S.A.* **76**, 3102-3106.

Taketo, A. (1975). *Z. Naturf.* **30**, 520-522.

Tsuchiya, T. and Rosen, B. P. (1975a). *J. biol. Chem.* **250**, 7687-7692.

Tsuchiya, T. and Rosen, B. P. (1975b). *J. biol. Chem.* **250**, 8409-8415.

Tsuchiya, T. and Rosen, B. P. (1976a). *J. biol. Chem.* **251**, 962-967.

Tsuchiya, T. and Rosen, B. P. (1976b). *J. Bacteriol.* **127**, 154-161.

Tsuchiya, T. and Takeda, K. (1979). *J. Biochem., Tokyo* **85**, 943-951.

van Alphen, L., Verkleij, A., Leunissen-Bijvelt, J. and Lugtenberg, B. (1978). *J. Bacteriol.* **134**, 1089-1098.

Young, E. G., Begg, R. W. and Pentz, E. I. (1944). *Archs Biochem.* **5**, 121-136.

Zahorchak, R. J., Charnetzky, W. T., Little, R. V. and Brubaker, R. R. (1979). *J. Bacteriol.* **139**, 792-799.

Zimniak, P. and Barnes, E. M. (1980). *J. supramolec. Struct. Suppl.* **4**, 82.

6

Sodium – Calcium Exchange:
Its Role in the Regulation of Cell Calcium

MORDECAI P. BLAUSTEIN and MARK T. NELSON

*Department of Physiology, University of Maryland School of Medicine,
Baltimore, Maryland, USA*

INTRODUCTION

The pioneering studies of Heilbrunn and Wiercinski (1947; and see Heilbrunn, 1940; Kamada and Kinosita, 1943) first established the fact that intracellular calcium ions control muscle contraction. Subsequently, intracellular Ca^{2+} ions were shown to trigger or modulate numerous other physiological processes including secretion (e.g. Katz and Miledi, 1967, 1969; Miledi, 1973; Douglas, 1968, 1975), phototransduction (e.g. Gold and Korenbrot, 1980; Yoshikami *et al.*, 1980), egg fertilization (e.g. Steinhardt *et al.*, 1977; Baker *et al.*, 1980) and cell reproduction (e.g. Hazelton *et al.*, 1979). The field is expanding at a very rapid rate, and additional Ca-regulated phenomena are continually being recognized.

Most of the Ca-regulated processes that have thus far been investigated appear to be controlled by intracellular ionized Ca^{2+} concentrations ($[Ca^{2+}]_i$) in the range of $10^{-7}-10^{-5}$ M (e.g. muscle contraction: see Costantin, 1977; secretion: see Baker and Knight, 1980). Indeed, direct measurements in a variety of cell types, with Ca-selective microelectrodes, indicate that $[Ca^{2+}]_i$ is normally in the range of about $10^{-7}-5 \times 10^{-7}$ M in resting cells (e.g. Marban *et al.*, 1980; Alvarez-Leefmans *et al.*, 1981; and see Ashley and Campbell, 1979). The concentration of Ca^{2+} in extracellular fluids ($[Ca^{2+}]_o$) is much higher: about 10^{-3} M in vertebrates, and about $3 \times 10^{-3}-4 \times 10^{-3}$ M in marine invertebrates (see Blaustein, 1974). Intracellular Ca^{2+} plays a fundamental role in so many physiological processes that it is essential to learn how cells regulate their $[Ca^{2+}]_i$.

Most animal cells, particularly excitable cells such as nerve and

muscle fibres, have resting membrane potentials (V_m), of $-50-$
$-90\,mV$ (cytoplasm negative). Therefore, cations such as Ca^{2+}, if
distributed according to the electrical gradient, should be present in
much higher concentration in the cytoplasm than in the extracellular
fluid. The Ca equilibrium potential, E_{Ca}, is given by the Nernst
relationship:

$$E_{Ca} = \frac{RT}{2F} \ln \frac{[Ca^{2+}]_o}{[Ca^{2+}]_i} \qquad (1)$$

where R, T and F are the gas constant, absolute temperature and
Faraday's number, respectively. If Ca^{2+} was distributed according to
the electrical gradient across the plasma membrane, E_{Ca} should equal
V_m. However, taking typical values for $[Ca^{2+}]_o$ and $[Ca^{2+}]_i$, 10^{-3}
and 10^{-7} M, respectively, we find that $E_{Ca} \simeq +120\,mV$, far different
from the aforementioned range of resting membrane potentials.
^{45}Ca tracer studies indicate that most cells are permeable to Ca^{2+};
thus, the large electrochemical gradients for Ca $(V_m - E_{Ca})$ must be
maintained by mechanisms that can extrude Ca^{2+} from the cells.

Another feature of cell Ca metabolism is also pertinent to this
discussion. In some types of cells, most of the Ca required to trigger
a physiological response comes from intracellular stores. For example,
in vertebrate skeletal muscle the rise in $[Ca^{2+}]_i$ that triggers contraction is due to Ca release from the sarcoplasmic reticulum, and
then this Ca is resequestered, causing relaxation (see Costantin,
1977). On the other hand, much of the Ca required to trigger contraction in vertebrate cardiac muscle (see Costantin, 1977; Chapman,
1979; Winegrad, 1979) and transmitter release at nerve terminals
(e.g. Katz and Miledi, 1967, 1969; Blaustein, 1975) comes from the
extracellular fluid during excitation.

Reduction of $[Ca^{2+}]_i$ may then be required to terminate the
physiological response. This reduction of $[Ca^{2+}]_i$ can be accomplished
by a variety of mechanisms. Initially, the Ca^{2+} may be rapidly buffered by cytoplasmic high-affinity Ca binding proteins (cf. Baker and
Schlaepfer, 1978), such as the parvalbumins in skeletal muscle sarco-
plasm (see Nockolds et al., 1972; Parello et al., 1978; Somlyo et al.,
1981). This rapid buffering would allow time (perhaps on the order
of tens of milliseconds) for the slower intracellular Ca-sequestering
organelles such as sarcoplasmic reticulum in muscle (see Costantin,
1977) and endoplasmic reticulum in other cells (see Blaustein et al.,
1980), to take up the Ca. Mitochondria may also be involved in Ca
sequestration (see Lehninger, 1970), although the physiological significance of the mitochondrial role in intracellular Ca buffering has

recently been questioned (e.g. Blaustein *et al.*, 1978, 1980; Brinley, 1980; Henkart, 1980).

The Ca-storing capacities of these intracellular systems are, of course, limited. In those cells in which Ca enters during activity, the Ca must be extruded during periods of rest, to avoid progressive Ca loading. Ca "pumps" in the plasmalemma must be the primary regulators of $[Ca^{2+}]_i$ over the long term, even though transfer of Ca between the cytosol and intracellular organelles may modulate $[Ca^{2+}]_i$ transiently.

PARADIGMS FOR CALCIUM TRANSPORT SYSTEMS

Calcium-dependent ATPase and Calcium Transport

In human erythrocytes, $[Ca^{2+}]_i$ is regulated by a Ca + Mg-dependent ATPase (Schatzmann and Vincenzi, 1969; Schatzmann and Burgin, 1978), that closely resembles the Na + K-dependent ATPase involved in cell Na and K regulation (see Glynn and Karlish, 1975). All of the energy for Ca extrusion from the red cells is apparently derived from ATP hydrolysis: one mole of ATP is hydrolysed by the Ca + Mg-dependent ATPase for every mole of Ca^{2+} extruded (Schatzmann, 1973). However, there is no evidence that the exiting Ca^{2+} is exchanged for any other ions, in contrast to the Na + K dependent ATPase, which normally mediates an exchange of K^+ (entering) for Na^+ (exiting).

Sodium–Calcium Counter-transport

In many other types of cells, however, such as nerve, muscle, epithelia and secretory cells, Ca^{2+} extrusion appears to depend upon the presence external $Na^+(Na_o)$ (see Blaustein, 1974). Available evidence indicates that the exit of this Ca^{2+} is tightly coupled to the entry of Na^+ (Blaustein and Russell, 1975). Because entering Na^+ moves down its steep electrochemical gradient, a number of workers have suggested that this (Ca coupled) influx of Na^+ may provide some or all of the energy needed to extrude Ca^{2+} against its large electro-chemical gradient (Reuter and Seitz, 1968; Baker *et al.*, 1969a; Blaustein and Hodgkin, 1969; and see Blaustein, 1974).

Unfortunately, despite the incontrovertible evidence for Na–Ca counter-transport in many types of cells (see below), the question of whether or not the Na electrochemical gradient serves as the sole source of energy for maintaining the Ca gradient has not been resolved. There are two complicating factors. One is that ATP influences the kinetics of Na–Ca exchange: this effect has been clearly demonstrated in squid axons and barnacle muscle fibres (see later). The second factor is that, in squid axons with very low $[Ca^{2+}]_i$ ($< 0.1 \mu M$), ATP promotes a small Ca efflux that apparently does not depend upon external Na. This flux has been termed an ATP-dependent "uncoupled" Ca efflux (see DiPolo and Beauge, 1979, 1980; and see later). DiPolo and Beauge's results raise the possibility that two different, independent plasma membrane Ca transport systems may operate in parallel in the same cell. One system, powered directly by ATP, may be analogous to the erythrocyte Ca pump; the second system may be the Na–Ca exchange system powered by energy from the Na electrochemical gradient. This possibility will be considered below.

THE THERMODYNAMICS OF SODIUM–CALCIUM EXCHANGE

If the downhill movement of Na^+ into cells provides energy for the (uphill) extrusion of Ca^{2+} via a coupled counter-transport system, we would expect the maximal Ca electrochemical potential gradient ($\Delta\bar{\mu}_{Ca}$) to be given by:

$$\Delta\bar{\mu}_{Ca} = n\Delta\bar{\mu}_{Na} \qquad (2)$$

In this expression, $\Delta\bar{\mu}_{Na}$ is the electrochemical potential gradient for Na^+, and n is the stoichiometry of the transport reaction (n Na^+ ions are exchanged for one Ca^{2+}). By expanding this expression, we find that the limiting maximal Ca^{2+} concentration gradient, $[Ca^{2+}]_o/[Ca^{2+}]_i$, obtainable with this system is:

$$\frac{[Ca^{2+}]_o}{[Ca^{2+}]_i} = \left(\frac{[Na^+]_o}{[Na^+]_i}\right)^n \exp -\frac{(n-2)V_m F}{RT} \qquad (3)$$

In several preparations for which data are available, n has a value of about 3: i.e. three Na^+ ions appear to be exchanged for one Ca^{2+} (see later).

The fundamental assumption that underlies equations (2) and (3) is that the Na–Ca exchange system is symmetrical or reversible. In other words, depending upon the prevailing electrochemical gradients for Na and Ca, the exchange mechanism should be capable of mediating either the entry of Na in exchange for exiting Ca (e.g. Blaustein and Hodgkin, 1969) or, conversely, the entry of Ca in exchange for exiting Na (e.g. Baker *et al.*, 1969a).

Thermodynamic considerations (see equation (3)) indicate that, if an Na–Ca exchange mechanism is responsible for the regulation of cell Ca^{2+}, there should be a net gain of Ca, and a rise in $[Ca^{2+}]_i$ when the Na electrochemical gradient $(\Delta\bar{\mu}_{Na} = E_{Na} - V_m)$ is reduced (where $E_{Na} = (RT/F) \ln ([Na^+]_o/[Na^+]_i)$). Indeed, this prediction is born out by experiments in a large variety of tissues (but not in human red blood cells: e.g. see Schatzmann and Vincenzi, 1969). The earliest observations along these lines were made by Wilbrandt and Koller (1948). They noted that the strength of contraction of cardiac muscle increased when external Na was reduced: i.e. lowering $[Na^+]_o$ caused $[Ca^{2+}]_i$ to increase, as expected from equation (3). Consistent with this hypothesis is their observation that lowering $[Ca^{2+}]_o$ counteracted the effect of reducing $[Na^+]_o$ (see equation (3)). These findings were confirmed and extended by Lüttgau and Niedergerke (1958; and see Niedergerke, 1963), who found that there was a large Ca influx, and net Ca gain when frog heart ventricles were incubated in media with reduced $[Na^+]_o$; there was prompt net loss of Ca and relaxation when the ventricles were replaced in control (Na-rich) Ringer's solution (also see Vassort, 1973).

Very similar observations on the induction of contractures by reduced $[Na^+]_o$ were also made in various types of smooth muscles (e.g. Hinke and Wilson, 1962; Judah and Willoughby, 1964; and see Blaustein, 1974 for additional references). In slow (tetanic) skeletal muscle, too, sustained contractures were observed when $[Na^+]_o$ was reduced (Schaechtelin, 1961).

Further evidence consistent with the Na–Ca exchange hypothesis (see equation (3)) is that twitch-type skeletal muscle (Cosmos and Harris, 1961) and smooth muscle (e.g. Leonard, 1956) show evidence of Ca gain under conditions that promote the net gain of Na — such as cooling, exposure to K-free media, or treatment with the Na pump inhibitor, ouabain. However, some caution must be exercised in interpreting these smooth muscle data because the depolarization that may result from these treatments could increase Ca entry through voltage-regulated Ca channels; this possibility does not appear to be a problem in twitch-type skeletal muscle. In this context, it is also important to note that ouabain, itself, does not appear to have any direct effect on the Ca transport system in these tissues (see Curtis, 1966; Blaustein, 1974).

EXPERIMENTAL EVIDENCE FOR SODIUM–CALCIUM EXCHANGE

The explanation for the observations reviewed in the preceding section first became apparent in the late 1960s as a result of Na-dependent Ca flux measurements carried out simultaneously by researchers in Mainz, Germany and in Plymouth, England. The Mainz group (Reuter and Seitz, 1968) observed that the rate of Ca efflux from ^{45}Ca-loaded mammalian cardiac muscle was, in part, dependent upon the presence of Na^+ in the bathing medium.

The Plymouth group made similar observations on squid giant axons (Blaustein and Hodgkin, 1968, 1969). In addition, they found that raising $[Na^+]_i$ and/or lowering $[Na^+]_o$ promoted inward Ca movement across the plasma membrane (Baker *et al.*, 1967b, 1969a). Moreover, a fraction of the Na efflux from squid axons depended upon external Ca (Baker *et al.*, 1967a, b). None of these Na and Ca fluxes was affected by concentrations of ouabain (10^{-5} M) that were sufficient to block the Na pumps in squid axons completely (Baker *et al.*, 1969b).

As a result of these observations, both the Mainz and Plymouth groups speculated that a carrier-mediated exchange of Na for Ca might play an important role in the regulation of cell Ca in both mammalian cardiac muscle (Reuter and Seitz, 1968) and squid axons (Blaustein and Hodgkin, 1968, 1969; Baker *et al.*, 1969a). They suggested that the large Ca electrochemical gradients in these cells might be attained by energy derived from the downhill movement of Na into the cells, rather than from ATP, directly. The Na gradients would have to be maintained by the ATP-dependent, ouabain-sensitive Na–K exchange pumps, operating in parallel with the Na–Ca exchange carriers. This type of scheme is illustrated in Fig. 1.

In the ensuing decade, similar observations about the interrelationship between Na and Ca movements were made in numerous other laboratories. Na–Ca exchange has now been implicated in the regulation of cell Ca in a large variety of tissues. A few examples, other than the ones listed above, include:

1. Invertebrate muscle (e.g. DiPolo, 1973; Russell and Blaustein, 1974; Ashley *et al.*, 1974).
2. Vertebrate neurons (Blaustein and Oborn, 1975; Blaustein and Ector, 1976; and cf. Blaustein 1974).
3. Photoreceptors of invertebrates (e.g. Lisman and Brown, 1972) and vertebrates (Gold and Korenbrot, 1980; Yoshikami *et al.*, 1980; Yoshikami and Hagins, 1981; Yau *et al.*, 1981).

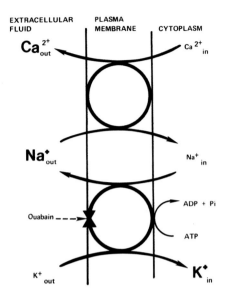

Fig. 1. *Diagram showing the relationship between the parallel Na–K exchange (Na pump) and Na–Ca exchange transport systems in the plasma membrane. As discussed in this review, energy from the hydrolysis of ATP can be stored in the Na electrochemical gradient, and then harnessed to drive Ca extrusion against a large electrochemical gradient. The Na–Ca exchange appears to be a symmetric transport system that is shown operating in the "forward" mode (with entering Na exchanging for exiting Ca). If the Na electrochemical gradient is reduced (e.g. by blocking the Na pump with ouabain) this transport system can operate in "reverse", and can extrude Na in exchange for entering Ca. Further details are given in the text.*

4. Various endocrine secretory tissues (e.g. Herchuelz *et al.*, 1980; and cf. Blaustein, 1974).
5. Intestinal, renal, and other epithelia (Blaustein, 1974; Grinstein and Erlij, 1978; Taylor and Windhager, 1979; Lee *et al.*, 1980).
6. Bone cells (Krieger and Tashjian, 1980).

This list is merely a sampling of published observations that indicates the widespread occurrence of Na–Ca exchange in animal tissues. It seems likely that similar Na–Ca exchange mechanisms will be elucidated in other types of cells as well, when they are tested appropriately. Indeed, the human erythrocyte is one of the few cell types in which careful study has failed to turn up evidence of Na–Ca exchange (see Schatzmann and Vincenzi, 1969). However dog erythrocytes do appear to have an Na-coupled Ca transport system, although its properties may differ somewhat from those of the Na–Ca exchange systems in the tissues listed above (see Parker, 1980).

STOICHIOMETRY OF SODIUM–CALCIUM EXCHANGE

The evidence outlined in the preceding sections shows that Na–Ca exchange mechanisms are present in many types of cells; the obvious implication is that Na–Ca exchange plays a role in the transport of Ca across the plasma membranes of many types of cells. But we have not yet tackled two critical questions: (1) Is there sufficient energy in the Na electrochemical gradient to maintain, via Na–Ca exchange, the Ca gradient (and low $[Ca^{2+}]_i$) that prevails in most animal cells? (2) even if there is sufficient energy available, is $[Ca^{2+}]_i$ actually regulated by Na–Ca exchange in the various cells mentioned above?

In order to answer the first question, we need information about the stoichiometry of the exchange (i.e. n in equation (3)). Indeed, as recognized early on, an electroneutral two-Na^+-for-one-Ca^{2+} exchange could not reduce $[Ca^{2+}]_i$ to the level that prevails in squid axons (see Blaustein and Hodgkin, 1969); an exchange of at least a three Na^+ for one Ca^{2+} was required. Considerable information has accrued, since 1969, concerning the stoichiometry of Na–Ca exchange in various preparations. For ease of presentation, the data for the "foward mode" of exchange (i.e. Na entry coupled to Ca exit) are considered first, followed by the data for the "reverse mode" (i.e. Ca entry coupled to Na exit).

Stoichiometry of "Forward Mode" of Sodium–Calcium Exchange

An easy, but indirect way to obtain information about the molecularity of a chemical reaction is to study its activation kinetics. Using this approach, experiments on barnacle muscle fibres (Russell and Blaustein, 1974) and squid axons (Blaustein et al., 1974; Blaustein, 1977a) indicate that Ca efflux is a sigmoid function of the external Na concentration. These activation curves can be fitted to the Hill equation with a Hill coefficient of 2.6–3 (but see Baker and McNaughton, 1976b); the implication is that three (or more) external Na^+ ions may be required to activate the efflux of one Ca^{2+}. Direct measurements of internal Ca-dependent Na influx and external Na-dependent Ca efflux in squid axons also provide evidence for a three-for-one stoichiometry (Blaustein and Russell, 1975).

If the exchange is, indeed, $\geqslant 3:1$, and not electroneutral, there are several interesting consequences: (i) the rate of exchange will be influenced by the membrane potential, and (ii) the exchange will be associated with the (net) flow of ionic current across the plasma membrane. Such a current could be detected directly (with voltage clamp methods), or indirectly as a voltage drop (current × resistance, or $I \times R_m$ drop) across the membrane resistance. In squid axons, Na_o-dependent Ca efflux is *slowed* by membrane depolarization (Blaustein *et al.*, 1974; Mullins and Brinley, 1975), and speeded up by hyperpolarization (Mullins and Brinley, 1975); the implication is that the exchange involves the net entry of positive charge. Unfortunately, these data do not provide explicit information on the number of Na^+ ions that exchange for one Ca^{2+}; they merely indicate that *n* has a value $\geqslant 3$.

A voltage $(I \times R_m)$ drop across the plasma membrane of giant barnacle muscle fibres that appears to correlate with Na–Ca exchange has recently been measured (Nelson and Blaustein, 1981b). In fibres with high $[Ca^{2+}]_i$ $(0.5-1.0 \times 10^{-5}$ M), the activation of Ca efflux by external Na is accompanied by a parallel depolarization. A Na_o-dependent Ca efflux of about 10 pmol/cm² s is associated with an inward current that depolarizes the muscle by ~ 2 mV. With a membrane resistances of about $2 \, K\Omega.cm^2$ (Nelson and Blaustein, 1980) these Ca flux and membrane potential data imply that the 10 pmol/cm² s Ca efflux is coupled to a Na influx of about 30 pmol/cm² s (i.e. a $3 \, Na^+ : 1 \, Ca^{2+}$ exchange); thus, there is a net inward current corresponding to 10 pmol Na/cm² s.

The data for vertebrate cardiac muscle are somewhat less clear cut. Early experiments on the interrelationship between $[Na^+]_o$, $[Ca^{2+}]_o$, Ca efflux and tension were interpreted in terms of a Na–Ca exchange carrier with a stoichiometry of two Na^+ for one Ca^{2+} (see Reuter and Seitz, 1968; Glitsch *et al.*, 1970; Jundt *et al.*, 1975; Chapman and Ellis, 1977). However, the expected level of $[Ca^{2+}]_i$ obtainable under these circumstances $(> 10^{-6}$ M) would cause considerable contractile activation (see Costantin, 1977; Winegrad, 1979). Recently, Horackova and Vassort (1979) and Chapman and Tunstall (1980; and see Chapman, 1979) presented evidence that, in frog heart, Na–Ca exchange is electrogenic, with a stoichiometry of about 3:1. Also, Sheu and Fozzard (1981) calculated a value of 2.6:1 from their observations on $[Na^+]_i$ and $[Ca^{2+}]_i$ in sheep ventricular muscle.

Stoichiometry of "Reverse Mode" of Sodium–Calcium Exchange

The Na–Ca exchange carriers also appear to mediate the entry of Ca in exchange for exiting Na, and an effort has been made in several laboratories to determine the stoichiometry of this exchange. The activation of Ca influx by $[Na^+]_i$ has been studied in mammalian cardiac muscle (Glitsch *et al.*, 1970); these data indicate that Ca influx is approximately proportional to $[Na^+]_i^2$, which may be consistent with a stoichiometry of at least $2:1$. Direct measurements of Ca influx and Ca_o-dependent Na efflux indicate that the influx of one Ca^{2+} ion is coupled to the exit of 3–5 Na^+ ions in squid axons (Baker *et al.*, 1969a) and about 3 Na^+ in *Myxicola* axons (Sjodin and Abercrombie, 1978). Further evidence that this exchange is electrogenic comes from the observations that Ca_o-dependent Na efflux (i) is *inhibited* by hyperpolarization in squid axons (Baker and McNaughton, 1976b), and (ii) is associated with a membrane hyperpolarization in barnacle muscle (Nelson unpublished data).

A recent development has been the use of plasma membrane vesicles from cardiac muscle homogenates for studies of Na–Ca exchange. The data from several laboratories demonstrate that Ca influx is promoted by intravesicular Na, while Ca efflux is stimulated by external Na (Reeves and Sutko, 1979; Carone and Carafoli, 1980; Pitts, 1979). [We must, of course, bear in mind the caveat that some of these vesicles may be inside-out, with the cytoplasmic surface facing the external medium (see Bers *et al.*, 1980). Thus, forward and reverse modes of Na–Ca exchange may be confused here.]

Pitts (1979) measured both Na influx and Ca efflux in these vesicles, and concluded that the stoichiometry of the Na–Ca exchange was $3:1$ (presumably in the "forward mode"). A similar conclusion was reached by Kadoma *et al.* (1981), who studied the activation of Ca efflux from cardiac sarcolemmal vesicles by external Na. Reeves and Sutko (1980) and Caroni *et al.* (1980) studied the redistribution of lipophilic cations when Na–Ca exchange was activated. Both groups found that the vesicle interiors became negative as Ca moved in and Na moved out; this is consistent with the hypothesis that more than two Na^+ exchange for one Ca^{2+} (presumably in the "reverse mode"). Further evidence that Na–Ca exchange is electrogenic comes from the observation that Na_i-dependent Ca influx in cardiac sarcolemmal vesicles is stimulated by inside-positive potentials, and inhibited by inside-negative potentials (Philipson and Nishimoto, 1980; Bers *et al.*, 1980). These data also suggest that three or more Na^+ are exchanged for one Ca^{2+}.

In sum, while not yet conclusive, the accumulating data appear to favour an Na–Ca exchange stoichiometry of about $3:1$ for both "forward" and "reverse" modes, in both vertebrate and invertebrate preparations. The apparent invariance of the coupling ratio may, of course, imply that the Na–Ca exchange mechanism has a fixed stoichiometry, and has not changed much during the course of evolution of the higher animals.

Energetics of Sodium–Calcium Exchange

The foregoing discussion indicates that the Na–Ca exchange system may be symmetric, since the stoichiometry for both the "forward" *and* "reverse" modes is about 3 : 1. This conclusion carries with it several important implications. In the first place, Na and Ca concentration data from several tissues including nerve and muscle (both vertebrate and invertebrate) indicate that sufficient energy may be available from the Na electrochemical gradient, alone, to maintain the observed $[Ca^{2+}]_i$ via Na–Ca exchange, in accord with equation (3) (see Blaustein, 1974).

The idea that $[Ca^{2+}]_i$ is regulated by Na–Ca exchange is supported by the evidence that Na_o-dependent Ca efflux is observed in metabolically poisoned nerve axons (Blaustein and Hodgkin, 1968, 1969) and muscle fibres (Nelson and Blaustein, 1981a). In fact, in poisoned nerve axons (Requena *et al.*, 1979) and muscle fibres (Baker and Singh, 1980) an appropriate Na electrochemical gradient is necessary and sufficient to power net Ca (or Sr) extrusion. In order for this net Ca extrusion to take place, $\Delta\bar{\mu}_{Ca}$ must be less than $3\Delta\bar{\mu}_{Na}$ (see equation (2)). An important corollary is that, with $\Delta\bar{\mu}_{Ca} > 3\Delta\bar{\mu}_{Na}$, energy from the Ca electrochemical gradient may be used to drive Na out of the cells. This type of effect (Ca_o-dependent net Na extrusion) has, in fact, been observed in cardiac muscle (Deitmer and Ellis, 1978).

The electrophysiological implications also deserve reiteration in this context. According to equation (3), with $n = 3$, changes in V_m, alone should directly affect net Ca movements (as noted above; and see Blaustein *et al.*, 1974; Mullins and Brinley, 1975): depolarization should promote net Ca gain, while hyperpolarization should effect Ca loss. This voltage dependency may have particularly interesting consequences in tissues such as cardiac muscle (see Chapman, 1979; Wiengrad, 1979) and secretory cells (see Blaustein, 1974) that require a large influx of Ca to trigger a physiological response. During depolarization, Ca will enter the cell cytoplasm by voltage-regulated Ca-selective channels (see Reuter, 1973; Hagiwara, 1975), by release from intracellular stores, and by the Na–Ca exchange system (Horackova and Vassort, 1979; Chapman and Tunstall, 1980). Conversely, when the cells repolarize, Ca efflux will be favoured and this change in V_m will help to restore the resting $[Ca^{2+}]_i$.

The electrogenic current that is generated as a result of Na–Ca exchange (see above) may also have interesting implications in, for example, cardiac muscle (see Mullins, 1979). In cardiac muscle, all of the Ca that enters during systole must normally be extruded during the subsequent diastole. If most, or all of this Ca extrusion is mediated by the Na–Ca exchange system, there may be a substantial inward current associated with the Ca efflux. The currents carried by the Na–Ca exchange system could, conceivably, be confused with the ionic currents that move through "classical" ion-specific channels; this would result in misinterpretation of electrophysiological data. For example, Yoshikami and Hagins (1981) have suggested that the inward "dark current" carried by Na^+ in retinal rods may actually be Na^+ that is exchanging for Ca^{2+}.

THE ROLE OF ATP IN SODIUM–CALCIUM EXCHANGE

The data described above suggest that, in many types of cells, the Ca electrochemical gradient may be maintained by energy stored in the Na electrochemical gradient, in accord with equations (2) and (3). The obvious implication is that ATP does not play a direct role in Ca extrusion from these cells. However, a number of observations on squid axons and barnacle muscle fibers indicate that ATP *does* influence Na–Ca exchange.

Baker and Glitsch (1973) first called attention to the fact that Ca efflux, including the Na_o-dependent fraction, was reduced when the cytoplasmic ATP concentration was lowered in squid axons in which $[Ca^{2+}]_i$ did not increase appreciably. Their evidence (and see Baker and McNaughton, 1976a) indicated that the apparent affinity of the Na–Ca exchange carriers for external Na was reduced when the ATP level was lowered. This finding was corroborated by DiPolo (1974) and Blaustein (1977a) in experiments on dialysed squid axons. In addition, lowering cytoplasmic ATP appeared to reduce the affinity of the Na–Ca exchange carriers for cytoplasmic Ca^{2+} without affecting the maximal rate of Na_o-dependent Ca efflux at high $[Ca^{2+}]_i$ (Blaustein, 1977).

Requena's (1978) data indicate that ATP promotes Ca efflux only in axons with a relatively high $[Na^+]_i$. He therefore suggested (Requena, 1978) that the primary effect of ATP is to "relieve the inhibition" of Ca efflux produced by intracellular Na (see Blaustein and Russell, 1975). But, in contrast to the observations by Requena, ATP *does* promote Na_o-dependent Ca efflux in axons (Blaustein, 1977) and barnacle muscle fibres (Nelson and Blaustein, 1981a) with low $[Na^+]_i$ (10–15 mM). At these low $[Na^+]_i$ levels the inhibitory effect of internal Na should be minimal.

The preceding discussion indicates that there is still considerable confusion about the role of ATP in Ca extrusion from squid axons and other excitable (and nonexcitable) cells. As already mentioned, ATP affects the kinetics of Na–Ca exchange. This influence may involve a specific phosphorylation reaction because, of 12 nucleotides tested, only the hydrolysable ATP analogues, 2β-methylene ATP, and 2-deoxy-ATP, were able to replace ATP in promoting Na_o-dependent Ca efflux (DiPolo, 1977). Moreover, this efflux can be inhibited by vanadate (Nelson and Blaustein, 1981a; cf. DiPolo *et al.*, 1979), an anion known to inferfere with several other phosphorylation reactions in which ATP serves as a substrate (e.g. Cantley *et al.*, 1977; Bond and Hudgins, 1980).

Of course, ATP may not be hydrolysed during each cycle of the Na–Ca exchange carrier. An alternative possibility is that phosphorylation of the carrier merely alters the affinities of the carrier for one or both of the transported ion species, thereby facilitating transfer in *both* directions across the plasma membrane. The latter possibility appears to be consistent with the evidence that, in cells with Na–Ca exchange systems (see above), small maintained changes in the Na electrochemical gradient result in Ca redistribution and alteration of the steady-state Ca concentration gradient (in accord with equations (2) and (3)). Moreover, because the exchange probably operates close to the equilibrium indicated by equation (2), these effects appear to be readily reversible: i.e. reducing the Na gradient causes the Ca gradient to fall, while increasing the Na gradient causes the Ca gradient to rise. In ATP-dependent transport systems where ATP serves as the immediate source of energy for ion translocation, there is significant loss of energy in the process, so that the transport reactions are not readily reversible (see Blaustein, 1977a): a classic example of a not easily reversible ATP-dependent transport system is the Na pump (see Caldwell, 1969).

OTHER CALCIUM EXTRUSION MECHANISMS

The discussion in the preceding section is based on the assumption that the Na–Ca exchange system observed in poisoned cells is the same entity as the ATP-modulated Na–Ca exchange system (see Blaustein, 1977a). However, the possibility that ATP-dependent and ATP-independent Na–Ca exchanges are two different and distinct transport systems, mediated by independent carrier molecules, cannot be excluded.

Adding to the confusion about the role of ATP in Ca extrusion is the evidence for a second ATP-dependent Ca transport system in the plasma membranes of squid axons. This system may be similar to the human red blood cell Ca extrusion mechanism mediated by a Ca+Mg-dependent ATPase, that does not require external Na, Ca or Mg, (Schatzmann and Vincenzi, 1969).

Recently, Baker and McNaughton (1978) noted that a substantial fraction (50–90%) of the Ca efflux from unpoisoned squid axons persists in the absence of external Na, Ca and Mg. This "uncoupled" Ca efflux was investigated further in dialysed squid axons (DiPolo, 1978; DiPolo and Beauge, 1979). DiPolo and Beauge, concluded (1979) that, "the 'uncoupled' Ca efflux operates in the physiological range of Ca^{2+} concentration (in squid axons: 50–200 nM), thus playing a fundamental part in the regulation of intracellular ionized calcium". However, before accepting this conclusion, several important factors must be considered. The first concerns the selectivity of the Na–Ca exchange mechanism for external Na. The Na–Ca exchange carrier monovalent cation binding sites may not bind Na exclusively. Other cations including Li, K and perhaps even choline or Tris, may be able to substitute for Na, albeit with very much lower affinity, to promote Ca efflux. Secondly, some large cells such as squid axons (Frankenhauser and Hodgkin, 1956) and barnacle muscle fibres (Hoyle et al., 1973) have a small surface to volume ratio and restricted extracellular space. In these cells, when all the Na in the bathing medium is replaced, $[Na^+]_o$ close to the plasma membrane may remain quite high, due to Na efflux from the cytoplasm. The apparent affinity of the Na–Ca exchange system for external Na is quite high in ATP fuelled cells (Baker and Glitsch, 1973, and see Blaustein, 1977a). Therefore, the $[Na^+]_o$ at the plasma membrane may be sufficient to activate a substantial fraction of the Na–Ca exchange even when $[Na^+]_o$ in the bulk solution is nominally zero.

THE ROLE OF SODIUM–CALCIUM EXCHANGE IN THE REGULATION OF $[Ca^{2+}]_i$

The preceding discussion indicates that some important questions about the transport of Ca across the plasma membrane still have not been resolved. Nevertheless, as we have already pointed out (and see below), in many types of cells small changes in the Na electrochemical gradient affect $[Ca^{2+}]_i$ as predicted from equations (2) and (3) (if $[Ca^{2+}]_o$ remains constant). This may be taken as *prima facie* evidence that Na–Ca exchange plays a critical role in the regulation of $[Ca^{2+}]_i$.

Despite this evidence, several authors have questioned the

physiological role of Na–Ca exchange (e.g. Droogmans and Casteels, 1979; DiPolo and Beauge, 1980). However, it may not always be possible to interpret experimental observations in terms of equations (2) and (3). For example, upon complete removal of external Na, $[Na^+]_i$ may fall to a very low level, so that $\Delta\bar{\mu}_{Na}$ cannot be defined. Moreover, other cations in the intra- and extracellular fluids may substitute for Na, albeit poorly, on the Na–Ca exchange carrier. In other circumstances, too (e.g. when external K is removed), $[Na^+]_i$ may not be known (see Droogmans and Casteels, 1979) unless it is directly measured.

Of course, participation of a second Ca transport system, comparable to the "uncoupled" transport in squid axons, in the lowering of $[Ca^{2+}]_i$ may also occur. The presence of Ca-dependent ATPase activity in "plasma membrane fractions" of tissue homogenates has often been construed as evidence for the existence of a Ca transport system, independent of Na–Ca exchange, in the plasma membranes (e.g. Hurwitz et al., 1973; Leslie and Borowitz, 1975; Godfraind et al., 1976; Janis et al., 1976; Gill et al., 1981). However, some authors fail to take into account the substantial contamination by endoplasmic reticulum or sarcoplasmic reticulum Ca-ATPases that may be present in these preparations (see Blaustein et al., 1980).

At present, the available evidence leads us to conclude, in contrast to DiPolo and Beauge (1979), that Na–Ca exchange is the *primary* mechanism responsible for controlling the steady level of $[Ca^{2+}]_i$ in squid axons and barnacle muscle fibres, and probably in many other types of cells as well (see Blaustein, 1974).

THE PHYSIOLOGICAL AND PATHOPHYSIOLOGICAL CONSEQUENCES OF SODIUM–CALCIUM EXCHANGE

As noted in the Introduction, intracellular Ca serves as a "second messenger" for the initiation or modulation of numerous physiological processes. Thus, if Na–Ca exchange plays a key role in the regulation of $[Ca^{2+}]_i$, small alterations in the Na electrochemical gradient across the plasma membranes of appropriate cells should affect the physiological activities that are mediated by intracellular Ca. Indeed, numerous examples appear to bear out this expectation. The initiation or modulation of tension in various types of muscle, by alteration of the trans-sarcolemmal Na gradient has already been mentioned. Several physiological implications follow directly.

Inhibition of Na transport by cardiotonic digitalis steroids should promote Na retention and, *pari passu*, Ca retention by cardiac muscle fibres; the cardiotonic effect could then be attributed to the increased Ca that is made available during activity (Baker *et al.*, 1969a; and see Blaustein, 1974, 1977b). Also, there is substantial evidence (see Blaustein, 1977b; Haddy and Overbeck, 1976; DeWardener and MacGregor, 1980) that a Na pump inhibitor circulates in the blood of many patients with essential hypertension. As a result Na and consequently Ca may be elevated in vascular smooth muscle cells; this elevation of $[Ca^{2+}]_i$ could explain the increased vascular tone, and could provide a key to the well-documented relationship between Na and hypertension (Blaustein, 1977b).

Calcium also plays a central role in stimulus–secretion coupling. Alteration of the Na gradient in secretory cells, including nerve terminals, has been shown to modulate secretion (see Blaustein, 1974). For example, the progressive accumulation of Na that occurs with repetitive stimulation of nerve terminals may contribute to post-tetanic potentiation by raising $[Na^+]_i$, thereby causing Ca retention, and making more Ca available to trigger transmitter release (Atwood *et al.*, 1975; Charlton *et al.*, 1980). In this way, Na–Ca exchange may play an important role in the modulation of synaptic transmission.

In both vertebrate and invertebrate photoreceptors, Ca plays a central role in phototransduction. The Na gradient in photoreceptor cells appears to modulate the response to light stimulation. This is exemplified in the *Limulus* (horseshoe crab) eye, where reduction of the Na gradient inhibits adaptation of the visual response, presumably by slowing Ca extrusion (Lisman and Brown, 1972). Furthermore, as noted above, the photoreceptor dark current in vertebrate retinal rods may be the manifestation of electrogenic Na–Ca exchange.

Net transport of Ca across intestinal, renal and other epithelia includes the "active (uphill) extrusion of Ca across the basolateral membranes of the epithelial cells. This "active" step is Na-dependent and probably involves an Na–Ca exchange mechanism (Blaustein, 1974; Taylor and Windhager, 1979, Lee *et al.*, 1980). Grinstein and Erlij (1978), and Taylor and Windhager (1979) have suggested that Na permeability at the brush border (apical membrane) of toad bladder and renal tubule cells may be modulated by intracellular Ca. They have shown that changing Ca extrusion by modifying the Na gradient across the basolateral border alters apical Na permeability.

The mechanism of Ca transport in bone cells has long been an enigma. But recently Kreiger and Tasjian (1980) showed that parathyroid hormone-stimulated bone resorption is inhibited by the cardiotonic steroid, ouabain, or by reduction of extracellular Na.

They therefore suggested that Ca extrusion from the bone cells is mediated by an Na–Ca exchange mechanism. It seems plausible to ask whether Na–Ca exchange also is involved in bone mineral deposition.

The aforementioned are only a few examples of processes in which Na–Ca exchange, by regulating $[Ca^{2+}]_i$, plays a key role. The data from epithelia and bone demonstrate that Na–Ca exchange is not limited to "excitable" cells. Moreover, a number of the observations mentioned above demonstrate that, in many types of cells, Na–Ca exchange helps to control $[Ca^{2+}]_i$ in the physiological range. These findings reinforce the conclusion enunciated in the previous section: namely, that Na–Ca exchange is the *primary* mechanism responsible for $[Ca^{2+}]_i$ regulation in many types of cells.

ACKNOWLEDGEMENTS

We thank Ms C. Revis for preparing the typescript. Supported by NIH grant NS-16106, NSF grant PCM 7911704, and a grant from the Muscular Dystrophy Association. MTN was supported by a fellowship from the Baltimore Chapter of the American Heart Association.

REFERENCES

Alvarez-Leefmans, F. J., Rink, T. J. and Tsien, R. Y. (1981). *J. Physiol., Lond.* **315**, 531-548.

Ashley, C. C. and Campbell, A. K. (Eds) (1979). "Detection and Measurement of Free Ca^{2+} in Cells." Elsevier/North-Holland Biomedical Press, Amsterdam.

Ashley, C. C., Ellory, J. C. and Hainaut, K. (1974). *J. Physiol., Lond.* **242**, 255-272.

Atwood, H. L., Swenarchuck, L. E. and Gruenwald, C. R. (1975). *Brain Res.* **100**, 198-204.

Baker, P. F. and Glitsch, H. G. (1973). *J. Physiol., Lond.* **233**, 44-46P.

Baker, P. F. and Knight, D. E. (1980). *J. Physiol., Paris* **76**, 497-504.

Baker, P. F. and McNaughton, P. A. (1976a). *J. Physiol., Lond.* **259**, 103-144.

Baker, P. F. and McNaughton, P. A. (1976b). *J. Physiol., Lond.* **260**, 24-25P.

Baker, P. F. and McNaughton, P. A. (1978). *J. Physiol., Lond.* **276**, 127-150.

Baker, P. F. and Schlaepfer, W. W. (1978). *J. Physiol., Lond.* **276**, 103-125.

Baker, P. F. and Singh, R. (1980). *J. Physiol., Lond.* **305**, 10-12P.

Baker, P. F., Blaustein, M. P., Hodgkin, A. L. and Steinhardt, R. A. (1967a). *J. Physiol., Lond.* **192**, 43-44P.

Baker, P. F., Blaustein, M. P., Manil, J. and Steinhardt, R. A. (1967b). *J. Physiol., Lond.* **191**, 100-102P.

Baker, P. F., Blaustein, M. P., Hodgkin, A. L. and Steinhardt, R. A. (1969a). *J. Physiol., Lond.* **200**, 431-458.

Baker, P. F., Blaustein, M. P., Keynes, R. D., Manil, J., Shaw, T. I. and Steinhardt, R. A. (1969b). *J. Physiol., Lond.* **200**, 459-496.
Baker, P. F., Knight, D. E. and Whittaker, M. J. (1980). *Proc. R. Soc. Lond. B* **207**, 149-161.
Bers, D. M., Philipson, K. D. and Nishimoto, A. Y. (1980). *Biochim. biophys. Acta* **601**, 358-371.
Blaustein, M. P. (1974). *Rev. physiol. Biochem. Pharmacol.* **70**, 33-82.
Blaustein, M. P. (1975). *J. Physiol., Lond.* **247**, 617-655.
Blaustein, M. P. (1977a). *Biophys. J.* **20**, 79-111.
Blaustein, M. P. (1977b). *Am. J. Physiol.* **232**, C165-C173.
Blaustein, M. P. and Ector, A. C. (1976). *Biochim. biophys. Acta* **419**, 295-308.
Blaustein, M. P. and Hodgkin, A. L. (1968). *J. Physiol., Lond.* **198**, 46-48P.
Blaustein, M. P. and Hodgkin, A. L. (1969). *J. Physiol., Lond.* **200**, 497-527.
Blaustein, M. P. and Oborn, C. J. (1975). *J. Physiol., Lond.* **247**, 657-686.
Blaustein, M. P. and Russell, J. M. (1975). *J. Membr. Biol.* **22**, 285-312.
Blaustein, M. P., Russell, J. M. and DeWeer, P. (1974). *J. Supramolec. Struct.* **2**, 558-581.
Blaustein, M. P., Ratzlaff, R. W. and Kendrick, N. C. (1978). *Ann. N.Y. Acad. Sci.* **307**, 195-212.
Blaustein, M. P., McGraw, C. F., Somlyo, A. V. and Schweitzer, E. S. (1980). *J. Physiol., Paris* **76**, 459-470.
Bond, G. H. and Hudgins, P. M. (1980). *Biochim. biophys. Acta* **600**, 781-789.
Brinley, F. J., Jr (1980). *Fedn Proc. Fedn Am. Socs exp. Biol.* **39**, 2778-2782.
Caldwell, P. C. (1969). *Curr. Topics Bioenerg.* **3**, 251-278.
Cantley, L. C., Jr, Josephson, L., Warner, R., Yanagisawa, M., Lechenne, C. and Guidotti, G. (1977). *J. biol. Chem.* **252**, 7421-7423.
Caroni, P. and Carafoli, E. (1980). *Nature, Lond.* **283**, 765-767.
Caroni, P., Reinlib, L. and Carafoli, E. (1980). *Proc. natn. Acad. Sci. U.S.A.*, **77**, 6354-6358.
Chapman, R. A. (1979). *Prog. Biophys. molec. Biol.* **35**, 1-52.
Chapman, R. A. and Ellis, D. (1977). *J. Physiol., Lond.* **272**, 331-354.
Chapman, R. A. and Tunstall, J. (1980). *J. Physiol., Lond.* **305**, 109-123.
Charlton, M. P., Thompson, C. S., Atwood, H. L. and Farnell, B. (1980). *Neurosci. Lett.* **16**, 193-196.
Cosmos, E. E. and Harris, E. J. (1961). *J. gen. Physiol.* **44**, 1121-1130.
Costantin, L. L. (1977). *In* "Handbook of Physiology" (Ed. E. R. Kandel) Vol. 1, 215-259. American Physiological Society, Bethesda, MD.
Curtis, B. A. (1966). *J. gen. Physiol.* **50**, 255-267.
Deitmer, J. W. and Ellis, D. (1978). *J. Physiol., Lond.* **277**, 437-453.
De Wardener, H. E. and MacGregor, G. A. (1980). *Kidney International* **18**, 1-9.
DiPolo, R. (1973). *Biochim. biophys. Acta* **298**, 279-283.
DiPolo, R. (1974). *J. gen. Physiol.* **64**, 503-517.
DiPolo, R. (1977). *J. gen. Physiol.* **69**, 795-813.
DiPolo, R. (1978). *Nature, Lond.* **274**, 390-392.
DiPolo, R. and Beauge, L. (1979). *Nature, Lond.* **278**, 271-273.
DiPolo, R. and Beauge, L. (1980). *Cell Calcium* **1**, 147-169.
DiPolo, R., Rojas, H. R. and Beauge, L. (1979). *Nature, Lond.* **281**, 228-229.
Douglas, W. W. (1968). *Br. J. Pharmacol.* **34**, 457-474.
Douglas, W. W. (1975). *In* "Handbook of Physiology" (Eds H. Blaschiko, G. Sayers and A. D. Smith) 367-388. American Physiology Society, Washington, DC.

Droogmans, G. and Casteels, R. (1979). *J. gen. Physiol.* **74**, 57-70.
Frankenhaeuser, B. and Hodgkin, A. L. (1956). *J. Physiol., Lond.* **131**, 341-376.
Gill, D. L., Grollman, E. F. and Kohn, L. D. (1981). *J. biol. Chem.* **256**, 184-192.
Glitsch, H. G., Reuter, H. and Scholz, H. (1970). *J. Physiol., Lond.* **209**, 25-43.
Glynn, I. M. and Karlish, S. J. D. (1975). *A. Rev. Physiol.* **37**, 13-55.
Godfraind, T., Sturbois, X. and Verbecke, N. (1976). *Biochim. biophys. Acta* **455**, 254-268.
Gold, G. H. and Korenbrot, J. I. (1980). *Proc. natn. Acad. Sci. U.S.A. – Biol. Sci.* **77**, 5557-5561.
Grinstein, S. and Erlij, D. (1978). *Proc. R. Soc. Lond.* B **202**, 353-360.
Haddy, F. J. and Overbeck, H. W. (1976). *Life Sci.* **19**, 935-948.
Hagiwara, S. (1975). *In* "Membranes. A Series of Advances" (Ed. G. Wisenman) 359-381. Marcel Dekker, New York.
Hazelton, B., Mitchell, B. and Tupper, J. (1979). *J. Cell Biol.* **83**, 487-498.
Heilbrunn, L. V. (1940). *Physiol. Zool.* **13**, 88-94.
Heilbrunn, L. V. and Wiercinski, (1947). *J. cell comp. Physiol.* **29**, 15-32.
Henkart, M. (1980). *Fedn Proc. Fedn Am. Socs exp. Biol.* **39**, 2783-2789.
Herchuelz, A., Sener, A. and Malaisse, W. J. (1980). *J. Membr. Biol.* **57**, 1-12.
Hinke, J. A. M. and Wilson, M. L. (1962). *Am. J. Physiol.* **203**, 1161-1166.
Horackova, M. and Vassort, G. (1979). *J. gen. Physiol.* **73**, 403-424.
Hoyle, G., McNeill, P. A. and Selverston, A. I. (1973). *J. Cell Biol.* **56**, 74-91.
Hurwitz, L., Fitzpatrick, D. F., Debbas, G. and Landon, E. J. (1973). *Science, N.Y.* **179**, 384-386.
Janis, R. A., Lee, E. Y., Allan, J. and Daniel, E. E. (1976). *Pflügers Arch. ges. Physiol.* **365**, 171-176.
Judah, J. D. and Willoughby, D. A. (1964). *J. cell. comp. Physiol.* **64**, 363-370.
Jundt, H., Portzig, H., Reuter, H. and Stucki, J. W. (1975). *J. Physiol., Lond.* **246**, 229-253.
Jundt, H. and Reuter, H. (1977). *J. Physiol., Lond.* **266**, 78-79P.
Kadoma, M., Froelich, J., Sutko, J. and Reeves, J. (1981). *Biophys. J.* **33**, (2, Pt. 2), 46a.
Kamada, T. and Kinosita, H. (1943). *Japan. J. Zool.* **10**, 469-493.
Katz, B. and Miledi, R. (1967). *J. Physiol., Lond.* **189**, 535-544.
Katz, B. and Miledi, R. (1969). *J. Physiol., Lond.* **203**, 459-487.
Kreiger, N. S. and Tashjian, A. H., Jr (1980). *Nature, Lond.* **287**, 843-845.
Lee, C. O., Taylor, A. and Windhager, E. E. (1980). *Nature, Lond.* **287**, 859-861.
Lehninger, A. L. (1970). *Biochem. J.* **119**, 129-138.
Leonard, E. (1957). *Am. J. Physiol.* **189**, 185-190.
Leslie, S. W. and Borowitz, J. L. (1975). *Biochim. biophys. Acta.* **394**, 227-238.
Lisman, J. E. and Brown, J. E. (1972). *J. gen. Physiol.* **59**, 701-719.
Lüttgau, H. C. and Niedergerke, R. (1958). *J. Physiol., Lond.* **143**, 486-505.
Marban, E., Rink, T. J., Tsien, R. W. and Tsien, R. Y. (1980). *Nature, Lond.* **286**, 845-850.
Miledi, R. (1973). *Proc. R. Soc. Lond.* B **1983**, 421-425.
Mullins, L. J. (1979). *Am. J. Physiol.* **236**, C103-C110.
Mullins, L. J. and Brinley, F. J. Jr (1975). *J. gen. Physiol.* **65**, 135-152.
Nelson, M. T. and Blaustein, M. P. (1980). *J. gen. Physiol.* **75**, 183-206.
Nelson, M. T. and Blaustein, M. P. (1981a). *Nature, Lond.* **289**, 314-316.
Nelson, M. T. and Blaustein, M. P. (1981b). *Biophys. J.* **33**, (2, pt 2), 61a.

Niedergerke, R. (1963). *J. Physiol., Lond.* **167**, 515-550.
Nockolds, C. E., Kretsinger, R. H., Coffee, C. J. and Bradshaw, R. A. (1972). *Proc. natn. Acad. Sci. U.S.A.* **69**, 581-584.
Parelloa, J., Lilja, H., Cave, A. and Lindman, B. (1978). *FEBS Lett.* **87**, 191-195.
Parker, J. C. (1978). *J. gen. Physiol.* **71**, 1-17.
Philipson, K. D. and Nishimoto, A. Y. (1980). *J. biol. Chem.* **255**, 6880-6882.
Pitts, B. J. R. (1979). *J. biol. Chem.* **254**, 6232-6235.
Reeves, J. P. and Sutko, J. L. (1979). *Proc. natn. Acad. Sci. U.S.A.* **76**, 590-594.
Reeves, J. P. and Sutko, J. L. (1980). *Science, N.Y.* **208**, 1461-1464.
Requena, J. (1978). *J. gen. Physiol.* **72**, 443-470.
Reuter, H. (1973). *Proc. Biophys. molec. Biol.* **26**, 1-43.
Reuter, H. and Seitz, N. (1968). *J. Physiol., Lond.* **195**, 451-470.
Russell, J. M. and Blaustein, M. P. (1974). *J. gen. Physiol., Lond.* **63**, 144-167.
Schaechtelin, G. (1961). *Pflügers Arch. ges. Physiol.* **273**, 164-181.
Schatzmann, H. J. (1973). *J. Physiol., Lond.* **235**, 551-569.
Schatzmann, H. J. and Burgin, H. (1978). *Ann. N.Y. Acad. Sci.* **307**, 125-146.
Schatzmann, H. J. and Vincenzi, F. F. (1969). *J. Physiol., Lond.* **201**, 369-395.
Sheu, S. S. and Fozzard, H. A. (1981). *Biophys. J.* **33**, (2, Pt 2), 11a.
Sjodin, R. and Abercrombie, R. (1978). *J. gen. Physiol.* **71**, 453-466.
Somlyo, A. V., Gonzales-Serratos, H., Shuman, H., McClellan, G. and Somlyo, A. P. (1981). *Biophys. J.* **33**, (2, Pt 2), 88a.
Steinhardt, R., Zucker, R. and Schatten, G. (1977). *Dev. Biol.* **58**, 185-196.
Taylor, A. and Windhager, E. E. (1979). *Am. J. Physiol.* **236**, F505-F512.
Vassort, G. (1973). *Pflügers Arc. ges. Physiol.* **298**, 23-30.
Wilbrandt, W. and Koller, H. (1948). *Helv. Physiol. Pharmacol. Acta.* **6**, 208-221.
Winegrad, S. (1979). *In* "Handbook of Physiology" (Ed. R. M. Berne) Vol. 1, 393-428. American Physiology Society, Bethesda, MD.
Yau, K. -W., McNaughton, P. A. and Hodgkin, A. L. (1981). *Nature, Lond.* **292**, 502-505.
Yoshikami, S., George, J. S. and Hagins, W. A. (1980). *Nature, Lond.* **286**, 395-398.
Yoshikami, S. and Hagins, W. A. (1981). *Biophys. J.* **33** (2, Pt 2), 288a.

7

Intestinal Calcium Absorption and Transport

FELIX BRONNER

School of Dental Medicine, The University of Connecticut Health Center, Farmington, Connecticut, USA

GENERAL

Calcium absorption is the resultant of two processes: extraction of calcium from the food as the ingesta, mixed with intestinal juices, move from the stomach to the colon, and transmural calcium movement across the intestinal tissue into the blood and lymph. Extraction results from enzymatic degradation of the food and from mechanical churning due to the segmental movement of the intestine which also propels the intestinal contents in a distal direction. Transmural movement is either trans- or paracellular, and, if the former, may involve cellular processes that require input of metabolic energy. Direct regulation of calcium transport involves cellular organelles and must in some way bear a relationship to the regulation of intracellular calcium on which the cell expends considerable energy. Indirect regulation subsumes diverse events and situations, such as ageing, effect of nutrients and nutritional state, hormonal state, and the like. These are considered indirect regulators as they are thought to modify calcium transport only via other events or processes. For example, if ageing were to alter the intercellular distance between mucosal cells, this could lead to a modification of whatever paracellular calcium movement exists, but this effect would be unlikely to be specific for calcium. Vitamin D and its metabolites, on the other hand, act to modify calcium transport directly, probably via specific protein synthetic events and/or the modification of lipid membrane components.

In what follows we shall describe intestinal calcium absorption in terms of the methods that have been used to study it. In turn these have led to some understanding of mechanisms, at the tissue and also at the cellular level. A detailed description of the orderly subcellular events that permit calcium to enter and be transported across the intestinal cell is not yet possible, though certain aspects can be outlined.

METHODS AND MECHANISMS

The Balance Approach

1. Conceptually simple, *net intestinal absorption* is defined as the difference between the amount that enters and leaves the intestine. In that sense it is a function of time, i.e. a rate, and of necessity an average rate. In whole animal studies, it represents the difference between the daily calcium intake, v_i, and daily faecal excretion, v_F:

$$S_i = v_i - v_F \qquad (1)$$

where S_i = net intestinal absorption (units: mass/time, typically mg Ca/d).

Intake is readily measured by weighing the food eaten and determining or estimating its calcium content. To measure faecal output requires the use of a faecal marker, like chromium oxide or carmine red, which is ingested at the beginning and end of the experimental period, and whose concentration in the stools is determined either visually or chemically. To obtain meaningful measurements the length of the experimental period must exceed the time required for net absorption to be completed. The latter tends to proceed in 24-h cycles (Wrobel, 1981).

To study the absorption process, it is necessary to obtain S_i as a function of the input, v_i. In rats this relationship can be best described by an exponential-like function (Sammon *et al.*, 1970), but, as will be shown, the true underlying relationship is a more complicated one, as it is the resultant of segmental and transmural movements, with the latter in turn differing in intensity and type along the length of the intestine. The relationship between v_i and S_i is akin to an input–output function and is a useful base measure to be able to study the effect of different physiological or nutritional states. This information can serve as a guide to the effect of these states on underlying transport mechanisms.

2. Different parts of the intestine differ in their capacity to extract calcium. This can be demonstrated by a variety of methods,

which however do not necessarily measure the same process. The *in situ loop procedure* is an application of the balance procedure to intestinal segments. A segment is tied off, rinsed, and a calcium solution is instilled. The rate of loss from the loop may be described as a hyperbolic type of function which approaches a slow, constant rate that is not zero. When this constant rate is plotted as a function of the instilled calcium concentration, one obtains an input–output function that is the equivalent of the $v_i - S_i$ relationship described above. But whereas the $v_i - S_i$ relationship may be described by an exponential-type function, the relationships obtained in the three parts of the intestine – duodenum, jejunum, ileum – cannot be described by an exponential function. In the rat duodenum (Pansu *et al.*, 1981) the relationship between rate of net calcium absorption and luminal calcium concentration is best described by a combination of hyperbolic (or saturable) and linear functions, as theoretically predicted by Wasserman and Taylor (1969):

$$J_{ms} = \frac{J_{max}\,[Ca_L^{2+}]}{K_T + [Ca_L^{2+}]} + A\,[Ca_L^{2+}] \qquad (2)$$

where J_{ms} = total calcium flux from lumen (mucosal side) to the body fluids (serosal side) (units: mass/time, e.g. $\mu mol/h$ or $\mu mol/cm^2/h$)

J_{max} = the maximum saturable flux

$[Ca_L^{2+}]$ = luminal calcium concentration (units: mass/volume, e.g. μM)

K_T = the luminal calcium concentration at which $J_{max}/2$ is attained

A = diffusivity constant (units: volume/time, e.g. litres/h or litres/cm^2/h)

In the rat jejunum, the saturable process is virtually absent and in the ileum it is totally absent. The numerical value for A is the same in all three segments (Pansu *et al.*, 1981; and unpublished). Thus the transmural movement of calcium all along the intestine comprises a linear or diffusional process to which a hyperbolic or saturable process is added in the duodenum. In the colon there seems to exist in addition to the diffusional process a secretory process, i.e. a saturable process in the opposite direction (Favus *et al.*, 1980; Lee *et al.*, 1980).

In all likelihood there is incomplete anatomical and functional identity between different intestinal segments or between species. Whereas in the proximal part of the rat jejunum there may exist a saturable calcium transport process which is completely absent in the distal jejunum, one could also imagine dividing the small intestine into two parts: one with and one without saturable transport, the former being equivalent to the duodenum and part of the jejunum, the later to the jejunum plus ileum. Recent evidence (Schedl *et al.*, 1981) suggests that in the hamster the calcium secretory process associated with the colon in the rat already occurs in the ileum.

3. A variant of the balance technique is the *perfusion approach*. The whole animal balance technique is conceptually a flow technique in that one labels at one time what is perceived as an input stream and collects the labelled output from an output system. The *in situ* loop technique uses a batch approach that can theoretically be improved upon by a perfusion procedure in which the input concentration is invariant and which moreover can be varied at will in the same experimental animal. Both techniques, if properly analysed, yield comparable results (see Wasserman and Taylor, 1969).

The Gradient Approach

Everted intestinal sacs

To study the saturable process alone, one can prepare everted intestinal sacs from various segments of the intestine (Wilson and Wiseman, 1954). In this procedure the tissue is everted so that the mucosal aspect is on the outside and transport can occur from the outside to the inside, i.e. the serosal side. The same buffer solution is placed on the inside and outside of the sac, the solution is oxygenated and the increase in calcium concentration is measured on the inside, the outside solution being considered an infinite sink. An increase in the ratio inside/outside beyond unity is a measure of the transport capacity of the tissue. As might be expected, this capacity is relatively high in the proximal duodenum and decreases in the distal jejunum and ileum, with the I/O ratio approaching unity. It should also be apparent that the I/O ratio of a given tissue will vary inversely with the calcium concentration of the buffer solution, approaching unity as the calcium concentration becomes high. Theoretically the ratio should approach unity in an inverse hyperbolic fashion; in other words, the function should be the mirror image of the saturable component of equation (2). In practice the ratio I/O has been used widely to document the existence or the rate of appearance of the saturable component of tissue calcium transport, but has not been used for quantitative studies of J_{max} or K_T.

Ussing chamber technique

The Ussing chamber (Ussing and Zerahn, 1951) technique, where a piece of intestinal tissue is mounted in a lucite chamber between solutions in a way so as to eliminate both electrical and chemical gradients, can be used to evaluate active transport. However, the amount of current carried by the calcium ion is small when compared to the current due to the sodium ion whose concentration is 130 times greater than that of calcium. Therefore, instead of short-circuiting the electrical current, one can apply widely varying voltages across the epithelial surface and derive a relationship between the change in calcium flux and the applied voltage. Nellans and Kimberg (1979a, b) found that under these conditions the unidirectional calcium flux across the ileum conformed to a modified Goldman (1943) equation and could be described by a relationship functionally equivalent to equation (2), the sum of a voltage-dependent component. The former would be equivalent to the diffusional component. While Nellans and Kimberg, on the basis of earlier work (Nellans and Kimberg, 1978), thought they could equate the voltage-dependent component to paracellular calcium flux, the analysis of these data according to Frizzell and Schultz (1972) indicated that the voltage-dependent Ca flux included both cellular and paracellular components. Therefore the identification of a class of transport with a specific route through the tissue appears hazardous.

The principal advantage of using the Ussing chamber as compared to everted sacs or ligated loops is that the conditions can be controlled carefully and therefore studies intended to explore mechanisms can be carried out. Virtually no such studies have appeared, however, with most reports concerned with establishing whether or not calcium transport involves an "active" component, i.e. movement against an electrochemical gradient. Walling and Rothman (1968, 1969) provided evidence that calcium is absorbed against a transmural electrochemical gradient of 3–5 mV, but most subsequent studies were concerned with regulatory aspects, i.e. the effect of the nutritional or endocrine status of the animal on this carrier-mediated flux.

Isolated Cells and Organelles

Transcellular calcium movement must involve an orderly, directed transfer of calcium such that the intracellular free calcium concentration remains stable while fairly large quantities of calcium move through the cell. One can estimate the amounts involved from the

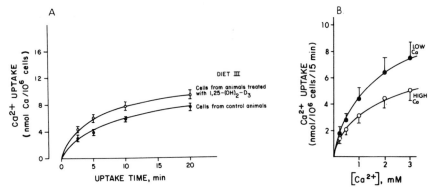

Fig. 1. *Cellular calcium uptake. A, uptake time course of cells isolated from vitamin D-repleted animals on a high calcium (1.5% Ca, 1.5% P) semisynthetic regimen (Bronner and Freund, 1975). One group of animals had been treated with 1,25-(OH)$_2$-D$_3$ (0.625 μg/100 g body wt) 3 h before sacrifice. The assay calcium concentration was 1 mM. B, calcium uptake as a function of assay calcium concentration. Animals had been placed at weaning on either high or low calcium (0.06% Ca, 0.2% P) diets. The uptake values were obtained 15 min after the initiation of the uptake assay. Cells were obtained by a modification of the procedure of Ueng and Bronner (1979). All values are shown as means with associated standard errors. (Lipton and Bronner, unpublished.)*

data of Pansu *et al.* (1981). In rats on a lactose-containing semi-synthetic diet that provides the equivalent of about 0.7% Ca, the luminal concentration, allowing for dilution by intestinal and other secretions, can reach 50 mM. At that concentration about 15 μmol/h are absorbed in the duodenum, with approximately half moving by the saturable, and the remainder by the nonsaturable route. Since the saturable route is certainly transcellular, at least 7×10^{-6} mol Ca move through the cells of the duodenum. The tissue probably contains 20×10^6 cells, equivalent to a volume of approximately 20×10^{-6} litres. The free calcium in these cells would amount to a maximum of 20×10^{-12} mol, so that the calcium throughput per hour is at least 3×10^5 times greater than the free calcium concentration. The mechanism available to handle and regulate this type of throughput must indeed be remarkable.

Isolated cells
Ca^{2+} uptake by isolated cells is a time- and concentration-dependent process (Fig. 1A) that differs in extent in cells obtained from animals on different calcium intakes (Fig. 1B). Cellular calcium uptake can also be modified by vitamin D status or exogenous vitamin D administration (Fig. 1A). Therefore uptake reflects the cellular history and state and is unlikely to constitute simple external binding. Rather

uptake must involve cellular calcium entry and therefore comprises a transport component. Since the total cellular calcium content following exposure to a calcium-containing solution is at least equal to that of the bathing medium – the true concentration depends on the true cellular volume – most of the cellular calcium must be bound. Cell viability, tested by trypan blue exclusion or the ability to synthesize protein, is not markedly depressed by exposure to 1 mM Ca^{2+} (Bronner, unpublished).

While isolated cells have been used to study calcium uptake and may shed light on some aspects of the mechanism of uptake, not enough studies have been done to relate cellular uptake to tissue uptake. Thus it is not yet known whether cellular uptake is strictly equivalent to saturable transport or comprises both saturable and nonsaturable processes. Preliminary work with isolated cells from newborn rats (Pansu and Bronner, unpublished) has revealed virtually no time-dependent or concentration-dependent calcium uptake, in contrast with what has been observed with cells from older animals. Studies with the *in situ* duodenal loops prepared in newborn rats have shown that in these animals calcium absorption proceeds without a saturable component (Bronner *et al.*, 1981a). Time-dependent and concentration-dependent cellular calcium uptake therefore appears to be related to the saturable component of tissue calcium transport, but the nature of this relationship is not known.

The relationship between cellular and tissue events has also been explored (Bronner *et al.*, 1981b) by kinetic studies of the response to exogenous 1,25-dihydroxyvitamin D_3, 1,25-$(OH)_2$-D_3, the most active metabolite of vitamin D (Norman, 1979; see also below). As a follow-up of earlier studies (Buckley and Bronner, 1980) that had shown a differential response in terms of calcium binding protein (CaBP) induction in vitamin D-replete animals on a high or a low-calcium diet, similar animals were treated with exogenous 1,25-$(OH)_2$-D_3 and their transport response evaluated in terms of an everted intestinal sac preparation and isolated intestinal cells. Both preparations showed an equivalent response: the peak of the response was reached in either sacs or cells (Fig. 6) from animals on a high calcium diet 3 h after drug adimiministration; the response thereafter dropped fairly rapidly to the preresponse level. In animals on a low-calcium diet, the peak response occurred at 7 h after dosing and thereafter diminished rapidly. These observations again show that a response comparable to the saturable transport response – tested by the everted sac preparation – also occurs in isolated cells, tested by an increase in calcium uptake.

Organelles
Calcium transport through the intestinal cell involves a series of steps – entry at the brush border membrane, movement inside the cell and possible interaction with major organelles, e.g. endoplasmic reticulum, the Golgi apparatus, mitochondria, and extrusion from the cell. Each of these stages or steps can be or has been studied with the aid of isolated organelles or membrane vesicles.

Brush border membrane vesicles (BBMV). The entry step has been studied most extensively with the aid of microvillus (brush border) membranes isolated either by an EDTA procedure (Forstner *et al.*, 1968; Hopfer *et al.*, 1973) or from a high calcium solution (Sigrist-Nelson *et al.*, 1975). Under appropriate conditions the membranes, largely stripped of their actin core, are harvested in the form of vesicles whose orientation is predominantly right-side out (Hasse *et al.*, 1978). These vesicles have been used for a variety of transport studies (Kinne, 1976; Hopfer, 1977; Sacktor, 1977), including calcium uptake (Rasmussen *et al.*, 1979; Mürer and Hildman, 1981; Wilson and Lawson, 1980; Miller and Bronner, 1981).

Calcium uptake by BBMV is a time- and concentration-dependent process (Fig. 3) and involves binding both at the inner and outer aspects of the membrane. A portion of the time-dependent uptake must therefore represent transport across the membrane into the intravesicular space.

It is generally thought that the calcium uptake process does not require metabolic energy, but is energetically downhill. In experiments using the lipophilic anions SCN^- and NO_3^-, or employing the ionophore valinomycin ($+ K^+$) and the uncoupler carbonyl cyanide p-triphenyl-fluoromethoxyphenyl-hydrazone (FCCP) it was shown that Ca^{2+} accumulation increased when the negativity inside the vesicle had increased (Miller and Bronner, 1981). These data could best be explained by significant binding to internal membrane sites and correlate with similar conclusions for the brush border from rat kidney cortex (Gmaj *et al.*, 1979). Rasmussen *et al.* (1979) feel that in their system most binding is completed within the first minute and that in all subsequent uptake only a small proportion involves binding at the internal binding. Wilson and Lawson (1980), using a chick membrane preparation similar to that used by Rasmussen *et al.* (1979), observed that at higher medium calcium concentrations uptake had not yet been completed by 50 min, whereas it had been at lower calcium concentrations. This indicates different classes of binding sites, with high and low affinities. Miller and Bronner (1981) have formally demonstrated the existence of two classes of binding sites and have suggested that the more numerous low affinity binding sites are the phosphate groups of membrane phospholipids. This is consistent with the observation of the inhibition of calcium binding by Na^+, K^+, La^{3+}, tetracaine and ruthenium red, and with binding studies done subsequent to extraction of the membrane by a lipid solvent (Miller, Li and Bronner, unpublished). Studies of high affinity sites suggest that these may be due to one or more calcium-binding proteins. The occurrence of a brush border membrane-related protein that binds calcium, is of relatively low molecular weight ($M_r \approx$ 18 500), and is vitamin D-dependent, has been reported (Miller et al., 1979). Kowarski and Schachter (1980) have reported the existence of a large particulate fraction isolated from rat intestinal mucosa. This complex is made up of three vitamin D-dependent activities, one of which appears to be a high molecular weight calcium-binding protein ($M_r \approx 200\,000$) which upon treatment with sodium dodecyl sulphate yields a monomer of low molecular weight ($\approx 20\,500$). The latter, as well as the brush border calcium-binding protein of Miller et al. (1979) can be clearly distinguished from the cytosolic, vitamin D-dependent calcium-binding protein ($M_r \approx 9000$) found in rat and other mammalian duodena (Wasserman and Feher, 1977), but whether the monomer and the protein of Miller et al. (1979) are the same is not known.

The initial rate of calcium uptake by the BBMV preparation is vitamin D-dependent (Rasmussen et al., 1979; Miller and Bronner, 1981) in that it is higher in BBMV from vitamin D-replete than -deficient animals and can be raised to the level found in preparations from replete animals by prior administration of a vitamin D metabolite to the deficient animal (Fig. 3).

The molecular nature of vitamin D regulation of the initial rate of calcium uptake is not known. It may involve synthesis of a channel or carrier site, as well as modification of the lipid composition of the membrane (Max et al., 1978; Norman et al., 1981). For more detailed discussion, see below.

Intracellular organelle and membrane preparations. A large literature exists on mitochondrial calcium uptake and release (Nicholls and Crompton, 1980), but the role of mitochondria in intestinal cells has not been studied widely. Bikle et al. (1980) found that isolated mitochondria from chick intestine accumulated calcium in two phases: an early, transient uptake into a Na^+-labile pool, followed by an ATP-dependent uptake into a Na^+-resistant pool. Uptake characteristics were similar to those of mitochondria from other tissues. For example, ruthenium red and dinitrophenol both blocked Ca^{2+} accumulation, and oligomycin depressed it. Atractyloside, an inhibitor of the adenine nucleotide translocase, had no effect. Similar results have been obtained with cardiac mitochondria (Asimakis and Sordahl, 1977). Vitamin D status of the animals from which the mitochondria were harvested had little effect on calcium accumulation by the organelles. Interestingly, 1,25-dihydroxyvitamin D_3 had a direct effect on mitochondrial calcium accumulation, as did much higher concentrations of vitamin D_3. These effects were quantitatively much less significant than the effect of higher calcium concentrations. Bikle et al. (1980) concluded that the effect of vitamin D_3 on Ca^{2+} release from kidney mitochondria, reported in early studies (DeLuca et al., 1962; Engstrom and DeLuca, 1962, 1964) was due to the vitamin D-induced increase in cytosolic calcium.

What role mitochondria play in intracellular calcium regulation and therefore in transcellular calcium movement is controversial. Some investigators attribute a major regulatory role to these organelles (Mela, 1977), while others (Somlyo et al., 1981) assert that the calcium-binding constant of mitochondrial membranes is so low in relation to the intracellular calcium concentration (10^{-6}–10^{-7} M) that they cannot regulate intracellular calcium.

Work from the laboratory of Weiser (Freedman et al., 1977, 1981; MacLaughlin et al., 1980; Weiser et al., 1981) has shown that calcium uptake by Golgi membrane vesicles from rat intestine was much greater than by microvillus or basolateral membrane vesicles and was vitamin D-dependent and responsive. Moreover, the level of uptake by the Golgi membrane vesicles from different regions of the intestine parralleled that found in everted sac preparations from those regions, i.e. duodenum > ileum > jejunum. This uptake was unaffected by ATP – in contrast with the effect of ATP on liver Golgi membranes (Hodson, 1978) – and was inhibited by cycloheximide administration. There was evidence for a H^+/Ca^{2+} and Mg^{2+}/Ca^{2+} countertransport system (Freedman et al., 1981) and Weiser et al. (1981) suggested there may exist a carrier-mediated calcium transport, although the amount of calcium binding in the system is very high.

How these findings relate to calcium transport by the intestinal cell is as yet entirely speculative. However, the possible existence of a vitamin D-dependent calcium-binding protein in the Golgi membrane is intriguing, since such a protein could constitute one in a series of intracellular binding proteins that help regulate intracellular calcium concentration under conditions of an increase in the saturable component of calcium transport.

Basolateral membrane studies. A number of methods for preparing basolateral membranes have been published (Douglas *et al.,* 1972; Fujita *et al.,* 1972; Mürer *et al.,* 1974; Scalera *et al.,* 1980), including a very detailed and elaborate procedure by Mircheff and Wright (1976). Two types of calcium transport system have been identified in the contraluminal membrane (Kinne, 1981; Mürer and Kinne, 1980): a Ca^{2+}-dependent ATPase system with high affinity for calcium, and a Na^+/Ca^{2+} countertransport system.

Evidence for calcium-dependent ATPase activity in rat intestine was provided by Martin *et al.* (1969), Mircheff and Wright (1976) and more recently by Ghijsen and van Os (1979; Ghijsen *et al.* 1980; van Os *et al.* 1981). The latter investigators have identified a protein ($M_r \approx 115\,000$) which comigrates with the purified Ca-ATPase from heart plasmalemma, is alkali-labile, sensitive to hydroxylamine treatment and whose phosphorylation is strongly inhibited by phenothiazine, but not by theophylline. The enzyme has a K_m for Ca of $0.3\,\mu M$ and a J_{max} of $1.2 \pm 0.4\,\mu mol\,P_i/h/mg$ protein. Van Os *et al.* (1981) conclude that the Ca-ATPase is clearly distinguishable from the theophylline inhibitable alkaline phosphatase, with the activity of which they associate two proteins ($M_r \approx 84\,000$ and $64\,000$). The Ca-ATPase seems exclusively associated with the basolateral membrane (Mircheff and Wright, 1976; van Os *et al.,* 1981).

Vitamin D dependence of the Ca-ATPase was already reported by Martin *et al.* (1969), but vitamin D dependence of the more purified enzymes remains to be demonstrated. In the red blood cell, calmodulin has been shown to stimulate calcium extrusion across the plasma membrane (Larsen and Vincenzi, 1979). In the intestinal cell, calmodulin stimulates the Ca-ATPase from the basolateral membrane (Nellans and Popovitch, 1981). Calmodulin does not appear to be vitamin D-dependent (Halloran *et al.,* 1980), but the vitamin D-dependent cytosolic calcium-binding protein, CaBP, is related in an evolutionary way to calmodulin (Goodman *et al.,* 1979) and may stimulate the intestinal ATPase (Freund, personal communication).

Mürer and Hildmann (1981) have shown that basolateral membrane vesicles prepared from rat intestine, because they are inside-out, can be stimulated to accumulate calcium by the addition of ATP, that the increased influx is markedly diminished when the calcium ionophore A23187 is present in the medium and that this ATP-stimulatable process is not ouabain-sensitive. These transport responses are consistent with mediation by the Ca-ATPase.

Although there is good evidence for the existence of a Na^+/Ca^{2+} exchange mechanism in basolateral membranes from the cells of the kidney proximal tubule (Gmaj *et al.,* 1979; for a discussion of the Na^+/Ca^{2+} exchange system on the peritubular side of the kidney cell, see Taylor and Windhager, 1979), direct evidence for a Na^+/Ca^{2+} exchange system in duodenal basolateral membranes has not yet been published (but is referred to by Mürer and Hildmann, 1981). Most recent reports (e.g. Windhager *et al.*, 1981) indicate a stoichiometric relationship of three Na^+ ions to one Ca^{2+} ion, i.e. an electrogenic mechanism, although one Na^+ may exchange with one Cl^- and the movement would then be electroneutral. Whether or not the Na^+/Ca^{2+} exchange mechanism is subject to regulation by vitamin D is not known.

REGULATORY ASPECTS

Calcium absorption varies with age, nutritional state, endocrine status, pregnancy and lactation, and is altered by dietary factors, such as intake of substances that modify the availability of dietary calcium. For example, under conditions of low calcium intakes, whether due to a low level of calcium in the food supply, as is true for many vegetarians, or to reduced bioavailability, as when phytate intake is high, the efficiency of calcium absorption will be elevated. Under conditions of high calcium intake, as when large amounts of dairy products are eaten, or when the bioavailability of calcium is increased, e.g. by the addition of large quantities of lactose to the diet, the efficiency of calcium absorption is diminished. Efficiency here refers to the fraction absorbed.

Available data show that in whole animal studies, calcium absorption is an exponential-type function of intake, Thus Sammon *et al.* (1970) have shown that the following relationship holds for young male rats (body wt ≈ 220 g) that were placed on three levels of calcium intake (0.06%, 0.5%, 1.5%) for about two weeks:

$$v_a = 84.0\,(1 - e^{-0.011v_i}) \tag{3}$$

where v_a = true calcium absorption, mg/day
v_i = calcium ingested, mg/day

Equation (3) shows that the quantity of calcium absorbed increases as intake increases, but v_a/v_i, i.e. the fraction absorbed, decreases, since v_a approaches a limit of 84. A probable upper limit of calcium intake in these rates is 300 mg, with a predicted $v_a = 81$ or calcium absorption efficiency of 27%.

Transmural movement in the absence of peristaltic movement, as described by equation (2), has no limit. This is because even in the total absence of saturable transport, absorption would still occur by way of the diffusional process and the efficiency of absorption would be represented by the value of the slope, i.e. the diffusivity constant. To reconcile equations (2) and (3) one must assume that the limit of v_a approached by a relationship such as depicted by equation (3) is equal to or smaller than the diffusional term represented by equation (2). If the value is smaller, peristaltic movement would represent a process that limits *in vivo* calcium absorption.

In terms of equation (3), regulation would modify the plateau value and/or the rate at which this value is approached. Few such data have been published. Sammon *et al.* (1970) showed that in parathyroidectomized rats the plateau value was lower, but was approached faster than in euparathyroid animals. In practice there was little difference between the two groups when v_a was expressed as a function of v_i. This would suggest that the steady state effect of parathyroidectomy on calcium absorption was minor.

In terms of equation (2), regulation would modify the saturable component of the duodenal transmural movement. Pansu *et al.* (1981) have shown that in three groups of rats on low, intermediate and moderately high calcium intakes the saturable component decreased, whereas, as expected, the nonsaturable component, i.e. the diffusivity constant was unchanged. Zornitzer and Bronner (1971) had previously shown that the diffusivity constant was unchanged in old and young rats, even though the saturable component was much higher in the young rats. Pansu *et al.* (1981) also showed that there was a direct linear relationship between the saturable component, J_{max}, and the duodenal level of the cytosolic, vitamin D-dependent calcium-binding protein, CaBP. Since CaBP represents a molecular expression of the action of vitamin D (Norman, 1979), this indicates that vitamin D might act directly on the component of transmural calcium movement. Regulation of calcium absorption will therefore be discussed in terms of direct regulators, i.e. substances shown to alter calcium transport directly, and indirect regulators, i.e. situations and states where calcium absorption is altered, but where the mechanism is either via a direct regulator or unknown.

Direct Regulators of Intestinal Calcium Transport

Vitamin D
Vitamin D is unquestionably the major direct regulator of intestinal calcium transport. During the past 15 years the metabolism of

vitamin D has been elucidated to the point where one may justifiably speak of a vitamin D endocrine system. For details, the reader is referred to the work by Norman (1979) and a recent review by DeLuca (1981) which also provides information on more detailed earlier reviews.

Vitamin D enters the body either via the food or as a precursor that is transformed to vitamin D_3 in the skin glands under the action of ultraviolet light. Vitamin D is then transported in the circulation to the liver, enters the hepatocytes by an unknown mechanism and by a microsomal enzyme system is converted to 25-hydroxyvitamin D_3. This compound enters the circulation and travels to the kidney where, at the inner aspect of the kidney mitochondrion, it is further hydroxylated at either the carbon 1 or carbon 24 position to form 1,25-dihydroxyvitamin D_3 (1,25-$(OH)_2$-D_3), or 24,25-dihydroxy-vitamin D_3 (24,25-$(OH)_2$-D_3). This conversion is physiologically regulated so that under conditions of low calcium intake and/or high parathyroid hormone secretion the levels of 1,25-$(OH)_2$-D_3 are relatively elevated, while under condition of high calcium intake and/or low parathyroid hormone secretion, the resulting levels of 1,25-$(OH)_2$-D_3 are relatively depressed. 1,25-$(OH)_2$-D_3 acts on intestinal cells to augment calcium transport. It also stimulates these cells to synthesize specific proteins, of which the cytosolic, vitamin D-dependent calcium binding protein (CaBP, $M_r \approx 9000$ in mammals, $M_r \approx 28\,000$ in birds) is the best studied.

In the whole animal, the effect of vitamin D on intestinal calcium absorption is exemplified by equations (4) and (5), derived by Hurwitz et al. (1969) for male rats raised from weaning on adequate calcium and phosphate intakes with (+ D) or without (− D) vitamin D in the diet:

$$v_a = 0.51\, v_i + 3.4 \quad + D \tag{4}$$

$$v_a = 0.55\, v_i - 8.6 \quad - D \tag{5}$$

where v_a = true calcium absorption, mg/day
v_i = daily calcium intake

The equations were derived from balance and radiocalcium studies on individual animals fed the same diet and apply to an intake range of 15–115 mg Ca/day. For this reason they are linear. The more complete description of absorption as an exponential function of intake (equation (3)) requires a wider range of calcium intake.

Equations (4) and (5) differ only in their intercepts and therefore reflect the diminished ability of vitamin D-deficient animals to absorb

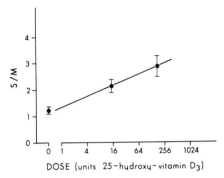

DOSE (units 25-hydroxy-vitamin D₃)

Fig. 2. *The effect of 25-hydroxyvitamin D_3 on saturable calcium transport. 15 male weaning Sprague-Dawley rats were placed on an intermediate calcium (0.5% Ca, 0.5% P) vitamin D-deficient semisynthetic regimen (Bronner and Freund, 1975). Approximately 3 weeks later, when the animals had become hypocalcemic (< 7 mg Ca/dl), some received 25-OH-D_3 (1 unit $\approx$ 0.025 μg) by i.p. injection. Thirty hours later, everted intestinal sacs were prepared from the proximal 5 cm of the duodenum according to Schachter and Rosen (1959). The calcium concentration was 10^{-5} M and measurements were made after 90 min incubation. S/M is the ratio of the serosal (S, inside) to the mucosal (M, outside) calcium concentration. Saturable transport has occurred when S/M > 1. From Bronner et al. (1976).*

calcium at a given level of intake. This is precisely what would happen if the saturable component of calcium absorption, equation (2), were suppressed. Wasserman and Kallfelz (1962) who measured mucosal–serosal fluxes in chicks, have reported that the diffusional component was diminished in vitamin D deficiency. Until the nature of the diffusional component is better understood, the reason for this discrepancy is unclear. In the rat, as pointed out above, the diffusional component appears to be the same throughout the duodenum and jejunum, with the saturable component present only in the proximal region. If the diffusional component of calcium transport were subjected to regulation by vitamin D, this would imply effects by vitamin D on the architecture of the intestinal cells and tissues. It is unlikely that this would be due to an acute effect of vitamin D, but might well be the result of a long-term effect of vitamin D deprivation.

The effect of vitamin D on the saturable component of calcium transport is best illustrated by the response of the everted sac system. In vitamin D-deficient animals the ratio of calcium concentration inside and outside the sac is unity, but rises in dose-dependent fashion following administration of an active vitamin D metabolite (Fig. 2).

The effect of vitamin D on cellular calcium uptake is illustrated in Fig. 1 which shows that cells from animals treated with 1,25-$(OH)_2$-D_3 exhibited increased calcium uptake compared with uptake

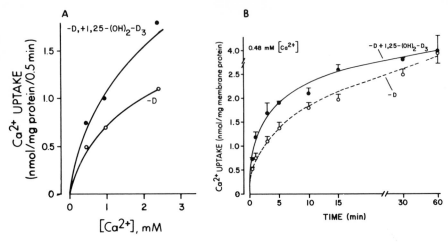

Fig. 3. *Effect of vitamin D on brush border membrane vesicle (BBMV) calcium uptake. BBMV were prepared from vitamin D-deficient rats on a low calcium (0.06% Ca, 0.2% P, no added vitamin D) semi-synthetic diet (Bronner and Freund, 1975). Some of the animals had received 0.5 μg 1,25-dihydroxyvitamin D_3 (1,25-(OH)$_2$-D_3) by intraperitoneal injection 16 h before sacrifice. Uptake was measured by a rapid filltration technique as a function of assay calcium at 0.5 min. A, or as a function of time at 0.48 mM Ca^{2+}. B. The solid circles refer to treated animals whose plasma Ca concentration had risen from 5.63 ± 0.14 (SE) to 7.68 ± 0.14 mg/dl. Note that treatment raised the initial rate of calcium uptake. From Miller and Bronner (1981).*

by cells from untreated controls. Figure 3 shows that brush border membrane vesicles (BBMV) harvested from 1,25-(OH)$_2$-D_3 treated animals also exhibit higher calcium uptakes than BBMV from untreated controls. It is quite likely that the calcium uptake process by cells and organelles involves a saturable component, but it is not possible to distinguish between saturable binding and saturable transport, unless it could be shown that one is very much larger than the other.

Figure 4 shows the close correlation observed in rats between intestinal Ca-ATPase activity and the saturable component of calcium transport as measured in the everted sac preparation. While this correlation was obtained under conditions of varying circulating levels of 1,25-(OH)$_2$-D_3, it may not reflect a direct response of the enzyme system to the hormone; rather it could have resulted from a 1,25-(OH)$_2$-D_3-mediated increase in cellular calcium throughput that in turn would have stimulated the extrusion system. Quite possibly the hormone may stimulate both calcium throughput and enzyme activity.

The molecular mechanisms of vitamin D action are not well

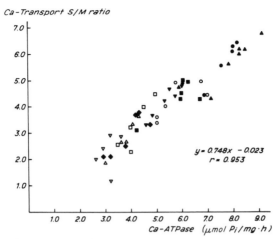

Fig. 4. *Correlation between saturable calcium transport and Ca-ATPase activity. Saturable calcium transport was evaluated by an everted sac technique; Ca-ATPase was evaluated according to Kowarski and Schachter (1973). Closed symbols refer to rats supplemented with 1,25-(OH)$_2$-D$_3$; open symbols refer to unsupplemented rats. △, ▲ uraemic rats, low-calcium diet; ◆ uraemic PTX animals; ▽, ▼ uraemic rats, normal calcium intakes. ○, ● sham-operated rats, low calcium diet; ◻, ■ sham-operated rats, normal calcium intake. From Schiffl and Binswager (1980).*

understood, but are likely to be complicated and may well involve more than a single mechanism. The existence of cytosolic receptors for 1,25-(OH)$_2$-D$_3$ in the 3.0–3.5 s class (Norman, 1979; Haussler and Brickman, 1982) and the 1,25-(OH)$_2$-D$_3$-induced synthesis of proteins such as CaBP have led to the proposal that 1,25-(OH)$_2$-D$_3$ acts like other steroid hormones. In other words the complex of 1,25-(OH)$_2$-D$_3$ and extranuclear receptor are thought to become associated with the chromatin fraction of the nucleus, leading ultimately to synthesis of CaBP or other specific proteins. This is a transcriptional phenomenon. In vitamin D-deficient animals treated with 1,25-(OH)$_2$-D$_3$ and simultaneously with actinomycin D, CaBP induction was reduced (Bikle *et al.*, 1978). If CaBP synthesis is not inhibited, it takes 6–8 h before a reliable increase in duodenal CaBP can be detected in vitamin D-deficient rats (Thomasset *et al.*, 1979a; Buckley and Bronner, 1980). The appearance of 1,25-(OH)$_2$-D$_3$-induced mRNA for chick CaBP prior to the detection of the translation product has also been reported (Spencer *et al.*, 1978). Cell-free translation of chick CaBP (Christakos and Norman, 1980) and of porcine CaBP (Mellersh *et al.*, 1980) has been accomplished.

Vitamin D-replete rats, i.e. animals raised on an ample supply of vitamin D, have CaBP levels that vary inversely with calcium intake, i.e. animals on a low calcium intake have high levels of CaBP and those on a high calcium intake have lower CaBP (Freund and Bronner, 1975; Buckley and Bronner, 1980; Pansu et al., 1981). Administration of 1,25-(OH)$_2$-D$_3$ to vitamin D-replete animals on a high calcium diet provokes a doubling of their duodenal CaBP level within ~ 1 h (Buckley and Bronner, 1980); this can be inhibited by simultaneous administration of cycloheximide, but not of actinomycin D (Buckley and Bronner, unpublished). These results suggest that the 1,25-(OH)$_2$-D$_3$-induced increase in CaBP in the vitamin D-replete animal on a high calcium diet is posttranscriptional or translational. This interpretation is reinforced by the fact that in vitamin D-replete animals on a low calcium diet exogenous 1,25-(OH)$_2$-D$_3$ does not provoke an increase in CaBP until ~ 7 h after treatment and that this increase can be inhibited by simultaneous administration of actinomycin D (Buckley and Bronner, unpublished). Thus 1,25-(OH)$_2$-D$_3$ can induce a molecular response – CaBP synthesis – which is either transcriptional or posttranscriptional, depending on prior calcium intake. Moreover, the posttranscriptional response can only occur in the vitamin D-replete animal. Figure 5 summarizes these events.

The situation in transport terms is similar. Figure 6 shows in kinetic terms the calcium uptake response of cells from three different groups of animals, + D high calcium, + D low calcium, and − D high calcium, treated with 1,25-(OH)$_2$-D$_3$. As can be seen, the peak transport response occured at 3 h, 7 h and 12 h following treatment; these times are comparable to the peak response times for CaBP synthesis. There are however two important differences: (a) the transport response decreases at times when CaBP levels still remain elevated; (b) in the + D high Ca animals, CaBP biosynthesis distinctly preceded the cellular uptake response.

Preliminary studies with everted intestinal sacs from vitamin D-replete animals on high and low calcium diets have shown that the tissue transport responses occur at about the same time as the cellular calcium uptake responses (Pansu and Bronner, unpublished).

These aspects of vitamin D regulation are being dwelt on because they bear on a current, as yet unresolved controversy. Does vitamin D stimulate calcium transport via its effect on protein synthesis or its effect on membrane composition and architecture, specifically its phospholipid composition (Max et al., 1978; Norman et al., 1981)? The over-all correspondence in terms of the kinetic response to 1,25-(OH)$_2$-D$_3$ stimulation between molecular and cellular responses depicted in Figs 5 and 6 may not apply to the tissue response

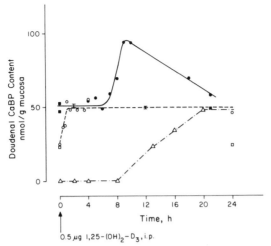

Fig. 5. *The effect of 1,25-dihydroxyvitamin D_3 on calcium binding protein in rat duodenum. 0.5 μg 1,25-(OH)$_2$-D_3 were adminstered by intraperitoneal injection to three groups of male Sprague-Dawley rats: vitamin D-replete animals on a high calcium (1.5% Ca, 1.5% P) semisynthetic diet (Bronner and Freund, 1975) (- - -); vitamin D-replete animals on a low calcium (0.06% Ca, 0.2% P) diet (———); and vitamin D-deficient animals on a high Ca (1.5% Ca, 1.5% P) vitamin D-deficient diet (- · -). The deficient animals had been on the regimen since weaning, the other for about 2 weeks. CaBP was analysed by an equilibrated column technique (Ueng et al., 1979). Square symbols represent CaBP levels in the untreated + D controls. Adapted from Buckley and Bronner (1980 and unpublished).*

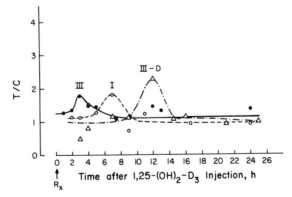

Fig. 6. *Time course of calcium uptake by duodenal cells. The cells were isolated from rats on three diet groups: high calcium (III), low calcium (I) and high calcium without vitamin D (III-D). See legend of Fig. 5 for further details. "Time after injection" refers to the time when the animals were killed after an intraperitoneal injection of 1,25-(OH)$_2$-D_3 (0.625 μg/100 g body wt). The ordinate refers to the ratio of calcium uptake by cells from treated animals (T) to that from controls (C). Calcium uptake was measured at 5 min (Fig. 1A). (Lipton and Bronner, unpublished.)*

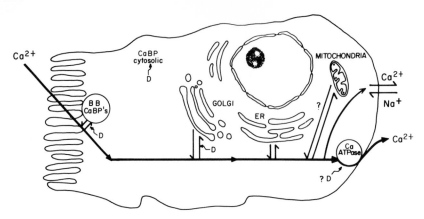

Fig. 7. *Scheme of transcellular calcium movement in the intestinal cell. Calcium is thought to move down its electrochemical gradient from lumen to cell, with brush border calcium binding proteins playing a gating role. Some of these proteins are vitamin D-dependent. Interaction with Golgi membrane and endoplasmic membrane proteins may serve as further gating and directional steps. The role of mitochondria in intracellular calcium regulation is in dispute. Calcium extrusion against an electrochemical gradient occurs by two routes: a Ca-dependent ATPase and a Na^+-Ca^{2+} exchanger. The later may come into play only when the Ca-ATPase is at its maximum capacity. Vitamin D dependence of the ATPase is uncertain (see Fig. 4). The role of the cytosolic, vitamin D-dependent CaBP in calcium transport is not known, but its response seems to parallel the effect of vitamin D on saturable calcium transport (see Figs 5 and 6).*

in vitamin D-deficient animals, where a response by the everted sac system has been reported to occur before CaBP biosynthesis (Spencer *et al.*, 1978; Thomasset *et al.*, 1979a). If that is the case, the relationship between transport response and protein synthesis becomes even more puzzling.

The possibility that $1,25\text{-}(OH)_2\text{-}D_3$ can act rapidly, perhaps by a posttranscriptional mechanism, is interesting because it could mean that vitamin D can participate in rapid regulation and that in the vitamin D-replete animal, at least, biosynthetic and functional events are not dissociated. If such dissociation is restricted to the vitamin D-deficient animal, it would be no less interesting, but would perhaps have less significance for normal function. On the other hand, Mac-Laughlin *et al.* (1980) have reported such a rapid response to $1,25\text{-}(OH)_2\text{-}D_3$ in the Golgi apparatus of the vitamin D-deficient rat that the question of the earliest event in response to $1,25\text{-}(OH)_2\text{-}D_3$ must still remain open.

At present it seems established (see Fig. 7) that $1,25\text{-}(OH)_2\text{-}D_3$ can directly stimulate calcium uptake by cells, brush border and Golgi membranes and can induce or stimulate saturable calcium transport by the duodenum, as evaluated by the everted sac preparation. An effect of vitamin D on nonsaturable transport is moot. Whether the biosynthetic events induced by $1,25\text{-}(OH)_2\text{-}D_3$ stimulation coincide with or are necessary for the transport response under all conditions or only in the vitamin D-replete state is as yet unresolved.

2. Parathyroid hormone

There is uncertainty in the literature whether acute parathyroidectomy is accompanied by depressed calcium absorption. Both Dowdle *et al.* (1960) and Rasmussen (1960) have reported that active calcium transport is depressed as a result of parathyroidectomy, whereas Kimberg *et al.* (1961) and Wasserman and Comar (1961) failed to find such an effect. Sammon *et al.* (1970) found little difference in calcium absorption of parathyroidectomized (PTX) as compared to euparathyroid animals, with the major effect at higher intake levels, when absorption was depressed in the ablated animals. Shah and Draper (1966), on the other hand, found calcium absorption was depressed in PTX animals at low calcium intake levels. Clark and Rivera-Cordero (1973) analysed the relationship between calcium absorption and intake not as an exponential function, as had Sammon *et al.* (1970), but as a linear regression. They found the slope of this relationship somewhat depressed in PTX animals, but the difference was not evaluated statistically. Moreover the authors interpreted their data as not reflecting differences over a wide range of calcium intake. It therefore seems that at the whole organism level the absence of parathyroid hormone has little effect on calcium absorption. Virtually no studies have been published on a direct effect of parathyroid hormone on calcium transport by cells or organelles. Since parathyroid hormone (PTH) is thought to act via the adenyl cyclase system, an effect of PTH should be reflected in variations of cyclic AMP. Again the evidence concerning a relationship between cyclic AMP and intestinal calcium transport is equivocal (see Nellans and Kimberg, 1979b, for a review). Until PTH receptors have been isolated from intestinal cells and a direct effect of PTH on cellular calcium uptake or efflux has been demonstrated, direct regulation of calcium transport by PTH must be considered unproven.

Part of the difficulty in interpreting the relationship between parathyroid hormone and calcium absorption stems from the trophic relationship between parathyroid hormone and the renal production of $1,25\text{-}(OH)_2\text{-}D_3$. Current evidence (reviewed by Haussler and Brickman, 1982) indicates that steady state circulating levels of $1,25\text{-}(OH)_2\text{-}D_3$ are depressed in parathyroidectomy and elevated in conditions of PTH excess. Saturable calcium absorption, in turn, varies positively and proportionately with body levels of $1,25\text{-}(OH)_2\text{-}D_3$ (see above). This explains why increased calcium absorption and hypercalciuria are common in hyperparathyroidism (Fischer, 1981). However, while PTH appears to regulate the steady state levels of circulating $1,25\text{-}(OH)_2\text{-}D_3$, PTX animals can still respond to a drop in calcium intake by increasing plasma $1,25\text{-}(OH)_2\text{-}D_3$ (Trechsel et al., 1980), raising duodenal CaBP (Thomasset et al., 1979b) and presumably saturable calcium absorption. Indeed, the data of Trechsel et al. (1980) indicate that the calcium intake-dependent rise in circulating $1,25\text{-}(OH)_2\text{-}D_3$ is proportionately the same in euparathyroid and PTX rats, though the absolute levels are higher in animals with normal parathyroid function. It thus appears that PTH can regulate plasma $1,25\text{-}(OH)_2\text{-}D_3$ by factors other than plasma calcium. In turn, calcium levels seem to regulate $1,25\text{-}(OH)_2\text{-}D_3$ production by means other than PTH. The question of a direct effect of PTH on calcium transport by the intestinal cell remains open.

Indirect Regulation of Intestinal Calcium Transport

Of the three major regulators of calcium metabolsim, parathyroid hormone, vitamin D, and calcitonin, only vitamin D has been demonstrated to have a direct effect on calcium transport of the intestinal cell; the direct effect of parathyroid hormone is uncertain, as just discussed. Calcitonin, whose only demonstrated role in plasma calcium regulation is that of a derivative control (Bronner, 1982), does not act on intestinal calcium absorption, either directly or indirectly (for reviews, see Levine et al., 1982). The effect of other hormones, many of which affect calcium or skeletal metabolism (Rude and Singer, 1982), is variable. Thus thyroxin has been reported to depress or to have no effect on intestinal calcium absorption (reviewed by Levine et al., 1982). In hypophysectomy, Kimberg et al. (1961) found a depression of saturable calcium transport, whereas Krawitt et al. (1977) found no effect.

As outlined earlier, calcium absorption varies in different nutritional and endocrine states, but whether or how these alter calcium transport by the intestinal cell has generally not been studied. Because of current interest in vitamin D, many of the observed changes have been explained in terms of $1,25\text{-}(OH)_2\text{-}D_3$ effects, but specific evidence is often lacking. Frequently cells and organelles retain calcium uptake behaviour characteristic of the tissue (see Figs 1–3). In that case, experiments testing a direct effect are possible, and if positive, may contribute significantly to an understanding of how the intestinal cell transports calcium.

ACKNOWLEDGMENTS

The author gratefully acknowledges support by the National Institutes of Health, the National Science Foundation and the University of Connecticut Research Foundation. Recent work has benefitted by NIH grants Nos. AM 26174 and AM 14261 and NSF grant No. PCM 782366.

REFERENCES

Asimakis, G. K. and Sordahl, L. (1977). *Archs. Biochem. Biophys.* **179**, 200–210.

Bikle, D. D., Askew, E. W., Zolock, D. T., Morrissey, R. L. and Herman, R. H. (1980). *Biochim. biophys. Acta* **598**, 561–574.

Bikle, D. D., Zolock, T., Morrissey, R. L. and Herman, R. H. (1978). *J. biol. Chem.* **253**, 484–488.

Bronner, F. (1982). *In* "Disorders of Mineral Metabolism" (Eds F. Bronner and J. W. Coburn) Vol. 2. Academic Press, New York. In press.

Bronner, F., Bellaton, C and Pansu, D. (1981a). *Fedn. Proc. Fedn. Am. Socs. exp. Biol.* **40** (3) 899 (abstract).

Bronner, F., Charnot, Y., Golub, E. E. and Freund, T. (1986). *In* "Proc. XIth European Symposium on Calcified Tissues" (Eds S. P. Mielsen and E. Hjorting-Hansen) 27–38. FADL, Copenhagen.

Bronner, F. and Freund, T. (1975). *Am. J. Physiol.* **229**, 689–694.

Bronner, F., Pansu, D., Buckley, M., Singh, R., Lipton, J. and Miller, A., III, (1981b). *In* "Calcium and Phosphate Transport Across Biomembranes" (Eds F. Bronner and M. Peterlick). Academic Press, New York. In press.

Buckley, M. and Bronner, F. (1980). *Archs. biochem. Biophys.* **202**, 235–241.

Christakos, S. and Norman, A. W. (1980). *Archs. biochem. Biophys.* **203**, 809–815.

Clark, J. and Rivera-Cordero, F. (1973). *Endocrinology* **92**, 62–72.

DeLuca, H. F. (1981). *A. Rev. Physiol.* **43**, 199–209.

DeLuca, H. F., Engstrom, G. W. and Rasmussen, H. (1962). *Proc. natn. Acad. Sci. U.S.A.* **48**, 1604–1609.

Douglas, A. P., Kerley, R. and Isselbacher, K. J. (1972). *Biochem. J.* **128**, 1329–1338.

Dowdle, E. B., Schachter, D. and Schenker, H. (1960). *Am. J. Physiol.* **198**, 269–274.

Engstrom, G. W. and DeLuca, H. F. (1962). *J. biol. Chem,* **237**, 974–974.

Engstrom, G. W. and DeLuca, H. F. (1964). *Biochemistry* **3**, 203–209.

Favus, M. J., Kathpalia, S. C. Coe, F. L. and Mond. E. (1980). *Am. J. Physiol.* **238**, G75–G78.

Fischer, J. (1982). *In* "Disorders of Mineral Metabolism" (Eds F. Bronner and J. W. Coburn). Vol. 2. Academic Press. New York. In press.

Forstner, G. G., Sabesin, S. M. and Isselbacher, K. J. (1968). *Biochem. J.* **106**, 381–390.

Freedman, R. A., Weiser, M. M. and Isselbacher, K. J. (1977). *Proc. natn. Acad. Sci. U.S.A.* **74**, 3612–2616.

Freedman, R. A., MacLaughlin, J. A. and Weiser, M. M. (1981). *Archs. Biochem. Biophys.* **206**, 233–241.

Freund, T. and Bronner, F. (1975). *Am. J. Physiol.* **228**, 861–869.

Frizzell, R. A. and Schultz, S. (1972). *J. gen. Physiol.* **59**, 318–346.

Fujita, M., Ohta, H., Kawai, K., Matsui, H. and Nakao, M. (1972). *Biochem. biophys. Acta* **274**, 336–347.

Ghijsen, W. E. J. M. and van Os, C. H. (1979). *Nature, Lond.* **279**, 802–803.

Ghijsen, W. E. J. M., DeJong, M. D. and van Os, C. H. (1980). *Biochim. biophys. Acta* **599**, 538–551.

Gmaj, P., Mürer, H. and Kinne, R. (1979). *Biochem. J.* **178**, 549–557.

Goldman, D. E. (1953). *J. gen. Physiol.* **27**, 37–60.

Goodman, M., Pechère, J-F., Haiech, J. and Gemaille, J. (1979). *J. molec. Evol.* **13**, 331–353.

Hasse, W., Schafer, A., Mürer, H. and Kinne, R. (1978). *Biochem. J.* **172**, 57–62.

Halloran, B. P., DeLuca, H. F. and Vincenzi, F. F. (1980). *FEBS Lett.* **114**, 89–92.

Haussler, M. and Brickman, A. (1982). *In* "Disorders of Mineral Metabolism" (Eds F. Bronner and J. W. Coburn). Vol. 2. Academic Press, New York, In press.

Hodson. S. (1978). *J. Cell Sci.* **30**, 117–178.

Hopfer, U. (1977). *Am. J. Physiol.* **233**, E445–E449.

Hopfer, U., Nelson. K., Perrotto, J. and Isselbacher, K. J. (1973). *J. biol. Chem.* **248**, 25–32.

Hurwitz, S., Stacey, R. E. and Bronner, F. (1969). *Am. J. Physiol.* **216**, 254–262.

Kessler, M., Acuto, O., Storelli, C., Mürer, H., Muller, M. and Semenza, G. (1978). *Biochim. biophys. Acta* **506**, 136–154.

Kimberg, D. V., Schachter, D. and Schenker, H. (1961). *Am. J. Physiol.* **200**, 1256–1262.

Kinne, R. (1976). *In* "Current Topics of Membranes and Transport" (Eds F. Bronner and A. Kleinzeller) Vol. 8, 209–267.

Kinne, R. (1981) *In* "Calcium and Phosphate Transport Across Biomembranes" (Eds F. Bronner and M. Peterlik). Academic Press, New York. In press.

Kowarski, S. and Schachter, D. (1973). *J. clin. Invest.* **52**, 2765–2773.

Kowarski, S. and Schachter, D. (1980). *J. biol. Chem.* **255**, 10 834–10 840.

Krawitt, E. L., Kunin, A. S., Sampson, H. W. and Bacon, B. J. F. (1977). *Am. J. Physiol.* **232**, E229–E233.

Larsen, F. L. and Vincenzi, F. F. (1979). *Science, N.Y.* **204**, 306–309.

Lee, D. B. N., Walling, M. W., Silis, V. and Coburn, J. W. (1980). *J. clin. Invest.* **65**, 1326–1331.

Levine, B. S., Walling, M. W. and Coburn, J. W. (1982). *In* "Disorders of Mineral Metabolism" (Eds F. Bronner and J. W. Coburn) Vol. 2. Academic Press, New York. In press.

MacLaughlin, J. A., Weiser, M. M. and Freedman, R. A. (1980). *Gastroenterology* **78**, 325–332.

Martin, D. L., Melancon, M. J. and DeLuca, H. F. (1969). *Biochem. biophys. Res. Commun.* **35**, 819–823.

Max, E. E., Goodman, D. B. and Rasmussen, H. (1978). *Biochim. biophys. Acta* **511**, 224–239.

Mela, L. (1977). *In* "Current Topics in Membranes and Transport" (Eds F. Bronner and A. Kleinzeller) Vol. 9, 321–366.

Mellersh, H., Tomlinson, S. and Pollock, A. (1980). *Biochem. J.* **185**, 601–607.

Miller, A. III, and Bronner, F. (1981). *Biochem. J*, **196** 391–401.

Miller, A. III, Ueng, T-H. and Bronner, F. (1979). *FEBS Lett.* **103**, 319–322.

Mircheff, A. K. and Wright, E. M. (1976). *J. Membr. Biol.* **28**, 309–333.

Mürer, H. and Hildmann, B. (1981). *In* "Calcium and Phosphate Transport Across Biomembranes" (Eds F. Bronner and M. Peterlik) Academic Press, New York. In press.

Mürer, H. and Kinne, R. (1980). *J. Membr. Biol.* **55**, 81–95.

Mürer, H., Hopfer, U. and Kinne, R. (1976). *Biochem. J.* **154**, 597–604.

Mürer, H., Hopfer, U. Kinne-Saffran, E. and Kinne, R. (1974). *Biochim. biophys. Acta* **345**, 179–179.

Nellans, H. N. and Kimberg, D. V. (1978). *Am. J. Physiol.* **236**, E726–E737.

Nellans, H. N. and Kimberg, D. V. (1979a). *Am. J. Physiol.* **236**, E474–E481.

Nellans, H. N. and Kimberg, D. V. (1979b). *In* "International Review of Physiology Gastrointestinal Physiology III" (Ed R. K. Crane) Vol. 19, 227–261. Park Press, Baltimore.

Nellans, H. N. and Popovitch, J. E. (1981). *Fedn. Proc. Fedn. Am. Socs. exp. Biol.* **40**(3), 369 (abstract).

Nicholls, D. G. and Crompton, M. (1980). *FEBS Lett.* **111**, 261–268.

Norman, A. W. (1979). "Vitamin D: The Calcium Homeostatic Steroid Hormone" 490 pp. Academic Press, New York.

Norman, A. W., Putkey, J. A. and Nemere, Ilka. (1981). *In* "Calcium and Phosphate Transport Across Biomembranes" (Eds F. Bronner and M. Peterlik). Academic Press, New York. In press.

Pansu, D., Bellaton, C. and Bronner, F. (1981). *Am. J. Physiol.* **240**, G32–G37.

Rasmussen, H. (1959). *Endocrinology* **65**, 517–519.

Rasmussen, H., Fontaine, O., Max, E. E. and Goodman, D. P. (1979). *J. biol. Chem.* **254**, 2993–2999.

Rude, R. and Singer, F. (1982). *In* "Disorders of Mineral Metabolism" (Eds F. Bronner and J. W. Coburn) Vol. 2. Academic Press, New York, In press.

Sacktor, B. (1977). *Curr. Topics Bioenerg.* **6**, 39–83.

Sammon, P. J., Stacey, R. E. and Bronner, F. (1970). *Am. J. Physiol.* **218**, 479–485.

Scalera, V., Storelli, C., Storelli-Joss, C., Hasse, W. and Mürer, H. (1980). *Biochem. J.* **186**, 177–181.

Schachter, D. and Rosen, S. M. (1959). *Am. J. Physiol.* **196**, 357–363.

Schedl, H. P., Miller, D. L. and Iskandarani, M. (1981). *In* "Calcium and Phosphate Transport Across Biomembranes" (Eds F. Bronner and M. Peterlik). Academic Press, New York. In press.

Schiffl, H. and Binswager, U. (1980). *Am. J. Physiol.* **238**, G424–G428.

Shah, B. G. and Draper, H. H. (1966). *Am. J. Physiol.* **211**, 963–966.

Sigrist-Nelson, K., Murer, H. and Hopfer, U. (1975). *J. biol. Chem.* **250**, 5674–5680.

Somlyo, A. P., Somlyo, A. V., Shuman, H., Scarpa, A., Endo, M. and Inesi, G.

(1981). *In* "Calcium and Phosphate Transport Across Biomembranes" (Eds F. Bronner and M. Peterlick). Academic Press, New York. In press.

Spencer, R., Chapman, R., Wilson, R. P. and Lawson, D. E. M. (1978). *Biochem. J.* **170**, 93–101.

Taylor, A. and Windhager, E. (1979). *Am. J. Physiol.* **236**, F505–F512.

Thomasset, M., Cuisinier-Gleizes, P. and Matthieu, H. (1979a). *FEBS Lett.* **107**, 91–94.

Thomasset, M., Cuisinier-Gleizes, P. Matthieu, H., Golub, E. E. and Bronner, F. (1979b). *Calcif. Tiss. Int.* **29**, 141–145.

Trechsel, U., Eisman, J. A., Fischer, J. A., Bonjour, J-P. and Fleisch, H. (1980). *Am. J. Physiol.* **239**, E119–E124.

Ueng, T-H. and Bronner, F. (1979). *Archs. Biochem. Biophys.* **197**, 205–217.

Ueng, T-H., Golub, E. E. and Bronner, F. (1979). *Archs, Biochem. Biophys.* **196**, 624–630.

Ussing, H. H. and Zerahn, K. (1951). *Acta Physiol. Scand.* **23**, 110–127.

van Os, C., Ghijsen, W. and deJonge, H. (1981). *In* "Calcium and Phosphate Transport Across Biomembranes" (Eds F. Bronner and M. Peterlik). Academic Press, New York. In press.

Walling, M. W. and Rothman, S. S. (1968). *Fedn. Proc. Fedn. Am. Socs. exp. Biol.* **27**, 386 (abstract).

Walling, M. W. and Rothman, S. S. (1969). *Am. J. Physiol.* **117**, 1144–1152.

Wasserman, R. H. and Comar, C. L. (1961). *Endocrinology* **69**, 1074–1079.

Wasserman, R. H. and Feher, J. J. (1977). *In* "Calcium Binding Proteins and Calcium Function" (Eds R. H. Wasserman, R. A. Corradino, E. Carafoli, R. H. Kretsinger, D. H. MacLennan and F. L. Siegel) 293–302. Elsevier/North-Holland, New York.

Wasserman, R. H. and Kallfelz, F. A. (1962). *Am. J. Physiol.* **203**, 221–224.

Wasserman, R. H. and Taylor, A. N. (1969). *In* "Mineral Metabolism – An Advanced Treatise" (Eds C. L. Comar and F. Bronner) Vol. 3, 321–403. Academic Press, New York and London.

Weiser, M. M., Bloor, J. H., Dasmahapatra, A., Freedman, R. A and MacLaughlin, J. A. (1981) *In* "Calcium and Phosphate Transport Across Biomembranes" (Eds F. Bronner and M. Peterlik). Academic Press, New York. In press.

Wilson, P. W. and Lawson, D. E. M. (1980). *Pflüg. Arch.* **389**, 69–74.

Wilson, T. H. and Wiseman, G. (1954). *J. Physiol. Lond.* **123**, 116–125.

Windhager, E. E., Lee, C. O., Lorenzer, M. and Taylor, A. (1981). *In* "Calcium and Phosphate Transport Across Biomembranes" (Eds F. Bronner and M. Peterlik). Academic Press, New York. In press.

Wrobel, J. (1981). *In* "Calcium and Phosphate Transport Across Biomembranes" (Eds F. Bronner and M. Peterlik). Academic Press, New York. In press.

Zornitzer, A. E. and Bronner, F. (1971). *Am. J. Physiol.* **220**, 1261–1265.

Subject Index